*Advances in*
# PARASITOLOGY

VOLUME 23

## Editorial Board

**W. H. R. Lumsden** University of Dundee Animal Services Unit, Ninewells Hospital and Medical School, P.O. Box 120, Dundee DD1 9SY, UK

**A. Capron** Director, Centre d'Immunologie et de Biologie Parasitaire, Institut Pasteur, 15 Rue Camille Guérin, 59019 Lille Cedex, France

**P. Wenk** Tropenmedizinisches Institut, Universität Tübingen, D7400 Tübingen 1, Wilhelmstrasse 31, W. Germany

**C. Bryant** Department of Zoology, Australian National University, G.P.O. Box 4, Canberra, A.C.T. 2601, Australia

**E. J. L. Soulsby** Department of Clinical Veterinary Medicine, University of Cambridge, Madingley Road, Cambridge CB3 0ES, UK

**K. S. Warren** Director for Health Sciences, The Rockefeller Foundation, 1133 Avenue of the Americas, New York, N.Y. 10036, USA

**J. P. Kreier** Department of Microbiology, College of Biological Sciences, Ohio State University, 484 West 12th Avenue, Columbus, Ohio 43210, USA

**M. Yokogawa** Department of Parasitology, School of Medicine, Chiba University, Chiba, Japan

# Advances in
# PARASITOLOGY

Edited by

## J. R. BAKER

Culture Centre of Algae and Protozoa,
Cambridge, UK

and

## R. MULLER

Commonwealth Institute of Parasitology,
St. Albans, UK

VOLUME 23

1984

## ACADEMIC PRESS
(*Harcourt Brace Jovanovich, Publishers*)
London   Orlando   San Diego   San Francisco   New York
Toronto   Montreal   Sydney   Tokyo   São Paulo

ACADEMIC PRESS INC. (LONDON) LTD.
24/28 Oval Road,
London NW1 7DX

*United States Edition published by*
ACADEMIC PRESS INC.
(Harcourt Brace Jovanovich, Inc.)
Orlando, Florida 32887

Copyright © 1984 by
ACADEMIC PRESS INC. (LONDON) LTD.

*All Rights Reserved*
No part of this book may be reproduced in any form by photostat, microfilm, or any other means, without written permission from the publishers

*British Library Cataloguing in Publication Data*

ISBN 0-12-031723-0
ISSN 0065-308X
LCCCN 62-22124

Typeset by Bath Typesetting Ltd., Bath
and printed in Great Britain by
Thomson Litho Ltd., East Kilbride, Scotland

# CONTRIBUTORS TO VOLUME 23

**A. E. Butterworth** *Department of Pathology, Tennis Court Road, Cambridge CB2 1QP, England*

**Tag E. Mansour** *Department of Pharmacology, Stanford University School of Medicine, Stanford, California 94305, USA*

**Heinz Mehlhorn** *Lehrstuhl für Spezielle Zoologie und Parasitologie, Ruhr-Universität Bochum, Postfach 102148—Gebaude ND, 4630 Bochum 1, West Germany*

**Richard E. Reeves** *Department of Biochemistry and Department of Tropical Medicine and Medical Parasitology, Louisiana State University Medical Center, 1901 Perdido Street, New Orleans, Louisiana 70112-1393, USA*

**Eberhard Schein** *Institut für Parasitologie der Freien Universität Berlin, Altensteinstrasse 40, 1000 Berlin 33, West Germany*

# PREFACE

In spite of our opening comments in the Preface to Volume 22, this volume contains an equal mixture of protozoology and helminthology. We apologize to entomological readers for the omission of a contribution specifically related to their interests but hope that they will find at least the chapter by Dr Mehlhorn and Dr Schein rewarding. The occurrence of a sexual phase in the life-cycle of piroplasms seems now to be established beyond reasonable doubt and further emphasizes the close relationship between these organisms and the Haemosporina. The emergence of evidence that some form of genetic exchange occurs in at least some trypanosomes is exciting and leads one to wonder whether such events, perhaps differing considerably from classical concepts of sexual behaviour, may be more widespread among parasitic protozoa than is currently believed. Perhaps this is a topic which "Advances" should review in a few years' time.

Knowledge of other topics increases so rapidly that relatively frequent reviewing is justified—indeed, essential. One such subject, now in its log-phase of growth, is immunology, becoming increasingly recognized as being of major importance in an understanding of host–parasite relations. Dr Butterworth provides a lucidly written synthesis of current knowledge and ideas in this field—the rapidity of advance being emphasized by the report since this review was written of platelets as yet another cellular component of the immune system.

The remaining two contributions by Dr Mansour and Dr Reeves summarize advances in a subject in which knowledge has been steadily increasing over a number of years: parasite biochemistry and physiology. The relevance of this knowledge to the attainment of the ultimate goal (or dream?) of rational chemotherapy is obvious.

1984                                                                                              J. R. Baker
                                                                                                   R. Muller

# Contents

Contributors.................................................................... v
Preface ........................................................................ vii

## Serotonin Receptors in Parasitic Worms
### TAG E. MANSOUR

| | | |
|---|---|---|
| I. | Introduction................................................................ | 2 |
| II. | Motility in Parasitic Helminths ...................................... | 3 |
| III. | Acetylcholine Receptors in Parasitic Worms .................... | 4 |
| IV. | Effect of Serotonin on Motility in Parasitic Flatworms ....... | 4 |
| V. | Role of Serotonin on Migration Behaviour of Cestodes ....... | 7 |
| VI. | Occurrence of Serotonin in Trematodes ........................... | 8 |
| VII. | Serotonin and Regulation of Metabolism in Parasites ......... | 10 |
| VIII. | $^{31}$P Nuclear Magnetic Resonance Spectroscopy of *Fasciola hepatica* and the Action of Serotonin ............................ | 12 |
| IX. | Adenosine Cyclic 3′, 5′-Monophosphate and Cell Regulation in Parasitic Worms................................................... | 14 |
| X. | Hormone-Activated Adenylate Cyclase in Higher Organisms... | 18 |
| XI. | Serotonin-Activated Adenylate Cyclase in the Liver Fluke *Fasciola hepatica* .................................................... | 20 |
| XII. | Serotonin Receptors in Nematodes ................................. | 27 |
| XIII. | Effect of Serotonin on Glycogenolysis and Glycolysis in Mammals ................................................................ | 28 |
| XIV. | General Summary and Conclusions................................. | 28 |
| | Acknowledgements .................................................... | 29 |
| | References ............................................................... | 29 |

## The Piroplasms: Life Cycle and Sexual Stages
### HEINZ MEHLHORN AND EBERHARD SCHEIN

| | | |
|---|---|---|
| I. | Introduction................................................................ | 38 |
| II. | The Genera *Theileria* and *Babesia* .............................. | 38 |
| III. | Life Cycle of *Theileria* species .................................... | 45 |
| IV. | The Life Cycle of *Babesia* species................................. | 69 |
| V. | Conclusions ............................................................... | 96 |
| | References ............................................................... | 97 |

## Metabolism of *Entamoeba histolytica* Schaudinn, 1903
RICHARD E. REEVES

| | | |
|---|---|---|
| I. | Introduction | 106 |
| II. | Carbohydrate Metabolism | 108 |
| III. | Other Metabolic Capabilities | 118 |
| IV. | Effect of Associate Cells | 122 |
| V. | Growth Requirements | 124 |
| VI. | Regulation | 126 |
| VII. | Evolutionary Considerations | 127 |
| VIII. | Speculation | 128 |
| IX. | Conclusions | 131 |
| | Acknowledgement | 132 |
| | Appendix A. Quantitation | 132 |
| | Appendix B. Free Amino Acids in Axenic Growth Media | 133 |
| | References | 133 |

## Cell-Mediated Damage to Helminths
A. E. BUTTERWORTH

| | | |
|---|---|---|
| I. | Introduction | 144 |
| II. | Methods for Studying Cell-Mediated Damage to Helminths *in Vitro* | 145 |
| III. | Characteristics of Cellular Effector Mechanisms Active Against Helminths | 148 |
| IV. | Effector Mechanisms Active Against Particular Helminth Species | 182 |
| V. | The Role of Cellular Effector Mechanisms in Immunity *in Vivo* | 192 |
| VI. | Summary and Conclusions | 205 |
| | Acknowledgements | 207 |
| | References | 207 |

Subject Index ... 237

# Serotonin Receptors in Parasitic Worms

TAG E. MANSOUR

*Department of Pharmacology, Stanford University School of Medicine, Stanford, California, USA*

| | | |
|---|---|---|
| I. | Introduction | 2 |
| II. | Motility in Parasitic Helminths | 3 |
| III. | Acetylcholine Receptors in Parasitic Worms | 4 |
| IV. | Effect of Serotonin on Motility in Parasitic Flatworms | 4 |
| V. | Role of Serotonin on Migration Behaviour of Cestodes | 7 |
| VI. | Occurrence of Serotonin in Trematodes | 8 |
| VII. | Serotonin and Regulation of Metabolism in Parasites | 10 |
| VIII. | $^{31}$P Nuclear Magnetic Resonance Spectroscopy of *Fasciola hepatica* and the Action of Serotonin | 12 |
| IX. | Adenosine Cyclic 3′,5′-Monophosphate and Cell Regulation in Parasitic Worms | 14 |
| | A. Conservation of the Regulatory Role of Adenosine Cyclic 3′,5′-Monophosphates | 14 |
| | B. Role of Neurotransmitters in Development of Eukaryotes | 15 |
| | C. Changes in Serotonin-activated Adenylate Cyclase During Development of *Schistosoma mansoni* | 16 |
| X. | Hormone-Activated Adenylate Cyclase in Higher Organisms | 18 |
| XI. | Serotonin-Activated Adenylate Cyclase in the Liver Fluke *Fasciola hepatica* | 20 |
| | A. Serotonin Receptors in *Fasciola hepatica* | 21 |
| | B. Desensitization of Serotonin Receptors in *Fasciola hepatica* | 23 |
| | C. The Guanosine Triphosphate Regulatory Protein | 25 |
| | D. Serotonin-Activated Protein Kinase in *Fasciola hepatica* | 25 |
| | E. Cyclic 3′,5′-Nucleotide Phosphodiesterase | 27 |
| XII. | Serotonin Receptors in Nematodes | 27 |
| XIII. | Effect of Serotonin on Glycogenolysis and Glycolysis in Mammals | 28 |
| XIV. | General Summary and Conclusions | 28 |
| | Acknowledgements | 29 |
| | References | 29 |

ADVANCES IN PARASITOLOGY
VOLUME 23 ISBN 0-12-031723-0

Copyright © 1984 Academic Press London
All rights of reproduction in any form reserved.

## I. Introduction

Serotonin (5-hydroxytryptamine) (Fig. 1) is an indoleamine that is widely distributed throughout the animal kingdom. Although it was originally isolated and characterized from mammalian serum, it has been recognized as playing an important role as a neurotransmitter in many invertebrate species. We owe a great debt to Erspamer (1966) and Welsh (1970) who established the importance of serotonin as a neurotransmitter in molluscs. Very soon after serotonin was chemically identified and was commercially available, I investigated its role in neuromuscular activity and metabolism of the liver fluke *Fasciola hepatica* (Mansour, 1957, 1959a,b). Early papers from my laboratory concerning serotonin were first obscured from parasitologists, probably because the work was published in pharmacological and biochemical periodicals, but they were soon recognized and pursued, using several species of parasitic worms. The large number of publications during the last decade, mainly in parasitology literature, is a pleasing reflection of the importance now given by parasitologists to understanding the role played by serotonin in the life of many parasitic worms. In this review I shall discuss the evidence available which implicates serotonin or a related indoleamine in the regulation of motility and metabolism of these helminths.

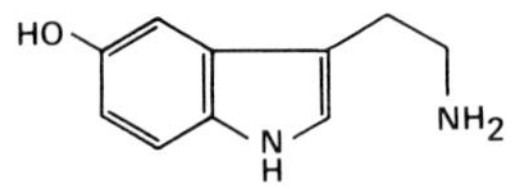


5-Hydroxytryptamine
(serotonin)

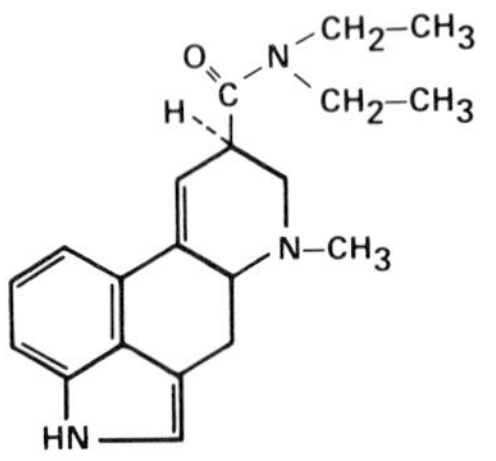


d-Lysergic acid diethylamide
(LSD)

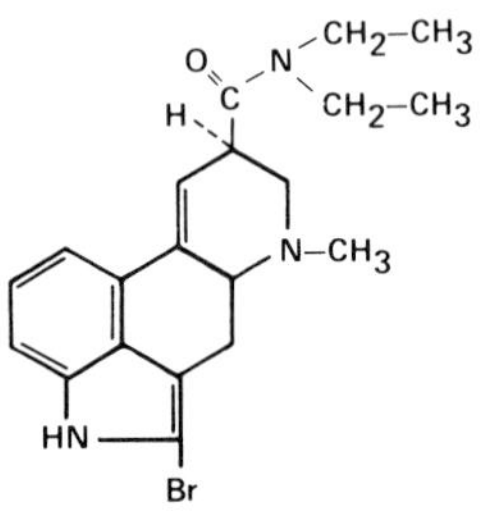


2-bromo-LSD
(BOL)

Fig. 1. Serotonin (5-hydroxytryptamine), lysergic acid diethylamide (LSD) and its 2-bromo derivative.

Occasionally I shall present some representative data from our own research to document certain recent aspects of the action of serotonin on parasitic worms. Since the direction of my work in this area has been mainly towards gaining more information on the molecular and biochemical effects of serotonin, there will be more emphasis given to this area. Although I have tried to present a comprehensive review of advances in this area, there is a chance that I may not have included all aspects relating to this problem. Omissions here should not be taken as an indication of lack of importance of the published work.

## II. Motility in Parasitic Helminths

The movement of parasitic worms is of particular interest because it is so important to their survival. These parasites maintain their site in the host through specific behaviour of their locomotory system. They are able to maintain themselves *in situ* in the face of motion of the host fluids, such as the gut content or bile in the case of intestinal or liver parasites, or the movement of blood or lymph in the case of some tissue parasites. When observed *in vitro*, many of these organisms show fast and well-coordinated rhythmical movement (Chance and Mansour, 1949, 1953). Presumably, the parasite is able to maintain itself in the host site through coordination of these movements. In addition, parasite motility must influence the movement of ingested food in the parasite intestinal caeca, its excretory and its egg laying systems. In addition to the motor system, parasitic worms are known to be endowed with sensory receptors that provide the organism with information concerning the environment of the host (Whitfield, 1979) and which are undoubtedly important for the specific migratory behaviour of the parasite in the host.

Parasitic worms have a well developed nervous system and musculature. Their muscle fibres are related to smooth rather than to striated muscle. They are differentiated to a specific arrangement such as outer circular, inner longitudinal and diagonal fibres. The nervous system is organized into ganglia, usually at the anterior of the parasite, peripheral ganglia that are more intermingled with the musculature, and sensory and motor nerves, with main trunks that run throughout the entire body of the worm. Information on the neuroanatomy and neurophysiology of these organisms is sparse. This has created a serious handicap to conducting modern experiments on the biochemistry and pharmacology of the neuromuscular system of some of these parasites. For example, some of the best classical research presently available on the neuroanatomy of these parasites was done during the late part of the nineteenth century with the limited technology available at the time (Looss, 1895; Bettendorf, 1897). It is regrettable that these well-documented investigations have not been pursued by modern biologists using the more sophisticated instrumentation that is now available.

The importance of motility in maintaining a successful parasitic life has provided an impetus for pharmacologists to use it as a biological parameter for the quick identification of new chemotherapeutic agents. Some investigators have used the isolated organ bath system that is used for mammalian tissues to study parasitic worm motility (Baldwin, 1943; Chance and Mansour, 1949). More elaborate systems of monitoring motility using instruments with multiphotocells (Hillman and Senft, 1973) or ultrasound (Chavasse *et al.*, 1979) have also been used successfully.

## III. Acetylcholine Receptors in Parasitic Worms

As in other multicellular organisms that possess nerves and ganglia, certain chemicals and neuromuscular regulators play a key role in the control of communication among nerve cells and from nerve to muscle cells. Acetylcholine, serotonin, epinephrine and dopamine have been implicated as putative neurohumoral transmitters in parasitic worms. Acetylcholine and the enzymes necessary for its synthesis and degradation have been identified in several species of parasitic worms and their larval stages (Bueding, 1952; Chance and Mansour, 1953; Probert and Durrani, 1977; Fripp, 1967; Pepler, 1958; Bruckner and Vage, 1974; Panitz and Knapp, 1967). Acetylcholine and its congeners caused neuromuscular preparations from *Ascaris* to contract (Baldwin and Moyle, 1949). These *Ascaris* preparations were more sensitive to nicotine than to acetylcholine. The receptors do not appear to be muscarinic but behave more like nicotinic receptors (sensitive to antagonism by d-tubocurarine but not by atropine). In the trematode *Fasciola hepatica*, acetylcholine, its congeners, and cholinesterase inhibitors relax the musculature, inhibit rhythmical movement and eventually produce paralysis (Chance and Mansour, 1949, 1953). These effects can be demonstrated even in preparations that have no central ganglia. Pharmacological evidence indicates that the acetylcholine receptors in trematodes may be different from those found in nicotinic or muscarinic mammalian synapses. For example, acetylcholine itself was not a very active effector on these receptors when compared with nicotine (Chance and Mansour, 1953). Furthermore, neither atropine nor d-tubocurarine, two of the most active cholinergic antagonists in mammals, affected the neuromuscular activity of these parasites to a significant degree.

## IV. Effect of Serotonin on Motility in Parasitic Flatworms

Shortly after the discovery of the presence of serotonin in several invertebrates, I initiated a study on the effect of this indoleamine and related

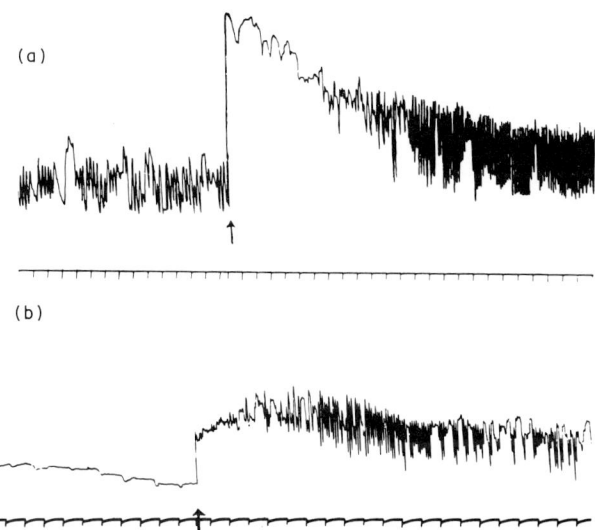

FIG. 2. Stimulatory action of serotonin on whole (a) and degangliated (b) preparations from *Fasciola hepatica*. Upward stroke of kymographic records represents contraction. Serotonin (0·5 mM) was added at the points indicated by the arrow. Time markings are in minutes. (Data from Mansour, 1957, by courtesy of the *British Journal of Pharmacology*.)

derivatives on motility of different parasites. The impetus for this investigation was our earlier observation that several synthetic amines stimulated the rhythmical movement of neuromuscular preparations of the liver fluke *F. hepatica* (Chance and Mansour, 1949). In 1957 I reported that motility of the liver fluke was stimulated by serotonin (Fig. 2), lysergic acid diethylamide (LSD) and related indoleamines (Mansour, 1957). The effect was peripheral and not mediated through the pharyngeo-oesophageal ganglia, the central ganglia of the parasite, because a degangliated preparation responded to the action of serotonin (Fig. 2). 2-Bromolysergic acid diethylamide, an analogue of LSD that has a bromine atom on the second position (Fig. 1), depressed rhythmical movement and antagonized the stimulant action of serotonin and of LSD. This was the first evidence that these parasites contain serotonin receptors that influence motility. This became even more evident when we later investigated the effect of several lysergic acid diethylamide and indoleamine derivatives on liver fluke neuromuscular preparations (Beernink *et al.*, 1963). Among these derivatives, LSD was the most potent compound. Surprisingly, this was in accord with data reported on the psychotomimetic action of this compound on man as reported by Isbell *et al.* (1959). In fact, a rank order correlation between the stimulatory effect of these derivatives on liver fluke motility correlated surprisingly well with the psychotomimetic effects of LSD derivatives on man (Beernink et al., 1963). This parallelism suggested that these compounds may act on receptors in the flukes which are similar to those in the central nervous system of man. Biochemical studies on these receptors may therefore give us a clue as to the biochemical effect of these agents on the complex mammalian nervous system.

The excitatory effect of serotonin was shown not to be unique to *F. hepatica*. Subsequent to our initial reports, we, as well as others, found that serotonin stimulates rhythmical movement in other species: *Schistosoma mansoni*, *Clonorchis sinensis*, and the cestode *Taenia pisiformis* (Mansour, 1964). These findings were further verified on *Schistosoma mansoni* by two other groups (Barker et al., 1966; Hillman and Senft, 1973) and on the cestode *Mesocestoides corti* (Hariri, 1974). Later, more research was done on the effect of indoleamines and their analogues on trematode motility. A list of different parasites that respond to serotonin is shown in Table 1.

TABLE 1
Role of serotonin in the regulation of motility in parasitic helminths

| Parasite | Action on motility | References |
|---|---|---|
| **Trematoda** | | |
| *Fasciola hepatica* | Stimulant | Mansour (1957); Beernink et al. (1963) |
| *Schistosoma mansoni* | Stimulant | Mansour (1964); Barker et al. (1966); Hillman and Senft (1973) |
| *Clonorchis sinensis* | Stimulant | Mansour (1964) |
| **Cestoda** | | |
| *Taenia pisiformis* | Stimulant | Mansour (1964) |
| *Mesocestoides corti* | Stimulant | Hariri (1974) |
| *Dipylidium caninum* | Stimulant | Terada et al. (1982a) |
| *Hymenolepis diminuta* | Stimulant | Mettrick and Cho (1981) |
| **Nematoda** | | |
| *Angiostrongylus cantonensis* | Paralysis | Terada et al. (1982b) |
| *Breinlia booliati* | No effect | Adaikan and Zaman (1982) |
| *Caenorhabditis elegans*[a] | Suppressed | Croll (1975) |

[a] Not a parasitic nematode.

Action of serotonin on parasite motion presumably was due to interaction with serotonin receptors. The effects on flukes as well as on cestodes could very well be explained in terms of the assumption that serotonin or a related chemical agent is a neurotransmitter in these trematodes.

Although early experiments on the action of serotonin on neuromuscular activity of parasitic worms were restricted to motility itself, some later reports examined the effect of serotonin on electrical stimulation of neuromuscular preparations of the parasites and on electrical activity emanating from the parasite. Pax *et al.* (1981) were able to record longitudinal muscle shortening of adult male *S. mansoni* produced by electrical stimulation. Neither serotonin nor its antagonists had an effect on the contraction of the muscle. It is possible therefore that the stimulatory effect of serotonin on rhythmical movement of schistosomes is due to an action at a site other than directly on the muscle. These authors suggested an effect on a neuromodulation system for motility. The effect of serotonin and other neurotransmitters on surface electrical activity from *S. mansoni* was examined by Semeyn *et al.* (1982). Recordings from control male worms were characterized by bi- and triphasic potentials occurring at a rate of about 60 to 70 $s^{-1}$ with amplitude ranging from greater than 1 mV down to values indistinguishable from background electrical noise. Serotonin caused a marked increase in frequency of larger potentials at concentrations as low as 0·1 μM. Dopamine, an amine that had no observable effect on motility, had a significant depressive effect on electrical activity, decreasing the frequency of the recorded potentials. Although this new methodology offers a new parameter that is more sensitive than motility for measuring serotonin-like agents, it does not elucidate the mechanism of action of serotonin. A better understanding of the nature of the measured potentials is needed before a definitive interpretation of these results can be made.

## V. Role of Serotonin on Migration Behaviour of Cestodes

Evidence that serotonin influences migration behaviour of flatworms was reported recently by Mettrick and Cho (1982) using *Hymenolepis diminuta* in the rat. Two distinct types of migration behaviour of this parasite in the rat small intestine are recognized. The first pertains to its life cycle (ontogenic) (Chandler, 1939; Holmes, 1962; Braton and Hopkins, 1969; Cannon and Mettrick, 1970) and the second is circadian (Read and Kilejian, 1969; Tanaka and MacInnis, 1975). The work of Read and Kilejian (1969), and more recently of Mettrick (1971, 1972), indicates that the circadian pattern of migration exhibited by *H. diminuta* is related to host feeding. Mettrick (1973) proposed that the activation of the helminth's migratory response was mediated via vagal stimulation of the gastrointestinal function (Mettrick, 1973; Mettrick and Cho, 1981). Stimulation of the vagal nerve induced anteriad migration of the worms in the small intestine. Since vagal stimulation mimics normal vagal response to host feeding, the circadian migratory behaviour of these worms is not caused by ingested nutrients *per se* but indirectly via the parasympathetic nerve stimulation of gastrointestinal function. Among the individual secretory responses under vagal control, serotonin was found to have the greatest effect on worm migration (Podesta and Mettrick, 1981). Cho and Mettrick (1982) reported that the circadian migration of *H. diminuta* in the small intestine of the rat correlated well with the circadian variation in serotonin levels in worm tissue, in the intestinal lumen, in the intestinal mucosa, and with the amount of food present in the small intestine. Serotonin administration parenterally or by mouth caused an anteriad migration by the worms (Mettrick and Cho, 1981). Methysergide, a serotonin antagonist, caused a 90% inhibition in the worm migratory response (Mettrick and Cho, 1982). These reports suggest the involvement of serotonin in the migration behaviour of these parasites in the host. It remains to be seen whether other more potent antiserotonin agents such as 2-bromolysergic acid diethylamide (Mansour, 1957) could influence the worm's behaviour without influencing the motility of the host intestine.

## VI. Occurrence of Serotonin in Trematodes

Serotonin was shown by Erspamer (1966) and by Welsh (1970) to be ubiquitous among invertebrate species. In spite of the demonstration that the indoleamine is found in a vast number of invertebrate species, its role as a neurohumoral transmitter has only been well-documented in molluscs. The different metabolic reactions for its synthesis and degradation have been elucidated by Udenfriend (1959) and his associates. L-Tryptophan, the natural amino acid, is taken up by the cells and converted into 5-hydroxytryptophan by a hydroxylase that is specific for L-tryptophan. This reaction requires the participation of oxygen. 5-Hydroxytryptophan is then converted into 5-hydroxytryptamine (serotonin) by a decarboxylase. In mammals, inactivation of serotonin occurs by the enzyme monoamine oxidase, a reaction that requires oxygen. Such a mechanism of inactivation has not been well-studied in invertebrates. Three major methods have been employed in the identification of serotonin in tissues. Early in its discovery the bioassay procedure was used for the indoleamine identification (Amin *et al.*, 1954). Subsequently, spectrophotofluorometric techniques, mass spectroscopy and fluorescence microscopy were used. In our laboratory, using a bioassay procedure (Mansour, 1964; Mansour *et al.*, 1957) and a spectrophotofluorometric method (Mansour and Stone, 1970), we reported that the indoleamine is present in the liver fluke *F. hepatica*. Andreini *et al.* (1970) published data suggesting that the indoleamine in the fluke is spectrophotofluorometrically different from serotonin. Chou *et al.* (1972) were unable to demonstrate the presence of the indoleamine in the liver fluke. This may be due to their use of unstarved parasites which contain large amounts of caecal contents including digested proteins. Lysine, a likely product of digestion, was reported by Tomosky-Sykes et al. (1977) to give a spectrophotofluorometric signal similar to that of serotonin. Obviously, many of these methods have their drawbacks, either in sensitivity or in selectivity. Serotonin has been reported to be present in other species of trematodes and cestodes. These include *S. mansoni, S. haematobium, S. japonicum, H. diminuta* and *H. nana* (Chou et al., 1972; Lee et al., 1978).

The biosynthetic machinery for the production of serotonin has been investigated ever since the effects of serotonin were shown on parasites. We reported that intact *F. hepatica* were capable of synthesizing the indoleamine from 5-hydroxytryptophan but not from tryptophan (Mansour, 1964; Mansour et al., 1957; Mansour and Stone, 1970). This finding indicated that, although the decarboxylase responsible for the synthesis of serotonin from 5-hydroxytryptophan was present, the hydroxylase that converts tryptophan into 5-hydroxytryptophan was absent. Since these organisms live in an environment that is predominantly anaerobic, it was not too surprising to find that the oxidative enzyme system tryptophan hydroxylase was not functional in these organisms. The presence of 5-hydroxytryptophan decarboxylase and its activation by pyridoxal phosphate was further demonstrated in cell-free extracts from whole flukes (Mansour, 1964; Mansour et al., 1957; Mansour and Stone, 1970). Subsequent to our findings with *F. hepatica*, similar data were reported in *S. mansoni* (Bennett and Bueding, 1973; Catto, 1981). The question of how these trematodes synthesize serotonin from the essential amino acid tryptophan will have to be resolved. The possibility exists that 5-hydroxytryptophan or serotonin itself may be provided to the parasite by the host. It has been reported that several essential constituents of schistosomes are not synthesized *de novo* by the parasite and are presumably taken from the host. These include adenine, steroids and fatty acids (Meyer et al., 1970; Senft, 1970; Jaffe et al., 1971). A good source of the indoleamine is the blood platelets in the hepatic vein where the flukes are known to feed. *Schistosoma mansoni* was reported to have an active system for the uptake of serotonin (Bennett and Bueding, 1973). Studies on the uptake mechanism for serotonin and the effect of chemical agents on this process are of great importance in a strategy for selecting new antiparasitic agents.

Except in *Schistosoma mansoni*, the inactivating enzyme systems for serotonin have not been studied. Nimmo-Smith and Raison (1968) assayed monoamine oxidase activity in *S. mansoni*. Male worms had much higher specific enzyme activity than females. Tryptamine appears to be a better substrate than serotonin. Indirect evidence reported by these authors suggests that there is more than one monoamine oxidase. More characterization of the enzyme is needed before the natural substrate can be proposed.

### VII. SEROTONIN AND REGULATION OF METABOLISM IN PARASITES

In many metazoan organisms regulation of metabolism is known to occur at two levels: through hormonal action and through allosteric kinetics of regulatory enzymes. The evidence available suggests that these parasites are influenced by mammalian hormones. For example, thyroxine (Dobson, 1966a) and progesterone (Dobson, 1966b) were reported to affect the size and number of parasites of *Amplicaecum robertsi* in mice infections and cortisol was reported to influence larval development of *Ostertagia circumcinta* in sheep (Dunsmore, 1961). On the other hand, Hutton et al. (1972) examined the direct effect of several mammalian hormones on whole liver flukes maintained *in vitro* as well as on fluke homogenates. No significant effects on the metabolism of this parasite were observed with thyroxine, epinephrine, norepinephrine, progesterone, testosterone and hydrocortisone. Although the information on this topic is scanty, the evidence we have from our work on the carbohydrate metabolism of trematodes indicates that control of metabolism in parasites may be mediated through chemical agents that are different from mammalian host hormones. In these parasites the adenylate cyclase system has been conserved and adenosine cyclic 3′,5′-monophosphate (cyclic AMP) acts as a second messenger for regulation of carbohydrate metabolism (Mansour, 1959a,b, 1967, 1970; Mansour et al., 1960; Higashi et al., 1973; Abrahams et al., 1976; Gentleman et al., 1976; Walter et al., 1974). Superimposed on these two levels of regulation is the control of several rate-limiting enzymes that are crucial for metabolism (Monod, 1966). Most of these enzymes have allosteric kinetics and act as regulators of metabolic pathways in addition to their function as catalysts of different reactions.

In my laboratory we have used the liver fluke *F. hepatica* as a model system to study the regulation of carbohydrate metabolism in trematodes. *Fasciola hepatica* is an anaerobic organism that metabolizes glucose at a high rate and converts it primarily into volatile fatty acids and carbon dioxide (Mansour, 1959c). A small amount of lactic acid is also formed. Serotonin, when added to fluke cultures, causes a marked increase in lactic acid production (Mansour, 1959c). This is accompanied by an increase in glycogen breakdown when glucose is omitted from the fluke medium. When glucose was made available in the medium, the uptake of this sugar by the parasites was markedly increased in the presence of serotonin. The indoleamine activates three enzymes that are involved in the regulation of carbohydrate metabolism: glycogen phosphorylase, phosphofructokinase and adenylate cyclase (Mansour, 1959b; Mansour et al., 1960). Thus, serotonin has an effect on the parasite similar to that seen with epinephrine in the host carbohydrate metabolism. Addition of epinephrine or norepinephrine did not affect the metabolism or motility. Because serotonin increases the motility of these parasites, we asked whether the effects on these enzyme systems are secondary to the increase in motility. This does not appear to be the case because serotonin, added directly to cell-free extracts from the flukes, activates glucose uptake, lactic acid production, phosphofructokinase and adenylate cyclase. Lahoud et al. (1971), using pyruvate as the substrate with intact F. hepatica, reported that serotonin causes an increase in the production of acetate, one of the metabolic products of the parasite. Whether the effect on acetate production is directly related to the action of serotonin or secondary to the effect of the indoleamine on motility has yet to be investigated.

TABLE 2
Role of serotonin in metabolism of parasitic helminths

| Parasite | Metabolic process | Serotonin action | References |
|---|---|---|---|
| Fasciola hepatica | Glycogenolysis | Increase | Mansour (1959b) |
|  | Glycolysis | Increase | Mansour (1959b) |
|  | Glycogen phosphorylase | Activate | Mansour et al. (1960) |
|  | Phosphofructokinase | Activate | Mansour and Mansour (1962) |
|  |  |  | Stone and Mansour (1967a) |
|  |  |  | Stone and Mansour (1967b) |
|  | Protein kinase | Activate | Gentleman et al. (1976) |
|  | Adenylate cyclase | Activate | Mansour et al. (1960) |
| Schistosoma mansoni | Glycogenolysis | Increase | Hillman et al. (1974) |
|  |  |  | Bueding and Fisher (1982) |
|  | Glycolysis | Increase | Hillman et al. (1974) |
|  |  |  | Bueding and Fisher (1982) |
|  | Protein kinase | Activate | Higashi et al. (1973) |
|  | Adenylate cyclase | Activate | Higashi et al. (1973) |
| Hymenolepis diminuta | Glycogenolysis | Increase | Mettrick et al. (1981) |
|  | Glycolysis | Increase | Mettrick et al. (1981) |
| Ascaris suum | Glycogenolysis | Increase | Donahue et al. (1981) |
|  | Glycolysis | Increase | Donahue et al. (1981) |
|  | Adenylate cyclase | Activate | Donahue et al. (1981) |

The effect of serotonin on glycolysis and on glycolytic enzymes does not seem to be restricted to *F. hepatica*. Subsequent to our finding, several investigators reported effects by serotonin similar to those we found in *F. hepatica* (Table 2). The indoleamine and its analogues methysergide and dihydroergotamine increase lactic acid production by schistosomes (Hillman et al., 1974). Recently Mettrick et al. (1981) reported stimulation of glycolysis by serotonin in *Hymenolepis diminuta*. Serotonin also activates adenylate cyclase from *S. mansoni* (Higashi et al., 1973; Kasschau and Mansour, 1982a,b). The data obtained thus far suggest that in the liver fluke and schistosomes serotonin has an effect on carbohydrate metabolism and on adenylate cyclase similar to that of epinephrine in mammalian hosts' muscle and liver.

## VIII. $^{31}$P NUCLEAR MAGNETIC RESONANCE SPECTROSCOPY OF *Fasciola* AND THE ACTION OF SEROTONIN

One of the most important methods that has been introduced recently in the field of metabolic regulation is nuclear magnetic resonance (NMR) spectroscopy. This technique relies on the fact that certain atomic nuclei have intrinsic magnetism. These nuclei include the phosphorus-31 nucleus ($^{31}$P) which is the nucleus of all phosphorus atoms that are included in natural metabolites such as adenosine triphosphate (ATP) and other related organic and inorganic phosphate compounds. Phosphorus atoms in a specific molecular site give a characteristic signal in an NMR spectrum. A good example that is relevant to cellular metabolism can be seen in the signals of the three phosphate atoms of ATP: $\alpha$, $\beta$ and $\gamma$ phosphate (Fig. 3). Since measurement in NMR spectroscopy depends on electromagnetic radiation in the radio-frequency part of the spectrum, intact organisms can be used for *in vivo* determination of many phosphorus-containing metabolic intermediates. A typical NMR spectrum of an intact mammalian tissue detects peaks from the three phosphorus atoms of ATP: creatine phosphate, inorganic phosphate and hexose phosphates. The technique has been of great value during the last 5 years to study *in vivo* changes of different metabolites in response to hormones or to different physiological changes (Schulman et al., 1979). We have recently utilized this technique to study changes in the metabolism of intact *F. hepatica* (Mansour et al., 1982). We found that this parasite provides good-quality $^{31}$P high-resolution NMR spectra for at least 6 hours under anaerobic conditions (Fig. 3). Signals for the $\alpha$-, $\beta$- and $\gamma$-phosphates and for inorganic phosphates that were seen in mammalian tissues were identified in NMR spectra of the parasites. There are significant differences between the $^{31}$P spectrum of mammalian heart or skeletal muscle and the

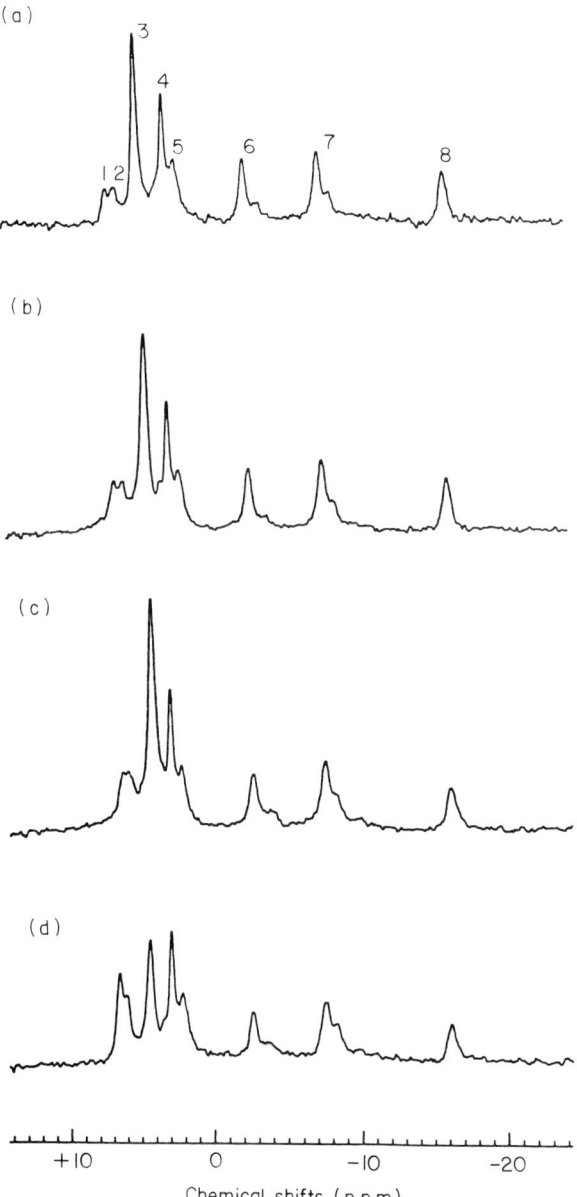

FIG. 3. The 81 MHz [31]P NMR spectrum of *Fasciola hepatica*: (a) in Krebs–Ringer solution, (b) in Krebs–Ringer solution containing 1 mM serotonin solution, collected during first 40 minutes after preparation, (c) as (b) but after 75 minutes, (d) in Krebs–Ringer solution containing 11 mM glucose to replace serotonin solution. The chemical shifts (p.p.m.) are referenced to phosphocreatine signal. The signal assignments are: 1,2, sugar phosphates; 3, inorganic phosphate; 4,5, phosphodiesters; 6,7,8, γ-, α-, β-phosphates of ATP. (Data from Mansour *et al.*, 1982, by courtesy of *Biochimica Biophysica Acta*.)

parasite. This includes the appearance of a large signal in the phosphodiester region in the fluke spectrum which was identified as L-α-glycerophosphoryl choline. There is also a distinct peak for the terminal phosphate of ADP observed just upfield of the γ-phosphate peak of ATP. This suggests that there is a significant amount of ADP present in the parasite in contrast to the findings in mammalian tissues. The most conspicuous difference from mammalian muscle tissues is the presence of large amounts of inorganic phosphates and the absence of phosphocreatine. In many invertebrates phosphoarginine is used as a reservoir of high-potential phosphoryl groups instead of phosphocreatine. The $^{31}$P spectrum showed no detectable signal for phosphoarginine. Serotonin, when added to intact parasites in the NMR spectrometer, caused an appreciable increase in the levels of sugar phosphates when the flukes were incubated in the absence of glucose. This is consistent with the known effects of serotonin in stimulating carbohydrate metabolism (see above). Additions of glucose to the incubation medium caused a marked increase in the sugar phosphate signals at the expense of the free phosphate signal. These experiments illustrate the potential of the $^{31}$P NMR technique for studies on the metabolism of parasites, and they show the effect of serotonin on the metabolism of the undisturbed parasites. Further experimentation is forthcoming, using probes other than $^{31}$P, such as $^{13}$C, to solve problems relevant to the physiology, biochemistry and pharmacology of serotonin.

## IX. Adenosine Cyclic 3′,5′-Monophosphate and Cell Regulation in Parasitic Worms

### A. Conservation of the Regulatory Role of Adenosine Cyclic 3′,5′-Monophosphate

Cyclic AMP plays a central role in cell regulation and communication. Many of the functions of cyclic 3′,5′-AMP as a cellular regulator have been conserved in a diverse variety of prokaryotic and eukaryotic cells. One function that has been conserved is the effect of cyclic AMP in promoting glycolysis and energy charge. In bacteria, cyclic AMP does it through catabolite repression and appears to be directly involved in transcriptional regulation (Rickenberg, 1974). In higher animal tissues, Sutherland and Robison (1966) showed that cyclic AMP acts as a second messenger, coupling hormone binding at the outer membrane surface to a variety of intracellular processes. Activation of protein kinase is one of these processes that could lead to a variety of biochemical effects. In lower eukaryotic cells, such as *Neurospora*, fungi and the slime moulds, the cyclic nucleotide has been implicated in metabolic and behavioural processes (Gerisch and Wick, 1975;

Pall, 1981). Adenylate cyclase in the slime mould *Physarum polycephalum* appears to have a mechanism of activation (Smith and Mansour, 1978) that is not coupled to a hormone-activated receptor.

In view of the importance of cyclic AMP in the regulation of metabolism, the enzymes involved in controlling its level in parasites deserve to be studied thoroughly. Such studies may pinpoint important differences between the parasite and the host revealing sites in the parasite which may be amenable to pharmacological manipulation in a strategy for the selection of new chemotherapeutic agents. More importantly, these studies may contribute to our understanding of this important regulatory mechanism. We have used *F. hepatica* as a prototype trematode to study the nature of adenylate cyclase. *Fasciola hepatica* probably has the most active adenylate cyclase in nature when the enzyme is measured after it is fully activated with NaF. The fluke cyclase seems to be related to motility and carbohydrate metabolism of the parasite. The fluke, therefore, may be an ideal model system to study the mechanism of action of a hormone like serotonin that may have double roles: as a putative transmitter and as a metabolic regulator.

### B. ROLE OF NEUROTRANSMITTERS IN DEVELOPMENT OF EUKARYOTES

It has been suggested that neurotransmitters such as norepinephrine, serotonin, dopamine and acetylcholine play an important role in the regulation of developmental processes including cell differentiation and morphogenesis (McMahon, 1974). These effects are presumably mediated through cyclic nucleotides. Intracellular concentrations of cyclic nucleotides, according to the hypothesis, can regulate the different morphogenic changes during development. Cyclic AMP has been implicated as a biochemical messenger in the development and differentiation of eukaryotic cells. Cyclic AMP induces changes in the morphology of fibroblasts, astrocytes and neurons (Hsie and Puck, 1971; Shapiro, 1973; Roisen *et al.*, 1972; Kirkland and Burton, 1972). Added cyclic AMP was reported to stimulate proliferation in certain cultured cell lines (Whitfield *et al.*, 1973; MacManus *et al.*, 1976). The involvement of neurotransmitters in the early development of metazoan organisms such as the sea urchin embryo has been reported by Buznikov and his associates (Buznikov *et al.*, 1968, 1970) as well as by Gustafson and Toneby (1970). Concentrations of serotonin and the catecholamines change with both gastrulation and cell division. Although the sensitivity of sea-urchin adenylate cyclase to these neurotransmitters has not as yet been established, the enzyme in some other invertebrates is sensitive to these neurotransmitters (Cedar and Schwartz, 1972; Nathanson, 1978; Nathanson and Greengard, 1974; Gentleman and Mansour, 1977). According to Buznikov's data, there is a sequence of changes of concentrations of serotonin, epinephrine and acetylcholine occurring in the early sea urchin embryo that is synchronized with the different phases of mitosis. According to McMahon's (1974) calculations of the projected concentrations of cyclic AMP in response to the neurotransmitter levels, the cyclic nucleotide could be responsible for producing the physiological and morphological changes of the cell during mitosis.

In addition to their possible role in development, the neurotransmitters have been implicated in the control of metabolism of eukaryotic cellular organisms that have no nervous system. Serotonin and epinephrine appear to stimulate the carbohydrate metabolism of *Tetrahymena pisiformis* (Blum, 1970). Both amines activate adenylate cylase of *Tetrahymena* (Rosenzweig and Kindler, 1972) and *Euglena* (Keirns et al., 1973). The effects of both serotonin and epinephrine on the metabolism of these organisms could be mediated through cyclic AMP.

C. CHANGES IN SEROTONIN-ACTIVATED ADENYLATE CYCLASE DURING DEVELOPMENT OF *Schistosoma mansoni*

Different developmental stages of *S. mansoni* can be obtained in experimentally infected laboratory animals and *in vitro* cultures. Because of the possible involvement of serotonin and cyclic AMP in the developmental process of eukaryotes, we examined the question whether adenylate cyclase in a trematode varies at different developmental stages. Recent experiments by Kasschau and Mansour (1982a,b) examined adenylate cyclase in adults and cercariae of *S. mansoni*. Total enzyme activity in membrane particles of adults can be obtained in the presence of saturating concentrations of NaF. As with *F. hepatica*, membrane particles of *S. mansoni* are rich in adenylate cyclase activity. On the other hand, the activity of the enzyme in the free-living cercariae of the parasite was one-tenth that of the adult (Fig. 4). The responsiveness of the particulate enzyme to serotonin was also examined in both developmental stages. Serotonin in the presence of guanosine triphosphate (GTP, a nucleotide that is important for the activation of all cyclases by different hormones) caused almost no activation of the cercarial enzyme, whereas it markedly stimulated adult adenylate cyclase (Fig. 5). Since there is such an obvious difference in total adenylate cyclase activity and serotonin-stimulated activity between adults and cercariae, the question was how rapidly does the enzyme system develop from low to high responsiveness to serotonin? We therefore studied the enzyme activity in the early intermediate stages of development from cercariae to adults. We examined this question utilizing schistosomules that were cultured at 36°C in a 5% $CO_2$ atmosphere, according to Basch's (1981) procedure. Although total activity of the enzyme (expressed as activity in the presence of NaF) was increased in the early stages of the schistosomules, the most significant change in the adenylate cyclase activity

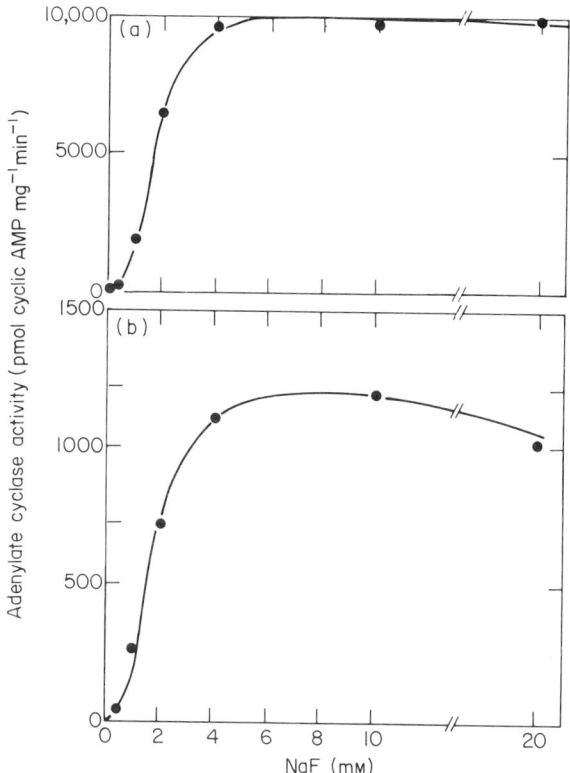

FIG. 4. Activity of adenylate cyclase in (a) adults and (b) cercariae of *Schistosoma mansoni* in the presence of different concentrations of NaF. Maximal enzyme activity can be seen in the presence of saturating concentrations of NaF. (Data from Kasschau and Mansour, 1982a, by courtesy of *Molecular and Biochemical Parasitology*.)

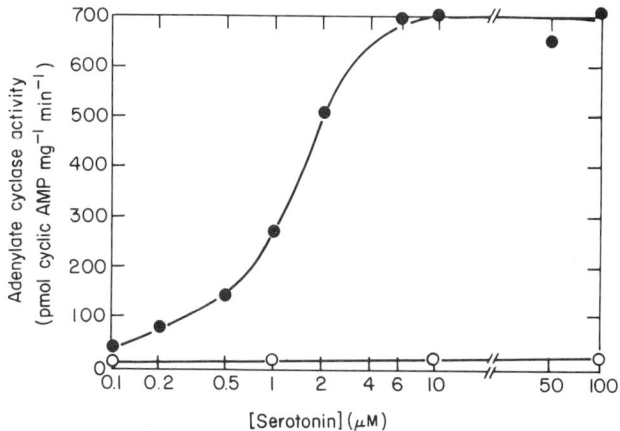

FIG. 5. Serotonin activation of adenylate cyclase from adult (●) and cercariae (○) particles. (Data from Kasschau and Mansour, 1982, by courtesy of *Nature*.)

was the marked increase in serotonin-activated enzyme on the second and fourth days. Adenylate cyclase activity of schistosomules that were maintained in culture for up to 10 days showed no further increase in sensitivity to serotonin activation. It is interesting to note from the results that cercariae, which have a very brief span of free-living life with growth and development at a standstill, have low total cyclase activity and almost no responsiveness to serotonin by the enzyme. On the other hand, adenylate cyclase becomes higher and its responsiveness to serotonin is markedly increased in the early stages of schistosomule development. The adult parasite shows an even greater increase in both total and serotonin-stimulated adenylate cyclase activity. Thus, as the parasite develops to maturity, it gains more adenylate cyclase that is sensitive to serotonin activation. The results may reflect an increase in the need for cyclic AMP early in the development so as to achieve full parasite maturation. The development of serotonin-sensitive cyclase comes at a time when the schistosomule is well-established in the mammalian host, where it encounters serotonin, presumably its natural activator.

Amplification of hormone receptors during differentiation in planarians has been reported by Csaba and Magda (1980). In these flatworms glucose uptake is enhanced by the catecholamines epinephrine and dopamine. Planarians in the early stage of regeneration, when treated with dopamine or epinephrine, showed great amplification of hormone action on glucose uptake.

Our findings with the increased serotonin-activated cyclase in the schistosomule and the results obtained in planarians and other invertebrates raise two main questions concerning amplification of hormone receptor action during development. The first is whether this phenomenon is due to a change in the kinetics of binding of the hormone to the receptor or to an increase in the coupling mechanism of the receptors to the cyclase or to other cell metabolic machinery. The second is whether the amplification in serotonin receptors is due to an increase in the number of hormone receptors resulting from an increase in their synthesis or a decrease in their degradation.

## X. Hormone-Activated Adenylate Cyclase in Higher Organisms

How does an extracellular hormone-like molecule communicate with the intracellular metabolic machinery of the organism? Extensive studies in mammals have indicated that many hormone signals are communicated across cell membranes by a receptor-linked adenylate cyclase. The following components of the membranous cyclase system were studied in our laboratory

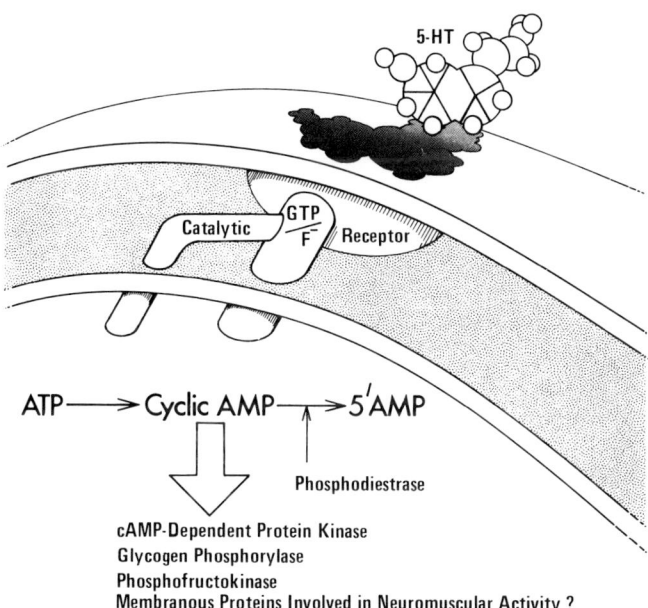

FIG. 6. Diagram depicting the different components of the adenylate cyclase system in the plasma membrane particles. The proposed physiological effects of a hormone (serotonin (5-HT) in this diagram) is initiated through binding of the indoleamine to the receptor that is part of a membranous adenylate cyclase. Such a binding activates the catalytic component of the enzyme through the participation of the regulatory protein (GTP/F$^-$). The catalytic component of the system catalyses the synthesis of cyclic AMP from ATP. Cyclic AMP (the second messenger) activates enzyme systems such as protein kinase, glycogen phosphorylase and phosphofructokinase, and increases phosphorylation of membranous proteins that may be involved in regulation of neuromuscular activity. (After Mansour, 1979, by courtesy of *Science*.)

and are shown diagrammatically in the accompanying figure depicting serotonin as the activator (Fig. 6): (1) the receptor that binds the hormone; (2) the regulatory component of the cyclase system (GTP/F$^-$) is a membranous protein that binds GTP which has the function of communicating the hormonal message from the receptor to the cyclase catalytic component. Activation of cyclase by serotonin depends on binding of GTP to this protein. Although the protein is highly specific for the guanine nucleotide, synthetic analogues of GTP which are poorly hydrolysed, such as guanylyl imidophosphate (GppNHp) and guanosine-5'-(3-*O*-thio)triphosphate (GTPγS), activate adenylate cyclase in the absence of serotonin; (3) the catalytic component of the cyclase is the enzyme that catalyses the cyclase reaction:

$$\text{ATP} \xrightarrow{Mg^{2+}} \text{Cyclic AMP} + \text{pyrophosphate}$$

This is the most highly conserved protein in the cyclase system. It has been shown to occur in bacteria as well as in higher organisms. It is the most difficult component to handle because of its instability; (4) cyclic AMP-dependent protein kinase (PK) that catalyses the following reaction:

$$\text{Protein} + \text{ATP} \xrightarrow{Mg^{2+}} \text{Protein phosphate} + \text{ADP}.$$

This enzyme has been reported in both vertebrates and invertebrates (Walsh et al., 1968; Greengard and Kuo, 1979), and more recently in eukaryotic microorganisms (Powers and Pall, 1980; Trevillyans and Pall, 1982). Protein kinase is composed of two catalytic (C) subunits and a regulatory dimer ($R_2$) to form a tetrameric holoenzyme ($C_2R_2$). When the catalytic (C) subunit is complexed with the regulatory dimer ($R_2$), the phosphotransferase activity of (C) is inhibited. Binding of cyclic AMP to the regulatory dimer causes dissociation of the (C) subunit from the regulatory dimer and activation of the PK activity.

$$R_2C_2 \text{ (inactive)} + 4 \text{ cyclic AMP} \longrightarrow (R \cdot 2 \text{ cyclic AMP})_2 + 2(C) \text{ (active)}$$

Evidence that the function of cyclic AMP in cell regulation in different eukaryotic cells has been evolutionarily conserved is seen from the fact that this regulatory subunit (R) has been isolated even from eukaryotic microorganisms such as *Neurospora* (Trevillyans and Pall, 1982).

## XI. SEROTONIN-ACTIVATED ADENYLATE CYCLASE IN THE LIVER FLUKE *Fasciola hepatica*

Adenylate cyclase in the liver fluke is a membranous enzyme with molecular functions that appear to have the same functions as those described above for higher organisms (Northup and Mansour, 1978a,b). Although basal enzyme activity (without serotonin or other activators) is low, it is markedly activated in the presence of serotonin. As in mammals, activation by the hormone is dependent on the presence of GTP. We have studied the serotonin receptors in the flukes in two different ways: first, by measuring (agonists* and antagonists); and secondly, by studying the ability of the receptors to directly bind serotonin and its analogues. Ultimately, we

* An agonist is a drug that has affinity for and stimulates physiological activity at cell receptors normally stimulated by naturally occurring substances.

wanted to compare the serotonin receptors from the flukes with those that have been studied from mammals.

### A. SEROTONIN RECEPTORS IN *Fasciola hepatica*

We confirmed the specificity of serotonin activation of the fluke cyclase by examining many analogues of serotonin on adenylate cyclase activity in membrane particles (Northup and Mansour, 1978a). Serotonin appears to be the best indoleamine to activate the adenylate cyclase (Fig. 7). Simple substitutions on the 5-hydroxytryptamine molecule decrease the efficacies of analogues as agonists of the serotonin receptor. Compounds lacking the 5-hydroxy group such as tryptamine, *N*-methyltryptamine, *N,N*-dimethyltryptamine and gramine (3-(dimethylaminomethyl)indole) are much poorer agonists and have decreased affinity compared with their 5-hydroxy congeners.

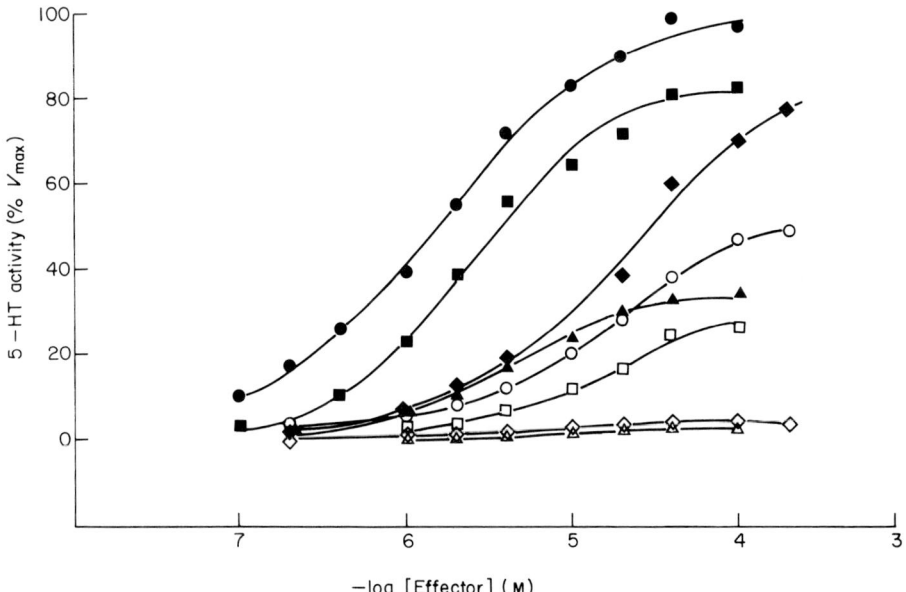

FIG. 7. Relative effect of serotonin analogues in activating adenylate cyclase in the liver fluke. Velocity was normalized as the fraction of maximum velocity ($V_{max}$) for serotonin (5-HT). Abscissa shows concentrations of different effectors. ●, 5-Hydroxytryptamine; ○, tryptamine; ▲, 5-hydroxy-*N,N*-dimethyltryptamine; △, *N,N*-dimethyltryptamine; ■, *N*-methyltryptamine; □, 5-hydroxy-*N*-methyltryptamine; ◆, 5-methoxytryptamine; ◇, 5-methoxy-*N,N*-dimethyltryptamine. (Data from Northup and Mansour, 1978, by courtesy of *Molecular Pharmacology*.)

Similarly, substitution of the 5-OH group by 5-methoxy as in 5-methoxytryptamine and 5-methoxy-*N*,*N*-dimethyltryptamine results in a poor agonist effect. Substitution on the ethylamine, as in the case of 5-hydroxy-*N*-methyltryptamime and *N*-methyltryptamine, causes reduction in both efficacy and affinity to the receptors. *N*,*N*-Dimethyltryptamine that lacks both 5-hydroxy and free amino groups was almost devoid of activity as an agonist.

D-Lysergic acid diethylamide is chemically and pharmacologically related to serotonin. In some isolated tissues it acts as an antagonist or a partial agonist. A partial agonist is an analogue which, although it mimics the effect of the full agonist, does not do so maximally. Partial agonists can also antagonize when the system is activated by a full agonist. I reported in the 1950s that, like serotonin, D-LSD stimulates motility (Mansour, 1957) and carbohydrate metabolism (Mansour, 1959b) of *F. hepatica*. It therefore appears to act as a serotonin agonist on intact flukes. D-LSD also mimics the effect of serotonin on activation of fluke adenylate cyclase. Its maximal effect is, however, only 25% that of serotonin. It also antagonizes the activation of the cyclase by serotonin, lowering it to the stimulatory level of LSD alone. LSD is therefore a partial agonist. LSD has a greater affinity to the receptors than serotonin itself (Northup and Mansour, 1978a). D-LSD was the most potent derivative, with a half-maximal activation at 46 nM compared to 2·1 μM for serotonin. Any substitution other than the diethyl on the amide nitrogen decreased the potency of the derivative. D-LSD and its bromo derivative, 2-bromo-LSD (Fig. 1), act as potent antagonists of serotonin activation. The effect of LSD is stereo-specific. D-LSD is much more potent than L-LSD, both as an activator and as an inhibitor of serotonin. It is interesting to note here that the hallucinogenic effect of LSD is also stereospecific for the D isomer. 2-Bromo-LSD, which is not an agonist, is the most potent antagonist of serotonin-activated adenylate cyclase, with half-maximal inhibition at 28 nM. It should be noted here that 2-bromo-LSD inhibits the motility of the flukes (see above). The antagonism of either LSD or 2-bromo-LSD is competitive with respect to serotonin. That all these agonists and antagonists interact at the same site, the serotonin receptor, was confirmed from kinetic studies (Northup and Mansour, 1978a).

In addition to studies on the receptors using the adenylate cyclase activity as a marker, we have studied the properties of the receptors to bind different ligands (McNall and Mansour, 1983). For these experiments we used D-[³H]LSD. We chose LSD over serotonin because it can be obtained with a high specific radioactivity and because it has a greater affinity to the receptors than serotonin. In order to measure specific binding to serotonin receptors we have used low concentrations of [³H]LSD (10 nM) and subtracted from the binding figure a blank containing the same concentration of [³H]LSD with an excess of unlabelled LSD. This allowed us to eliminate the non-specific binding of the radioligand to sites other than the specific serotonin receptors. Specific binding of radioactively labelled LSD has all the characteristics of binding to physiological receptors: specificity, stereospecificity and saturability. Mathematical analysis of specific binding data showed at least one set of sites with a high affinity to LSD with half-maximal binding at 39 nM. This can be compared with the binding constant of 46 nM when the cyclase assay was used (see above). Using the [$^3$H]LSD binding method we measured the effectiveness of different serotonin agonists and antagonists in displacing [$^3$H]LSD binding. The concentration of each compound which caused 50% inhibition of specific binding of [$^3$H]LSD was determined. The data for different derivatives obtained by the radioligand binding method corresponds well with data obtained using the adenylate cyclase assay procedure. Binding data obtained with fluke membrane fractions was compared with data obtained by Peroutka et al. (1981) on mammalian brain receptors using identical techniques. According to these authors, the brain has two distinct types of receptors that have been characterized by their binding to different agonists and antagonists of serotonin. They were designated 5-HT$_1$ and 5-HT$_2$. Striking differences between the fluke receptors and both mammalian receptors were observed. For example, the relative affinities of a series of serotonin antagonists to the fluke receptors were LSD = 2-bromo-LSD > methiothepin > cyproheptadine = metergoline > mianserin = serotonin > spiroperidol > ketanserin (McNall and Mansour, 1983; 1984a,b). In the mammalian system, of the compounds tested, serotonin has the highest affinity for the 5-HT$_1$ receptors whereas spiroperidol and ketanserin have the highest affinity for the 5-HT$_2$ receptors. Thus, the fluke has serotonin receptors that are different from the mammalian receptor. Differences between mammalian and fluke serotonin receptors may provide a new site for the selection of new chemotherapeutic agents against trematode infection.

### B. DESENSITIZATION OF SEROTONIN RECEPTORS IN *Fasciola hepatica*

Continued exposure of cells or tissues containing receptors that are linked to adenylate cyclase to effector hormones results in refractoriness of the cell or the tissue to hormone action (Lefkowitz et al., 1980). A typical effect is characterized by a rapid initial increase in intracellular concentration of cyclic AMP. Continued exposure of the cell to the hormone results in the cyclic AMP concentrations returning to near control values and failing to increase on subsequent hormonal stimulation. Teleologically, such a process might be viewed as a protective regulatory mechanism, allowing the organism to respond rapidly to the hormone and reducing the reactivity of the organism to continued stimulation. Hormone-induced desensitization appears to be a general phenomenon that involves cyclase-coupled systems as well as hormones not coupled to cyclase.

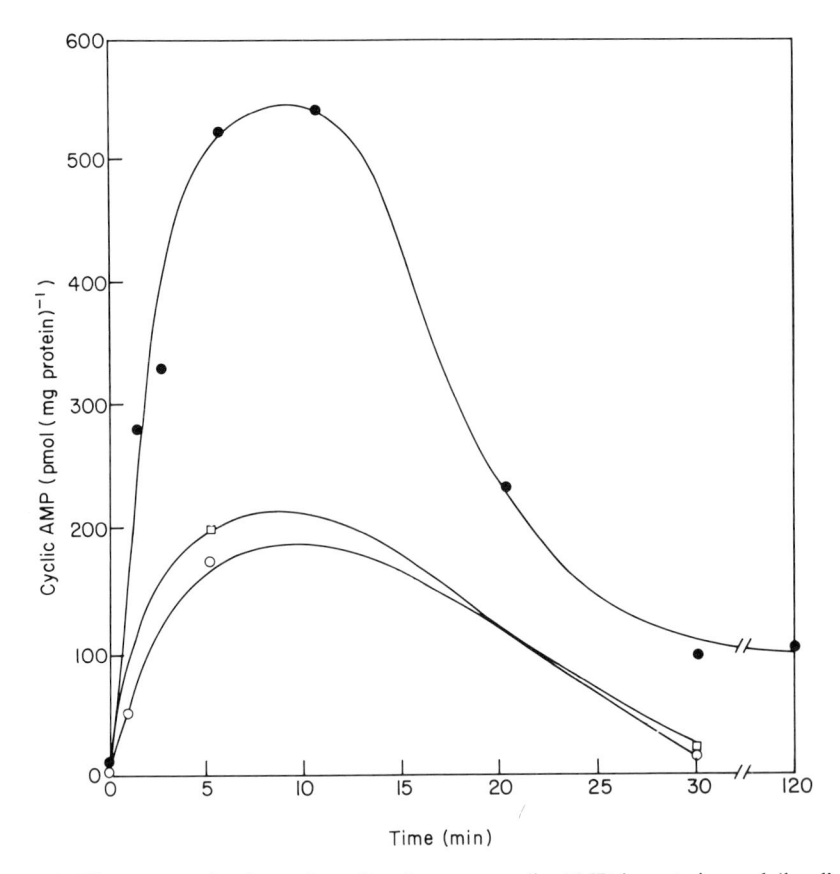

FIG. 8. Time course for formation of endogenous cyclic AMP in anterior end 'head', posterior part 'tail' of the parasite, and whole flukes. In this representative experiment, 20 heads (●), six tails (□) or six whole flukes (○) were incubated with 1 mM serotonin for the indicated times and then assayed for cyclic AMP. Experimental conditions were as described by Abrahams *et al.* (1976). (By courtesy of *Molecular Pharmacology*.)

A similar phenomenon was observed in our laboratory in *Fasciola hepatica* when it was exposed to serotonin (Abrahams *et al.*, 1976). Exposure of intact flukes or dissected portions of these parasites (heads or tails) to a fixed concentration of serotonin results in an immediate accumulation of cyclic AMP in the tissues of the organism (Fig. 8). Maximal accumulation of the cyclic AMP is attained within 5 to 10 minutes. After such an increase there was a rapid decrease to a plateau level slightly above control activity.

Although the decrease in cyclic AMP concentrations happens very soon after addition of serotonin, the physiological effects of the indoleamine on motility and carbohydrate metabolism persist for many hours after such a decrease. The decrease in cyclic AMP levels was not due to depletion of serotonin from the incubation medium. The desensitization process shown in the intact organisms can be reversed after overnight incubation in hormone-free medium. Although all our early experiments on the process of desensitization were done on intact organisms, we have been able to demonstrate the process in isolated membrane particles. We found that membrane particles prepared from intact flukes that were desensitized to serotonin show fewer specific binding sites. Adenylate cyclase in these membrane particles showed refractoriness to activation by serotonin and by LSD. Although our recent work illustrates that the serotonin receptors in the fluke can be desensitized like other hormone receptors, elucidating the molecular events involved in such a process awaits further experimentation.

### C. THE GUANOSINE TRIPHOSPHATE REGULATORY PROTEIN

In almost all receptor-linked adenylate cyclase systems, a membranous protein that binds GTP is required. That this is true in the liver fluke was first indicated by our finding that the guanine nucleotide or one of its analogues is needed for either serotonin or LSD activation of adenylate cyclase (Northup and Mansour, 1978b). The requirement for GTP is highly specific. ATP, UTP, CTP and xanthine triphosphate cannot be substituted for GTP, although inosine-5'-triphosphate was partially effective. Guanylyl imidophosphate (GppNHp), guanylyl methylene phosphate (GppCHp), $Cr^{+3}$-GTP and guanosine-5'-(3-$O$-thio)triphosphate (GTP$\gamma$S) all activated fluke adenylate cyclase in the absence of serotonin. They can also replace GTP to show activation of cyclase by serotonin. More direct evidence about the guanine nucleotide involvement in the liver fluke cyclase system was obtained from studies by Renart *et al.* (1979) in my laboratory. Utilizing a solubilized cyclase from the flukes, it was possible to separate a cell fraction that contains the guanine nucleotide component and reconstitute it with a fraction that has the catalytic activity. This technique makes it possible to assay each component of the serotonin-activated adenylate cyclase separately and is helpful for purification of the GTP regulatory protein.

### D. PROTEIN KINASE IN *Fasciola hepatica* ACTIVATED BY SEROTONIN

Many of the effects of hormones that are linked to adenylate cyclase are mediated through activation of protein kinases which phosphorylate specific proteins. In mammals the increase in glycogenolysis following administration of epinephrine offers a classical example of such a control mechanism. In this case, the enzyme involved is glycogen phosphorylase which is activated through a cascade of phosphorylation reactions initiated by cyclic AMP-dependent protein kinase. Our studies (Gentleman et al., 1976) established the presence of a protein kinase both in the membrane particulate and the supernatant fractions of *Fasciola hepatica*. Enzyme activity was 2–5 times greater in the anterior end ('head') of the fluke than in the posterior end. Cyclic AMP was very effective in activating the fluke protein kinase. Half-maximal activation of the enzyme either in the particles or in the soluble fraction can occur in the presence of the low concentration of cyclic AMP that can be produced in the fluke following activation of cyclase by serotonin. Incubation of the flukes with serotonin activated the protein kinase both in the particles and in the soluble fractions. Communication between the serotonin receptor and the cyclic AMP-dependent protein kinase was further established by the finding that LSD, under conditions that favour an antagonistic effect on the adenylate cyclase system in the flukes, was shown to antagonize serotonin activation of the protein kinase (Gentleman et al., 1976).

When extracts from the liver fluke were incubated with serotonin and [$\gamma$-$^{32}$P]ATP, several proteins were phosphorylated, suggesting that there are many substrates available for the protein kinase reaction. Although we have not as yet identified all these proteins, there are two enzymes that appear to be good candidates for this phosphorylation reaction. The first is the enzyme glycogen phosphorylase. We reported before that incubation of flukes with serotonin is accompanied by an increase in glycogenolysis and activation of glycogen phosphorylase (Mansour et al., 1960). It has already been shown that activation of the same enzyme in other tissues by catecholamines is accompanied by a conversion from the inactive dephosphorylated form into the active phosphorylated form. The second enzyme is phosphofructokinase, which was reported by us to be activated in response to serotonin and whose activation appears to be dependent on ATP and cyclic AMP (Mansour and Mansour, 1962). Our direct evidence that phosphofructokinase is phosphorylated has only been confirmed recently on the pure mammalian enzyme, which was shown to be phosphorylated by cyclic AMP-dependent protein kinase (Hofer and Furst, 1975). There are many other proteins that we found to be phosphorylated in response to the effect of serotonin on intact flukes or following incubation of the homogenates with ATP and cyclic AMP. The nature of these proteins should be determined in order to better understand the molecular events involved in the action of serotonin in these parasites.

E. CYCLIC 3',5'-NUCLEOTIDE PHOSPHODIESTERASE

The concentration of cyclic 3',5'-AMP—the second messenger—is determined by a balance between synthesis by adenylate cyclase and degradation of the cyclic nucleotide. Cyclic 3',5'-nucleotide phosphodiesterase is the only enzyme known to hydrolyse the cyclic nucleotide to 5'-AMP (see Fig. 6). This phosphodiesterase was identified in the fluke homogenates in our laboratory (Mansour and Mansour, 1977). The enzyme is involved in the regulation of the level of cyclic AMP in these organisms. This was indicated by the demonstration that incubation of parasite heads with isobutylmethylxanthine (IBMX), a potent inhibitor of the phosphodiesterase, amplified and slightly prolonged the increase in cyclic AMP caused by serotonin (Mansour and Mansour, 1977). We asked whether the effect of conventional phosphodiesterase inhibitors on fluke motility would be similar to that of compounds that activate adenylate cyclase. Indeed, we found that many of the phosphodiesterase inhibitors increase motility of the flukes in the same way as do serotonin and its analogues (Mansour and Mansour, 1977). Although the mechanism by which cyclic AMP affects fluke motility has not been elucidated, the participation of phosphodiesterase in the regulation of cyclic AMP levels in these parasites is indicated.

## XII. Serotonin Receptors in Nematodes

Muscle of *Ascaris* sp. has always been an attractive material to scientists who are interested in the regulation of carbohydrate metabolism. Besides having a high concentration of glycogen and an abundance of glycolytic enzymes, these muscles can be conveniently separated from other internal organs of the parasite. Although serotonin did not show any stimulant effect on the motility of nematodes (see above), recent experiments suggest that there are serotonin receptors that may be linked to glycogenolytic enzymes through adenylate cyclase. Donahue *et al.* (1981) have recently reported on the utilization of isolated segments of ascarid muscle which can be perfused with biogenic amines. Epinephrine, norepinephrine, synephrine, octopamine, dopamine and histamine have no effect on glycogen synthase or glycogen phosphorylase activities. Serotonin, on the other hand, caused a significant increase in active glycogen phosphorylase and a decrease in the affininty of glycogen synthase for its substrate, glucose 6-phosphate, both of which effects are induced in mammals by epinephrine and other related catecholamines. Serotonin also caused a concomitant threefold rise in the concentration of cyclic 3',5'-AMP in these muscles. Although the results support the concept that serotonin may function as a regulatory hormone of carbohydrate metabolism in *Ascaris*, there are several gaps in our knowledge that have to be filled. Most notable are the questions concerning the source of serotonin and the effect of cyclic AMP on these enzyme systems.

## XIII. Effect of Serotonin on Glycogenolysis and Glycolysis in Mammals

Although serotonin, when given parenterally to experimental animals, is not known to have a significant effect on glycogenolysis or glycolysis, recent reports suggest that it may have an effect when applied locally to certain tissues. Quach *et al.* (1982) reported that serotonin, when added to slices from mouse cerebral cortex, stimulated glycogenolysis. This effect is concentration dependent and can be shown with other indoleamine derivatives such as tryptamine or 5-methoxytryptamine and with LSD. *N,N*-Dimethyltryptamine acted as a competitive antagonist. Serotonin was also reported to induce phosphorylation of a protein in slices from facial motor nucleus in the mammalian brain at similar concentrations to those required for the glycogenolytic response (Dolphin and Greengard, 1981). A serotonin-sensitive adenylate cyclase has been shown to be present in rat brain (Enjalbert *et al.*, 1978a,b). The response to serotonin is weak in newborn animals and almost absent from adults. These reports suggest that there may be serotonin receptors that could have a localized effect on the carbohydrate metabolism in certain areas of the brain. More work is needed to characterize these receptors and establish their relationship to glycogenolysis via cyclic AMP.

## XIV. General Summary and Conclusions

It is evident from the above review that during the last two decades a great deal of interest in investigating the action of serotonin in parasitic worms has been shown by parasitologists as well as by scientists from several other disciplines. What we have initially reported concerning the effect of serotonin on motility and carbohydrate metabolism of *F. hepatica* has been pursued on several other parasitic worms. The studies so far indicate that serotonin stimulates motility of every species tested among the phylum Platyhelminthes. The indoleamine also stimulates glycogenolysis in the few flatworm parasites that have been investigated. The information in nematodes is scanty and the role of serotonin in these parasites is still open for experimentation.

Recent biochemical investigations on *F. hepatica* and *S. mansoni* demonstrated that serotonin and related compounds utilize a common class of receptors in plasma membrane particles which I designate as 'serotonin receptors'. These receptors are linked to an adenylate cyclase that catalyses the synthesis of the second messenger, cyclic 3′,5′-AMP. Serotonin and its congeners increase the concentration of cyclic AMP in intact parasites whereas antagonists inhibit such an effect. Cyclic AMP stimulates glycogenolysis, glycolysis and some rate-limiting glycolytic enzymes. It activates a protein kinase that may be involved in activation of glycogen phosphorylase and phosphofructokinase.

Serotonin-activated adenylate cyclase in *S. mansoni* is activated early in the life of the schistosomule. The possibility is discussed that the availability of cyclic AMP through serotonin activation in these parasites may be a prelude to the development processes that take place in the parasite.

The different components of the serotonin-activated adenylate cyclase in the parasite are the same as those that have been previously described for the host. Binding characteristics of the receptors indicate that the receptors in *F. hepatica* appear to be different from those that have been described in the host. The discovery of these receptors and their differences from those in the host offer a new site which is amenable to pharmacological manipulation. The search for new agents that influence serotonin receptors in these parasites could be included in a strategy for the development of new chemotherapeutic agents against these parasites.

## Acknowledgement

Some of the material in this review was presented as part of the Heath Clark Lectures delivered at the London School of Hygiene and Tropical Medicine in October, 1981. Financial support for the Heath Clark Lectureship by the London School of Hygiene and Tropical Medicine, London University, is gratefully acknowledged by the author. Some of the experiments reviewed herein were supported by Public Health Service research grants AI 16501 and MH 23464.

## References

Abrahams, S. L., Northup, J. K. and Mansour, T. E. (1976). Adenosine cyclic 3′,5′-monophosphate in the liver fluke, *Fasciola hepatica*. I. Activation of adenylate cyclase by 5-hydroxytryptamine. *Molecular Pharmacology* **12**, 49–58.

Adaikan, P. G. and Zaman, V. (1982). *In vitro* activity of anthelmintic drugs and biologically active substances on *Breinlia booliati*. *Southeast Asian Journal of Tropical Medicine and Public Health* **13**, 257–61.

Amin, A. H., Crawford, T. B. B. and Gaddum, J. H. (1954). The distribution of substance P and 5-hydroxytryptamine in the central nervous system of the dog. *Journal of Physiology (London)* **126**, 596–618.

Andreini, G. L., Beretta, C., Faustini, R. and Gallina, G. (1970). Spectrofluorometric and chromatographic characterization of a butanol extract from *Fasciola hepatica*. *Experientia* **26**, 166–167.

Baldwin, E. (1943). An *in vitro* method for the chemotherapeutic investigation of anthelmintic potency. *Parasitology* **35**, 89–111.

Baldwin, E. and Moyle, V. (1949). A contribution to the physiology and pharmacology of *Ascaris lumbricoides* from the pig. *British Journal of Pharmacology* **4**, 145–152.

Barker, L. R., Bueding, E. and Timms, A. R. (1966). The possible role of acetylcholine in *Schistosoma mansoni*. *British Journal of Pharmacology* **26**, 656–665.

Basch, P. F. (1981). Cultivation of *Schistosoma mansoni in vitro*. I. Establishment of cultures from cercariae and development until pairing. *Journal of Parasitology* **67**, 179–185.

Beernink, K. D., Nelson, S. D. and Mansour, T. E. (1963). Effect of lysergic acid derivatives on the liver fluke *Fasciola hepatica*. *International Journal of Neuropharmacology* **2**, 105–112.

Bennett, J. L. and Bueding, E. (1973). Uptake of 5-hydroxytryptamine by *Schistosoma mansoni*. *Molecular Pharmacology* **9**, 311–319.

Bettendorf, H. (1897). Über musculatur und sinneszellen der trematoden. *Zoologische Jahrbücher. Anatomie Band* **10**, 307–358.

Blum, J. J. (1970). Biogenic amines and metabolic control in *Tetrahymena*. In "Biogenic Amines as Physiological Regulators" (J. J. Blum, ed.), pp. 95–118. Prentice-Hall, New Jersey.

Braton, T. and Hopkins, C. A. (1969). The migration of *Hymenolepis diminuta* in the rat's intestine during normal development and following surgical transplantation. *Parasitology* **59**, 891–905.

Bruckner, D. A. and Vage, M. (1974). The nervous system of larval *Schistosoma mansoni* as revealed by acetylcholinesterase staining. *Journal of Parasitology* **60**, 437–446.

Bueding, E. (1952). Acetylcholinesterase activity of *S. mansoni*. *British Journal of Pharmacology* **7**, 563–566.

Bueding, E. and Fisher, J. (1982). Metabolic requirements of schistosomes. *Journal of Parasitology* **68**, 208–212.

Buznikov, G. A., Chudakova, I. V., Berdysheva, L. V. and Vyazmina, N. M. (1968). The role of neurohumors in early embryogenesis. II. Acetylcholine and catecholamine content in developing embryos of sea urchin. *Journal of Embryology and Experimental Morphology* **20**, 119–128.

Buznikov, G. A., Kost, A. N., Kucherova, N. F., Mndzhoyan, A. L., Suvarov, M. N. and Berdysheva, L. V. (1970). The role of neurohumors in early embryogenesis. III. Pharmacological analysis of the role of neurohumors in cleavage division. *Journal of Embryology and Experimental Morphology* **23**, 549–569.

Cannon, C. E. and Mettrick, D. F. (1970). Changes in the distribution of *Hymenolepis diminuta* (Cestoda: Cyclophyllidea) within the rat intestine during prepatent development. *Canadian Journal of Zoology* **48**, 761–769.

Catto, B. A. (1981). *Schistosoma mansoni:* decarboxylation of 5-hydroxytryptophan, L-Dopa, and L-histidine in adult and larval schistosomes. *Experimental Parasitology* 51, 152–157.

Cedar, H. and Schwartz, J. H. (1972). Cyclic adenosine monophosphate in the nervous system of *Aplysia californica.* II. Effect of serotonin and dopamine. *Journal of General Physiology* 60, 570–587.

Chance, M. R. A. and Mansour, T. E. (1949). A kymographic study of the action of drugs on the liver fluke (*Fasciola hepatica*). *British Journal of Pharmacology* 4, 7–13.

Chance, M. R. A. and Mansour, T. E. (1953). A contribution to the pharmacology of movement in the liver fluke. *British Journal of Pharmacology* 8, 134–138.

Chandler, A. C. (1939). The effects of number and age of worms on development of primary and secondary infections on *Hymenolepis diminuta* in rats, and an investigation into the true nature of "premunition" in tapeworm infections. *American Journal of Hygiene* 29, 105–114.

Chavasse, C. J., Brown, M. C. and Bell, D. R. (1979). *Schistosoma mansoni*: ultrasonic detectable motor activity responses to 5-hydroxytryptamine *in vitro.* *Annals of Tropical Medicine and Parasitology* 73, 363–367.

Cho, C. H. and Mettrick, D. F. (1982). Circadian variation in the distribution of *Hymenolepis diminuta* (Cestoda) and 5-hydroxytryptamine levels in the gastrointestinal tract of the laboratory rat. *Parasitology* 84, 431–441.

Chou, T.-C. T., Bennett, J. and Bueding, E. (1972). Occurrence and concentrations of biogenic amines in trematodes. *Journal of Parasitology* 58, 1098–1111.

Croll, N. A. (1975). Indolealkylamines in coordination of nematode behavioral activities. *Canadian Journal of Zoology* 53, 894–903.

Csaba, G. and Kadar, M. (1980). Durable sensitization of hormone receptors during differentiation in regenerating planarians by treatment with homologous or analogous hormone molecules. *Experimental Cell Biology* 48, 240–244.

Dobson, C. (1966a). The effects of thiouracil and thyroxine on the host-parasite relationship of *Amplicaecum robertsi* Sprent and Mines, 1960, in the mouse. *Parasitology* 56, 425–429.

Dobson, C. (1966b). The effects of pregnancy and treatment with progesterone on the host-parasite relationship of *Amplicaecum robertsi* Sprent and Mines, 1960, in the mouse. *Parasitology* 56, 417–424.

Dolphin, A. C. and Greengard, P. (1981). Serotonin stimulates phosphorylation of protein I in the facial motor nucleus of rat brain. *Nature* 289, 76–79.

Donahue, M. J., Yacoub, N. J., Michnogg, C. A., Masaracchia, R. A., and Harris, B. G. (1981). Serotonin (5-hydroxytryptamine): a possible regulator of glycogenolysis in perfused muscle segments of *Ascaris suum.* *Biochemical and Biophysical Research Communications* 101, 112–117.

Dunsmore, J. D. (1961). Effect of whole-body irradiation and cortisone on the development of *Ostertagia* spp. in sheep. *Nature* 192, 139–140.

Enjalbert, A., Bourgoin, S., Hamon, M., Adrien, J. and Bockaert, J. (1978a). Postsynaptic serotonin-sensitive adenylate cyclase in the central nervous system: Development and distribution of serotonin- and dopamine-sensitive adenylate cyclases in rat and guinea pig brain. *Molecular Pharmacology* 14, 2–10.

Enjalbert, A., Hamon, M., Bourgoin, S. and Bockaert, J. (1978b). Postsynaptic serotonin-sensitive adenylate cyclase in the central nervous system. II. Comparison with dopamine- and isoproterenol-sensitive adenylate cyclases in rat brain. *Molecular Pharmacology* 14, 11–23.

Erspamer, V. (1966). *In* "5-Hydroxytryptamine and Related Indolealkylamines". (*Handbook of Experimental Pharmacology XIX*) (V. Erspamer, ed.), pp. 132–181. Springer Verlag, New York.

Fripp, P. J. (1967). Histochemical localization of esterase activity in schistosomes. *Experimental Parasitology* **21**, 380–390.

Gentleman, S. and Mansour, T. E. (1977). Control of $Ca^{2+}$ efflux and cyclic AMP by 5-hydroxytryptamine and dopamine in abalone gill. *Life Sciences* **20**, 687–694.

Gentleman, S., Abrahams, S. L. and Mansour, T. E. (1976). Adenosine cyclic 3′,5′-monophosphate in the liver fluke, *Fasciola hepatica*. II. Activation of protein kinase by 5-hydroxytryptamine. *Molecular Pharmacology* **12**, 59–68.

Gerisch, G. and Wick, U. (1975). Intracellular oscillators and release of cyclic AMP from *Dictyostelium* cells. *Biochemical and Biophysical Research Communications* **65**, 364–370.

Greengard, P. and Kuo, J. F. (1979). On the mechanism of action of cyclic AMP. *Advances in Biochemical Psychopharmacology* **3**, 287–306.

Gustafson, T. and Toneby, M. (1970). On the role of serotonin and acetylcholine in sea urchin morphogenesis. *Experimental Cell Research* **62**, 102–117.

Hariri, M. J. (1974). Occurrence and concentration of biogenic amines in *Mesocestoides corti* (Cestoda). *Journal of Parasitology* **60**, 737–743.

Higashi, G. I., Kreiner, P. W., Keirns, J. J. and Bitensky, M. W. (1973). Adenosine 3′,5′-cyclic monophosphate in *Schistosoma mansoni*. *Life Sciences* **13**, 1211–1220.

Hillman, G. R. and Senft, A. W. (1973). Schistosome motility measurements: response to drugs. *Journal of Pharmacology and Experimental Therapeutics* **185**, 177–184.

Hillman, G. R., Olsen, N. J. and Senft, A. W. (1974). Effect of methylsergide and dihydroergotamine on *S. mansoni*. *Journal of Pharmacology and Experimental Therapeutics* **188**, 529–535.

Hofer, H. W. and Furst, M. (1975). Isolation of a phosphorylated form of phosphofructokinase from skeletal muscle. *FEBS Letters* **62**, 118–122.

Holmes, J. C. (1962). Effects of concurrent infections on *Hymenolepis diminuta* (Cestoda) and *Moniliformis dubius* (Acanthocephala). II. Effects on growth. *Journal of Parasitology* **48**, 87–96.

Hsie, A. W. and Puck, T. T. (1971). Morphological transformation of Chinese hamster cells by dibutyryl adenosine cyclic 3′,5′-monophosphate and testosterone. *Proceedings of the National Academy of Sciences USA* **68**, 358–361.

Hutton, J. C., Schofield, P. J. and McManus, W. R. (1972). Metabolic sensitivity of *Fasciola hepatica* to mammalian hormones *in vitro*. *Comparative Biochemistry and Physiology* **42**, 49–56.

Isbell, H., Miner, E. J. and Logan, C. R. (1959). Relationships of psychotomimetic to anti-serotonin potencies of congeners of lysergic acid diethylamide (LSD-25). *Psychopharmacologia* **1**, 20–28.

Jaffe, J. J., Meymarian, E. and Doremus, H. M. (1971). Antischistosomal action of tubercidin administered after absorption into red cells. *Nature* **230**, 408–409.

Kasschau, M. R. and Mansour, T. E. (1982a). Adenylate cyclase in adults and cercariae of *Schistosoma mansoni*. *Molecular and Biochemical Parasitology* **5**, 107–116.

Kasschau, M. R. and Mansour, T. E. (1982b). Serotonin-activated adenylate cyclase during early development of *Schistosoma mansoni*. *Nature* **296**, 66–68.

Keirns, J. J., Carritt, B., Freeman, J., Eisenstadt, J. M. and Bitensky, M. W. (1973). Adenosine 3′,5′-cyclic monophosphate in *Euglena gracilis*. *Life Sciences* **13**, 287–302.

Kirkland, W. and Burton, P. R. (1972). Cyclic adenosine monophosphate-mediated stabilization of mouse neuroblastoma cell neurite microtubules exposed to low temperature. *Nature (New Biology)* **240**, 205–207.

Lahoud, H., Prichard, R. K., McManus, W. R. and Schofield, P. J. (1971). The relationship of some intermediary metabolites to the production of volatile fatty acids by adult *Fasciola hepatica*. *Comparative Biochemistry and Physiology* **39**, 435–444.

Lee, M. B., Bueding, E. and Schiffler, E. L. (1978). The occurrence and distribution of 5-hydroxytryptamine in *Hymenolopis diminuta* and *H. nana*. *Journal of Parasitology* **64**, 257–264.

Lefkowitz, R. J., Wessells, M. R. and Stadel, J. M. (1980). Hormones, receptors, and cyclic AMP: Their role in target cell refractoriness. *Current Topics in Cellular Regulation* **17**, 205–230.

Looss, A. (1895). Zur anatomie und histologie der *Bilharzia haematobium* (Cobbold). *Archiv für Microskopische Anatomie und Entwicklungsgeschechte* **46**, 1–108.

MacManus, J. P., Whitfield, J. F., Boynton, A. L., Rixon, R. H. (1975). Role of cyclic nucleotides and Ca in the positive control of cell proliferation. *Advances in Cyclic Nucleotide Research* **5**, 719–734.

Mansour, T. E. (1957). The effect of lysergic acid diethylamide, 5-hydroxytryptamine, and related compounds on the liver fluke *Fasciola hepatica*. *British Journal of Pharmacology* **12**, 406–409.

Mansour, T. E. (1959a). Actions of serotonin and epinephrine on intact and broken cell preparations from the liver fluke, *Fasciola hepatica*. *Pharmacological Reviews* **11**, 465–466.

Mansour, T. E. (1959b). The effect of serotonin and related compounds on the carbohydrate metabolism of the liver fluke, *Fasciola hepatica*. *Journal of Pharmacology and Experimental Therapy* **126**, 212–216.

Mansour, T. E. (1959c). Studies on the carbohydrate metabolism of the liver fluke *Fasciola hepatica*. *Biochimica et Biophysica Acta* **34**, 456–464.

Mansour, T. E. (1964). The pharmacology and biochemistry of parasitic helminths. *Advances in Pharmacology* **3**, 129–165.

Mansour, T. E. (1967). Effect of hormones on carbohydrate metabolism of invertebrates. *Federation Proceedings* **26**, 1179–1185.

Mansour, T. E. (1970). Biogenic amines as metabolic regulators in invertebrates. In "Biogenic Amines as Physiological Regulators" (J. J. Blum, ed.), pp. 119–138. Prentice Hall, New Jersey.

Mansour, T. E. (1979). Chemotherapy of parasitic worms: new biological strategies. *Science* **205**, 462–469.

Mansour, T. E. and Mansour, J. M. (1962). Effects of serotonin (5-hydroxytryptamine) and adenosine $3',5'$-phosphate on phosphofructokinase from the liver fluke *Fasciola hepatica*. *Journal of Biological Chemistry* **237**, 629–634.

Mansour, T. E. and Mansour, J. M. (1977). Phosphodiesterase in the liver fluke *Fasciola hepatica*. *Biochemical Pharmacology* **26**, 2325–2330.

Mansour, T. E. and Stone, D. B. (1970). Biochemical effects of lysergic acid diethylamide on the liver fluke *Fasciola hepatica*. *Biochemical Pharmacology* **19**, 1137–1145.

Mansour, T. E., Lago, A. D. and Hawkins, J. L. (1957). Occurrence and possible role of serotonin in *Fasciola hepatica*. *Federation Proceedings* **16**, 319.

Mansour, T. E., Sutherland, E. W., Rall, T. W. and Bueding, E. (1960). The effect of serotonin on the formation of adenosine-$3',5'$-phosphate by tissue particles from the liver fluke *Fasciola hepatica*. *Journal of Biological Chemistry* **235**, 466–470.

Mansour, T. E., Morris, P. G., Feeney, J. and Roberts, G. C. K. (1982). A $^{31}$P NMR study of the intact liver fluke *Fasciola hepatica*. *Biochimica et Biophysica Acta* **721**, 336–340.

McMahon, D. (1974). Chemical messengers in development: A hypothesis. *Science* **185**, 1012–1021.

McNall, S. and Mansour, T. E. (1983). A novel serotonin receptor in the liver fluke *Fasciola hepatica*. *Federation Proceedings* **42**, 1876.

McNall, S. and Mansour, T. E. (1984a). Novel serotonin receptors in *Fasciola*: Characterization by studies on adenylate cyclase activation and [$^3$H]LSD binding. *Biochemical Pharmacology* (in press).

McNall, S. and Mansour, T. E. (1984b). Desensitization of serotonin-stimulated adenylate cyclase in the liver fluke *Fasciola hepatica*. *Biochemical Pharmacology* (in press).

Mettrick, D. F. (1971). Effect of host dietary constituents on intestinal pH and on the migratory behaviour of the rat tapeworm *Hymenolepis diminuta*. *Canadian Journal of Zoology* **49**, 1513–1525.

Mettrick, D. F. (1972). Changes in the distribution and chemical composition of *Hymenolepis diminuta*, and in the intestinal nutritional gradients of uninfected and parasitized rats following a glucose meal. *Journal of Helminthology* **46**, 407–429.

Mettrick, D. F. (1973). Competition for ingested nutrients between the tapeworm *Hymenolepis diminuta* and the rat host. *Canadian Journal of Public Health* **64**, 70–82.

Mettrick, D. F. and Cho, C. H. (1981). Migration of *Hymenolepis diminuta* (Cestoda) and changes in 5-HT (serotonin) levels in the rat host following parenteral and oral 5-HT administration. *Canadian Journal of Pharmacology* **54**, 281–286.

Mettrick, D. F. and Cho, C. H. (1982). Changes in tissue and intestinal serotonin (5-HT) levels in the laboratory rat following feeding and the effect of 5-HT inhibitors on the migratory response of *Hymenolepis diminuta* (Cestoda). *Canadian Journal of Zoology* **60**, 790–797.

Mettrick, D. F., Rahman, M. S. and Podesta, R. B. (1981). Effect of 5-hydroxytryptamine (5-HT; serotonin) on *in vitro* glucose uptake and glycogen reserves in *Hymenolepis diminuta*. *Molecular and Biochemical Parasitology* **4**, 217–223.

Meyer, F., Meyer, H. and Bueding, E. (1970). Lipid metabolism in the parasitic and free-living flatworms, *S. mansoni* and *Dugesia dorotocephala*. *Biochimica et Biophysica Acta* **210**, 257–266.

Monod, J. (1966). The Upjohn Lecture of the Endocrine Society on the mechanism of molecular interactions in the control of cellular metabolism. *Endocrinology* **78**, 412–425.

Nathanson, J. A. (1978). Octopamine receptors, adenosine 3',5'-monophosphate, and neural control of firefly flashing. *Science* **203**, 65–68.

Nathanson, J. A. and Greengard, P. (1974). Serotonin-sensitive adenylate cyclase in neural tissue and its similarity to the serotonin receptor: A possible site of action of lysergic acid diethylamide. *Proceedings of the National Academy of Sciences USA* **71**, 797–801.

Nimmo-Smith, R. H. and Raison, C. G. (1968). Monoamine oxidase activity of *Schistosoma mansoni*. *Comparative Biochemistry and Physiology* **24**, 403–416.

Northup, J. K. and Mansour, T. E. (1978a). Adenylate cyclase from *Fasciola hepatica*: 1. Ligand specificity of adenylate cyclase-coupled serotonin receptors. *Molecular Pharmacology* **14**, 804–819.

Northup, J. K. and Mansour, T. E. (1978b). Adenylate cyclase from *Fasciola hepatica*. 2. Role of guanine nucleotides in coupling adenylate cyclase and serotonin receptors. *Molecular Pharmacology* **14**, 820–833.

Pall, M. L. (1981). Adenosine 3′,5′-phosphate in fungi. *Microbiological Reviews* **45**, 462–480.

Panitz, E. and Knapp, S. E. (1967). Acetylcholine activity in *Fasciola hepatica* miracidiae. *Journal of Parasitology* **53**, 354.

Pax, R. A., Siefker, C., Hickox, T. and Bennett, J. L. (1981). *Schistosoma mansoni*: neurotransmitters, longitudinal musculature and effects of electrical stimulation. *Experimental Parasitology* **52**, 346–355.

Pepler, W. J. (1958). Histochemical demonstration of an acetylcholinesterase in the ova of *Schistosoma mansoni*. *Journal of Histochemistry and Cytochemistry* **6**, 139.

Peroutka, S. J., Lebovitz, R. M. and Snyder, S. H. (1981). Two distinct central serotonin receptors with different physical functions. *Science* **212**, 827–829.

Podesta, R. B. and Mettrick, D. F. (1981). A simple method for analyzing oral and intravenous treatment effects on biomass redistribution of *Hymenolepis diminuta* (Cestoda) in the rat intestine. *Canadian Journal of Zoology* **59**, 861–863.

Powers, P. A. and Pall, M. L. (1980). Cyclic AMP-dependent protein kinase of *Neurospora crassa*. *Biochemical and Biophysical Research Communications* **95**, 701–706.

Probert, A. J. and Durrani, M. S. (1977). *Fasciola hepatica* and *Fasciola gigantica*: Total cholinesterase, characteristics, and effects of specific inhibitors. *Experimental Parasitology* **42**, 203–210.

Quach, T. T., Rose, C., Duchemin, A. M. and Schwartz, J. C. (1982). Glycogenolysis induced by serotonin in brain: identification of a new class of receptor. *Nature* **298**, 373–375.

Read, C. P. and Kilejian, A. Z. (1969). Circadian migratory behaviour of a cestode symbiote in the rat host. *Journal of Parasitology* **55**, 574–578.

Renart, M. F., Ayanoglu, G., Mansour, J. M. and Mansour, T. E. (1979). Fluoride and guanosine nucleotide activated adenylate cyclase from *Fasciola hepatica*: reconstitution after inactivation. *Biochemical and Biophysical Research Communications* **89**, 1146–1153.

Rickenberg, H. V. (1974). Cyclic AMP in prokaryotes. *Annual Review of Microbiology* **28**, 353–369.

Roisen, F. J., Murphy, R. A., Pichichero, M. E. and Braden, W. G. (1972). Cyclic adenosine monophosphate stimulation of axonal elongation. *Science* **175**, 73–74.

Rosenzweig, Z. and Kindler, S. H. (1972). Epinephrine and serotonin activation of adenyl cyclase from *Tetrahymena pyriformis*. *FEBS Letters* **25**, 221–223.

Schulman, R. G., Brown, T. R., Ugurbil, K., Ogawa, S., Cohen, S. M. and den Hollander, J. A. (1979). Cellular applications of $^{31}P$ and $^{13}C$ nuclear magnetic resonance. *Science* **205**, 160–166.

Semeyn, D. R., Pax, R. A. and Bennett, J. L. (1982). Surface electrical activity from *Schistosoma mansoni*: a sensitive measure of drug action. *Journal of Parasitology* **68**, 353–362.

Senft, A. W. (1970). De novo or salvage pathway of purines in *S. mansoni*. *Journal of Parasitology* **56**, 314–315.

Shapiro, D. L. (1973). Morphological and biochemical alterations in foetal rat brain cells cultured in the presence of monobutyryl cyclic AMP. *Nature* **241**, 203–204.

Smith, D. L. and Mansour, T. E. (1978). An adenosine-3′,5′-monophosphate-activated adenylate cyclase in the slime mold *Physarum polycephalum*. *FEBS Letters* **92**, 57–62.

Stone, D. B. and Mansour, T. E. (1967a). Phosphofructokinase from the liver fluke *Fasciola hepatica*. 1. Activation by adenosine 3′,5′-phosphate and by serotonin. *Molecular Pharmacology* **3**, 161–176.

Stone, D. B. and Mansour, T. E. (1967b). Phosphofructokinase from the liver fluke, *Fasciola hepatica*. 2. Kinetic properties of the enzyme. *Molecular Pharmacology* **3,** 177–187.

Sutherland, E. W. and Robison, G. A. (1966). The role of cyclic 3′,5′-AMP in responses to catecholamines and other hormones. *Pharmacological Reviews* **18,** 145–161.

Tanaka, R. D. and MacInnis, A. J. (1975). An explanation of the apparent reversal of the circadian migration of *Hymenolepis diminuta*. *Journal of Parasitology* **61,** 271–280.

Terada, M., Ishii, A. I., Kino, H. and Sano, M. (1982a). Studies on chemotherapy of parasitic helminths. (VI) Effects of various neuropharmacological agents on the motility of *Dipylidium caninum*. *Japanese Journal of Pharmacology* **32,** 479–488.

Terada, M., Ishii, A. I., Kini, H. and Sano, M. (1982b). Studies on chemotherapy of parasitic helminths. (VIII) Effects of some possible neurotransmitters on the motility of *Angiostrongylus cantonensis*. *Japanese Journal of Pharmacology* **32,** 643–653.

Tomosky-Sykes, T. K., Jardine, I., Mueller, J. F. and Bueding, E. (1977). Sources of error in neurotransmitter analysis. *Analytical Biochemistry* **83,** 99–108.

Trevillyans, J. M. and Pall, M. L. (1982). Isolation and properties of a cyclic AMP binding protein from *Neurospora*. *Journal of Biological Chemistry* **257,** 3978–3986.

Udenfriend, S. (1959). Biochemistry of serotonin and other indoleamines. *Vitamins and Hormones* **17,** 133–154.

Walsh, D. A., Perkins, J. P. and Krebs, E. G. (1968). An adenosine 3′,5′-monophosphate-dependent protein kinase from rabbit skeletal muscle. *Journal of Biological Chemistry* **243,** 3763–3765.

Walter, R. D., Nordmeyer, J. P. and Konigk, E. (1974). Adenylate cyclase from *Trypanosoma gambiense*. *Hoppe-Seyler's Zeitschrift für Physiologische Chemie* **355,** 427–430.

Welsh, J. H. (1970). Phylogenetic aspects of the distribution of biogenic amines. *In* "Biogenic Amines as Physiological Regulators" (J. J. Blum, ed.), pp. 75–94. Prentice Hall, New Jersey.

Whitfield, P. J. (1979). *In* "The Biology of Parasitism: An Introduction to the Study of Associating Organisms", pp. 158–201. University Park Press, Baltimore.

Whitfield, J. F., Rixon, R. H., MacManus, J. P. and Balk, S. D. (1973). Calcium, cyclic adenosine 3′,5′-monophosphate, and the control of cell proliferation: A review. *In Vitro* **4,** 257–278.

# The Piroplasms: Life Cycle and Sexual Stages*

HEINZ MEHLHORN

*Lehrstuhl für Spezielle Zoologie und Parasitologie der Ruhr-Universität Bochum, West Germany†*

and

EBERHARD SCHEIN

*Institut für Parasitologie der Freien Universität Berlin, West Germany*

| | | |
|---|---|---|
| I. | Introduction | 38 |
| II. | The Genera *Theileria* and *Babesia* | 38 |
| III. | Life Cycle of *Theileria* Species | 45 |
| | A. Schizogony Inside Lymphocytes | 45 |
| | B. Development Inside Erythrocytes | 51 |
| | C. Sexual Stages in the Intestine of Ticks and *In Vitro* | 53 |
| | D. Development in Salivary Glands of Vector Ticks | 65 |
| IV. | The Life Cycle of *Babesia* Species | 69 |
| | A. *Babesia* Species with Intralymphocytic Schizonts | 69 |
| | B. Development in Erythrocytes | 72 |
| | C. Sexual Stages in the Intestine of Ticks and *In Vitro* | 75 |
| | D. Development in Various Organs of Ticks | 83 |
| | E. Development in Salivary Glands of Vector Ticks | 89 |
| V. | Conclusions | 96 |
| | References | 97 |

* Supported by DFG.
† Previously at Instiut für Zoologie, Universität Düsseldorf.

## I. Introduction

The piroplasms, comprising mainly two genera (*Theileria* and *Babesia*), are protozoa that are highly pathogenic to cattle, sheep, goats and occasionally even to man. The diseases they induce, known collectively as theilerioses and babesioses, cause fevers and lead to important economic losses in the tropics, subtropics and southern Europe. The parasites, which are situated in the blood cells of their vertebrate hosts, are transmitted by a broad spectrum of ixodid tick species (Riek, 1966, 1968; Friedhoff and Smith, 1981). Although the different *Theileria* and *Babesia* species and some of their vectors were already described at the turn of the century by famous investigators like Babès (1869, 1888), Smith and Kilborne (1893), Starcovici (1893), Koch (1898) and Theiler (1904, 1906), their life cycles were incompletely known, since these parasites are very small and morphological details remained undefinable by light microscopy. Thus only asexual reproduction within the salivary glands of vector ticks and within the blood cells of vertebrates could be identified with some certainty. This fact led to the doubtful systematic position of the piroplasms among the protozoa.

Laboratory transmission experiments and *in vitro* cultures in combination with electron microscopy by our group, starting in 1973–1974, led to unequivocal identification of a typical sporozoan life cycle for both genera of piroplasms comprising the three typical phases: schizogony, gamogony and sporogony.

This review is intended to comment briefly on the differences between the life cycles of *Babesia* and *Theileria* species with respect to their morphology, studied by means of light and electron microscopy.

## II. The Genera *Theileria* and *Babesia*

Among the piroplasms (so named because of their pear-shaped intraerythrocytic stages), the genera *Theileria* and *Babesia* are most important. Most species may be clearly differentiated by their appearance in erythrocytes (Fig. 1, a–m). In general, intraerythrocytic stages of *Theileria* species (Table 1) are smaller than 1–2 μm, whereas in the genus *Babesia* two groups of species occur: (1) those that are larger than 3 μm (Table 2) and (2) those that are smaller than 3 μm (Table 3). The vectors, geographic distribution and the diseases they cause are listed in Tables 1–3.

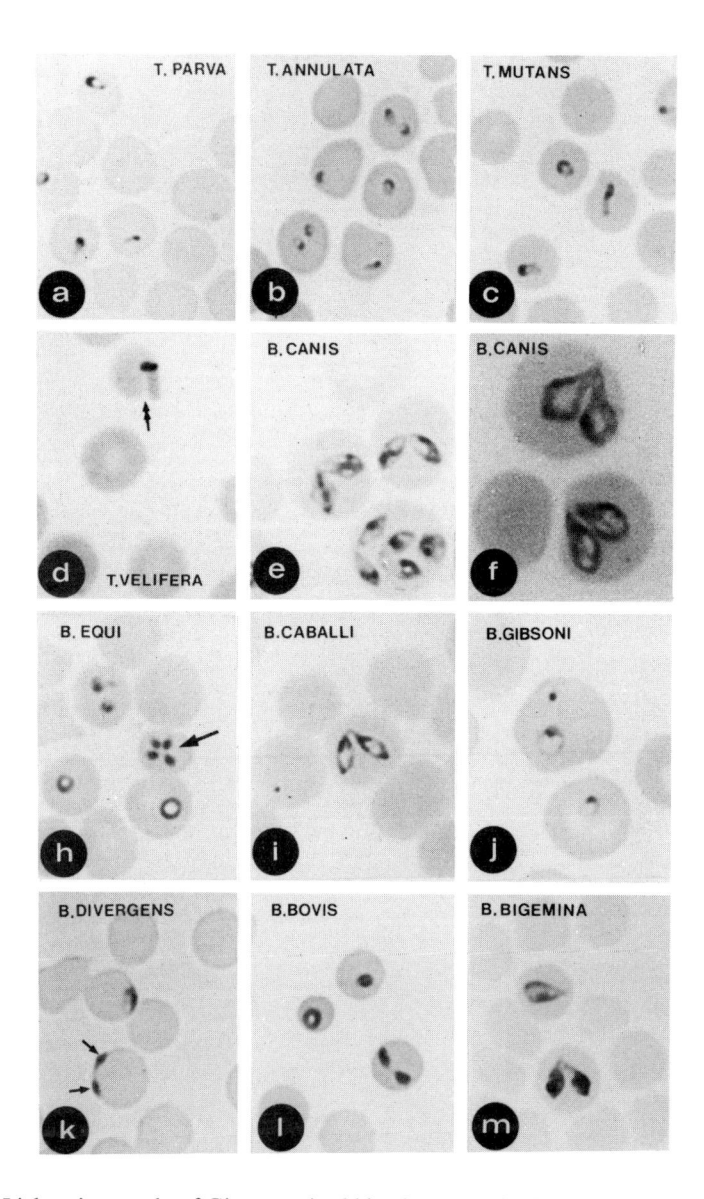

FIG. 1. Light micrographs of Giemsa-stained blood smears with intraerythrocytic stages of important *Theileria* and *Babesia* species. Magnifications: a–e, ×1600; f, 2080; h–m, 1600. Note different size of parasites in the genera *Babesia* and *Theileria*. Pecularities of some species are indicated by arrows: *T. velifera* (d) leads to a velum-like appearance of host cell; *B. equi* stages (h) often show a 'Maltese-cross' arrangement; *B. divergens* (k) parasites are mostly located at the margin of the erythrocyte. For hosts see Tables 1–3.

TABLE 1
Important species of Theileria

| Species | Vector | Vertebrate | Disease | Geographic distribution | References |
|---|---|---|---|---|---|
| *T. parva parva* | *Rhipicephalus appendiculatus*, *R.* spp. | Cattle, *Syncerus caffer* | East Coast fever | Africa | Koch (1906), Gonder (1910, 1911), Cowdry and Ham (1932), Reichenow (1937, 1938, 1940), Martin et al. (1964), Purnell and Joyner (1968), Schein et al. (1977a), Mehlhorn et al. (1978) |
| *T. parva lawrencei* | *Rhipicephalus appendiculatus*, *R.* spp. | Cattle, *Syncerus caffer* | Corridor disease | Africa | Schein et al. (1978) |
| *T. annulata* | *Hyalomma* spp. | Cattle, domestic water buffalo | Mediterranean or tropical theileriosis | Africa, Asia, South Europe | Sergent et al. (1936), Hadani et al. (1963, 1969), Dyakonow and Godzaev (1971), Schein (1975), Schein et al. (1975), Schein and Friedhoff (1978), Mehlhorn et al. (1975) |
| *T. mutans* | *Amblyomma* spp. | Cattle, *Syncerus caffer* | Benign African theileriosis I | Africa | Young et al. (1978a,b), Warnecke et al. (1980) |
| *T. velifera* | *Amblyomma* spp. | Cattle, *Syncerus caffer* | — | Africa | Warnecke et al. (1979) |

| | | | | | |
|---|---|---|---|---|---|
| *T. taurotragi* | *R. appendiculatus*, *R.* spp. | Cattle, *Taurotragus oryx* | Benign African theileriosis II | Africa | Martin and Brocklesby (1960), Young *et al.* (1980) |
| *T. sergenti*, *T. orientalis* | *Haemaphysalis* spp. | Cattle | Oriental theileriosis | Asia, Europe, Australia, Africa ? | — |
| *T. hirci* | *Hyalomma* spp. | Sheep, goats | Malignant ovine and caprine theileriosis | South Europe, Asia, Africa | — |
| *T. ovis* | ? | Sheep | — | Africa | Mehlhorn *et al.* (1979) |
| *T. separata* | *Rhipicephalus evertsi*, *R.* spp. | Sheep | — | Africa | — |
| *Theileria* sp. | ? | Sheep | — | Europe | — |

TABLE 2
*Important species of* Babesia

| Species | Vector Species | Stage | Vertebrate host | Size in erythrocytes (μm) | Geographic distribution | References |
|---|---|---|---|---|---|---|
| *B. bigemina* | *Boophilus* spp. | Nymphs, adults | Cattle, water buffalo, wild ruminants | 5 × 2 | South Europe, America, Africa, Asia, Australia | Koch (1906), Crawley (1916), Rosenbusch (1927), Dennis (1932), Regendanz (1936), Muratov and Kheisin (1959), Riek (1964), Potgieter and Els (1977), Friedhoff and Büscher (1976) |
| *B. bovis* | *Boophilus* spp., *Ixodes* spp.? *Rhipicephalus bursa* | Larvae | Cattle, water buffalo, wild ruminants | 2·5 × 1·5 | South Europe, America, Africa, Asia, Australia | Rieck (1966), Potgieter *et al.* (1976), Potgieter and Els (1976) |
| *B. divergens* | *Ixodes ricinus* | Larvae | Cattle, wild ruminants | 1·5 × 0·5 | Europe | Joyner *et al.* (1963), Donnelly and Peirce (1975) |
| *B. major* | *Haemaphysalis punctata* | Adults | Cattle | 3 × 1·5 | West and South Europe, Great Britain, North-west Africa | Bool *et al.* (1961), Morzaria *et al.* (1976, 1978) |

| | | | | | | |
|---|---|---|---|---|---|---|
| *B. motasi* | *Haemaphysalis* spp. | Adults | Sheep, goats | 4 × 2·5 | South Europe, Middle East, USSR, Africa, Asia | Uilenberg *et al.* (1980) |
| *B. ovis* | *Rhipicephalus bursa* | Adults | Sheep, goats | 2 × 1 | South Europe, Middle East, USSR, Africa, Asia | Friedhoff and Scholtyseck (1968), Friedhoff (1969), Friedhoff *et al.* (1972), Weber and Friedhoff (1977) |
| *B. taylori* | ? | ? | Goats | 2 × 1·5 | India | Sarwar (1935) |
| *B. caballi* | *Hyalomma* spp., *Dermacentor* spp., *Rhipicephalus* spp. | Adults | Horses, mules, donkeys, *Equus burchelli* | 4 × 2·5 | Europe, Asia, Africa, America, Australia | Enigk (1943, 1944), Holbrook *et al.* (1968) |
| *B. canis* | *Rhipicephalus sanguineus*, *Haemaphysalis leachi*, *Dermacentor reticulatus* | Nymphs adults | Dogs, *Vulpes vulpes*, other wild canines | 5 × 2·5 | Europe, Asia, Africa, America, Australia | Christophers (1907), Regendanz and Reichenow (1933), Shortt (1973), Schein *et al.* (1979) |
| *B. trautmanni* | *Rhipicephalus* sp. | ? | Pigs | 4 × 2·5 | South Europe, Africa | Knuth and Du Toit (1921) |
| *B. herpailuri* | ? | ? | *Felis sylvestris* | 3 × 2·2 | America | Dennig (1969) |
| *B. panthera* | ? | ? | *F. sylvestris*, *Panthera leo* | 2·5 × 1·5 | Africa | Dennig and Brocklesby (1972) |

TABLE 3
Babesia species of doubtful systematic position

| Species | Vector Species | Vector Stage | Vertebrate host | Size in erythrocytes ($\mu$m) | Geographic distribution | References |
|---|---|---|---|---|---|---|
| B. equi | Dermacentor spp., Hyalomma spp., Rhipicephalus spp. | Nymphs to adults | Horses, mules, donkeys, Equus burchelli | 2–3 | South Europe, Africa, Asia, America | Enigk (1943, 1944), Schein et al. (1981), Moltmann et al. (1983a,b) |
| B. gibsoni | Rhipicephalus sanguineus, Haemaphysalis bispinosa | ? | Canidae, including dogs, Vulpes vulpes and other wild canines | 1·5–2 | Asia, Africa | Groves and Dennis (1972) |
| B. microti group | Ixodes spp. | Larvae to nymphs | Rodents | 1·5–2 | Europe, America | Rudzinska et al. (1979), Weber and Walter (1980) |
| B. felis | Haemaphysalis leachi ? | ? | Felidae, including Panthera leo, Felis sylvestris | 1·5–2 | Africa | Davis (1929) |

## III. Life Cycle of *Theileria* Species

Infection of vertebrate hosts with *Theileria* parasites starts with the feeding of infected ticks (Fig. 2). However, ticks are unable to transmit parasites immediately after attachment of their mouthparts. This is due to a biological speciality of ixodid ticks, the salivary gland of which becomes reduced during moult. After attachment to a new host, alveolae of the salivary glands grow in a few days and only then do the piroplasms, which undergo multiplication inside the salivary gland cells (see page 65; Figs 10; 14–15; 29–32), have a chance to develop. Thus infectious sporozoites of piroplasms are often not transmitted earlier than 3–5 days after first attachment of ticks to their hosts.

### A. SCHIZOGONY INSIDE LYMPHOCYTES

*Theileria* sporozoites (see Fig. 15 b, d, e) that have been injected with the saliva into a host rapidly enter lymphoid cells as shown by *in vitro* studies (Fawcett *et al.*, 1982b). This penetration is completed within 10 minutes. At the site of initial contact, the membranes of parasite and host cell come into very close apposition. As the overlapping of the membranes spreads laterally, the sporozoite sinks into a progressively deepening recess in the surface of the host cell membrane until the rim of the invagination closes and fuses over the parasite. Sporozoites may enter in any orientation and not only after attachment of the apical complex, as is necessary with other sporozoan parasites such as sporozoites or merozoites of the genera *Toxoplasma*, *Sarcocystis*, *Eimeria* etc. Owing to this peculiar mode of entry into the host cell, theilerian sporozoites are at first surrounded by a membrane in the cytoplasm. This, however, disintegrates within 24 hours of entry and is not retained as a parasitophorous vacuole as it is, for example, in *Eimeria* or *Plasmodium* spp. As in the two latter genera, the sporozoite reduces the two inner membranes of its pellicle and remains bound by a single membrane during further development in lymphoid cells (Schein *et al.*, 1978). At 72 hours, the parasite has enlarged considerably to about 2 μm in diameter, apparently by ingested host cell cytoplasm via several typical sporozoan micropores at its surface. Simultaneously with the growth of the parasite, nuclear division starts; it consists of binary fission with the appearance of a typical intranuclear spindle-apparatus at two poles of the nucleus. By constant repetition of this procedure, large schizonts of about 10–15 μm in diameter are formed containing specific numbers of nuclei measuring 1·5–2·0 μm (Fig. 3 a–c; SCK). These schizonts are apparently identical with the 'Koch'sche Kugeln', 'Koch's bodies', 'blue bodies' or 'macroschizonts' seen by light microscopy. The schizonts stimulate the host cells, which—if lymphocytes—normally do not divide, to undergo cell division. During this division the schizont also

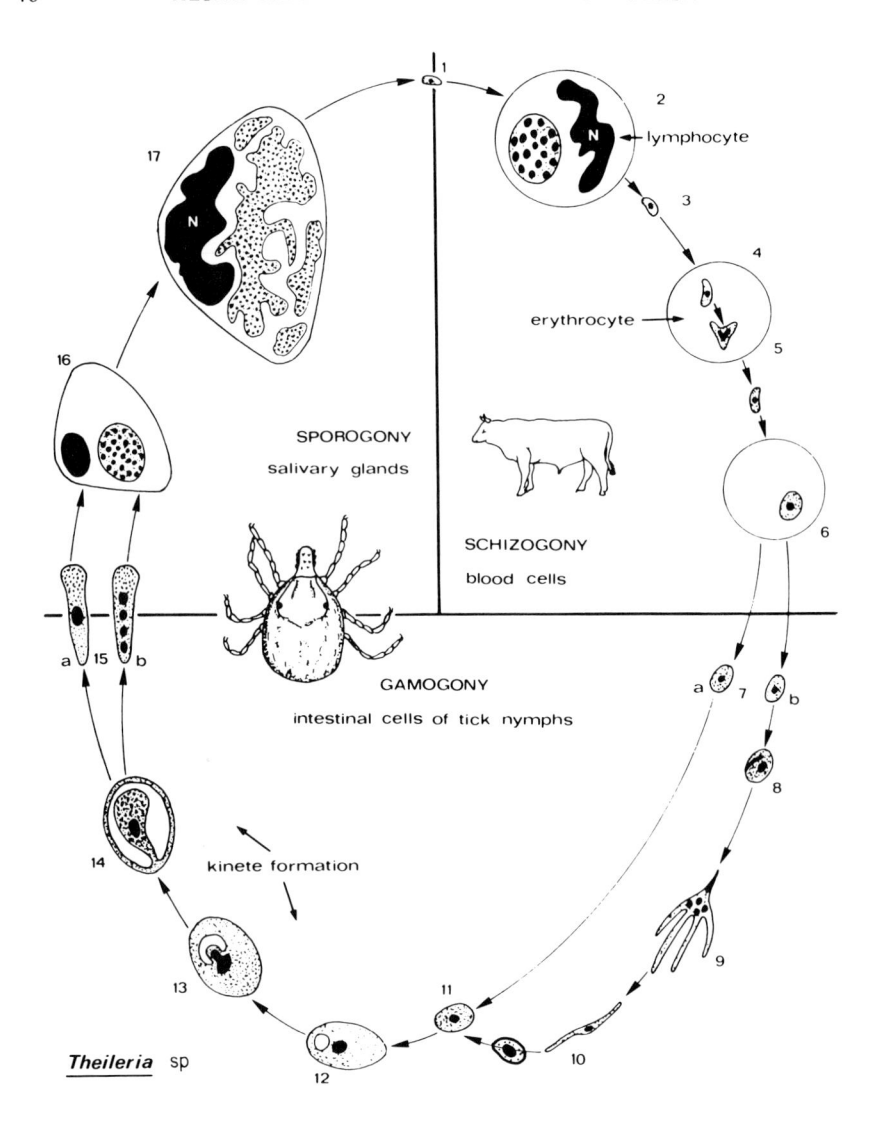

FIG. 2. Diagram of life cycle of *Theileria* species (for hosts see Table 1). **1.** Sporozoite in saliva of a feeding tick. **2.** Schizont (Koch's body) inside newly divided lymphocyte, forming merozoites (N, nucleus). **3.** Merozoite. **4–5.** Binary fission inside erythrocyte. **6.** Ovoid stage inside erythrocyte. **7a–b.** Ovoid stages free in blood masses in tick gut. **8–10.** Formation of microgamonts (**9**) and microgametes (**10**). **11.** Macrogamete. **12.** Zygote. **13–15.** Formation of motile kinete from ovoid stationary zygote. **15b.** In *T. parva* division of the nucleus may start in kinetes before leaving the intestinal cells of ticks. **16.** Kinete that has entered a salivary gland cell and started asexual reproduction by growth and nuclear division. **17.** Enlargement of host cell and its nucleus. Inside this giant cell thousands of sporozoites are formed.

divides and is thus distributed by the action of the host cell spindle apparatus to the daughter cells. Infected cells therefore displace uninfected lymph node tissue and often, within a few days, bring about symptoms that are identical to those of leucosis. *In vitro* culture of such infected cells leads in a short time to culture flasks overfilled with cells harbouring schizonts. These schizonts are morphologically similar in all *Theileria* species, variations occurring only in the number of nuclei (13–50); the numbers, however, are not sufficiently constant to be used for specific diagnosis. *In vivo*, the spherical schizonts start forming merozoites 8–10 days after infection in *T. annulata*, or 12–14 days in *T. parva*. This process, which (for unknown reasons) occurs only very seldom *in vitro*, may be completed in two different ways.

(1) When only two schizonts at most are situated within a host cell they both become spherical and relatively large, measuring often up to 6–10 μm in diameter. Their nuclei are arranged at the periphery and, during the last nuclear division, a merozoite-anlage is observed above each nuclear pole. Thus the schizonts appear rosette-like during this phase of merozoite-formation (Fig. 3 a, d). At the apical pole of such structures, an inner pellicular complex becomes visible, forming a polar ring, to which subpellicular microtubules are anchored. Into the opening of the polar ring three or four rhoptries extend, each with a diameter of 0·1–0·15 μm and a length of about 0·25 μm (Fig. 3d; R). Also, a very few micronemes, about 0·05 μm in size, are scattered at the apical pole of the structure. Finally it appears that these merozoites become free, more or less simultaneously, from the residual body, which, in *T. parva* and *T. annulata*, often has a diameter of 3 μm. This residual body still contains 1 or 2 very large nuclei (Fig. 3a; white N). As, in smear preparations as well as in ultrathin sections, residual bodies are never observed alone outside a host cell, it may be suggested that finally they are used up by repeated merozoite formation.

(2) When the schizonts are relatively small, as they always are in multiply infected host cells, merozoite formation is observed around large wrinkled nuclei. Above the different poles of these nuclei, up to six typical merozoite-structures are observed in a section in the plane of a single nucleus; thus the absolute number may be still higher. The cytoplasm of these smaller schizonts seems to become completely absorbed during daughter cell formation without leaving a residual body. The structure of the merozoites formed here is, however, identical with that described above.

Beginning with the phase of merozoite formation, the schizonts appear as microschizonts by light microscopy (Fig. 3c; SCM). The small size of both types of schizonts led to different interpretations and to speculation as to their fate during the life cycle (see Büttner, 1967b). Following electron microscopic observations in recent years, the terms 'macroschizont' and 'microschizont' should no longer be used, because they could lead to mis-

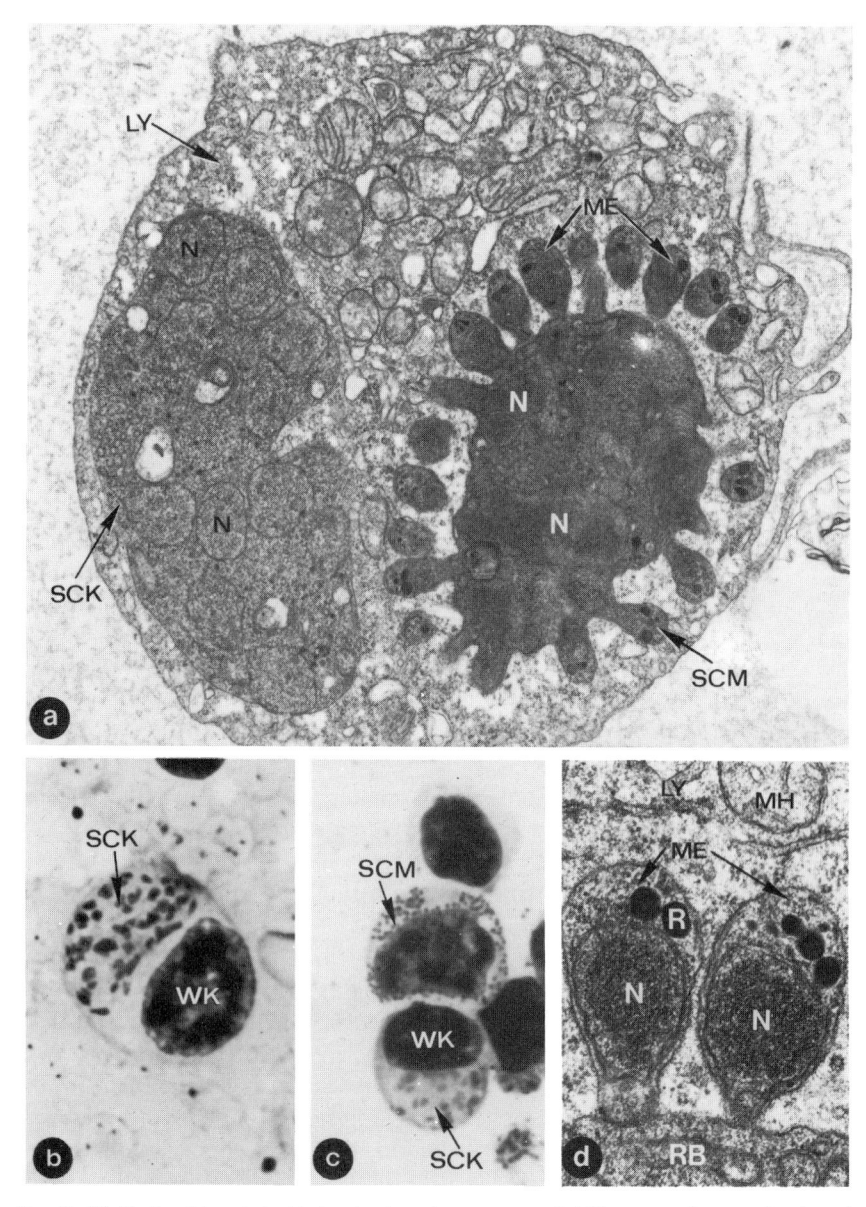

FIG. 3. *Theileria* schizonts inside bovine lymphocytes. a and d Electron micrographs; b and c light micrographs of whole cells. a. Section through a lymphocyte containing two schizonts (Koch's bodies) of *T. annulata* in different developmental phases (SCK, SCM). Magnification ×11 200. b. Lymphocyte containing a schizont of *T. annulata* during the phase of reproduction of the nuclei. Magnification ×1400. c. Bovine lymphocytes containing schizonts (SCM, SCK) of *T. parva*. Magnification ×1400. d. Section through the periphery of a schizont of *T. annulata* with differentiated merozoites. The parasite is surrounded directly by the cytoplasm of the host cell, with no parasitophorous vacuole or intervening membrane. Magnification ×31 500.

## Abbreviations used on Figures

| | | | |
|---|---|---|---|
| APK | Apical complex of kinetes | LST | Limiting membrane of ray bodies |
| BB | Bright body | LY | Lymphocyte |
| BC | Body cavity | MB | Single membrane-bound vacuole |
| BG | Blood masses during digestion | ME | Merozoite |
| BM | Basal membrane | MH | Mitochondrion of host cell |
| CH | Chromatin | MI | Mitochondrion |
| CY | Cytomere (uni- or multinucleate) | MN | Microneme |
| D | Dense inclusion | MP | Micropore (Cytostome) |
| DH | Degenerated host cell cytoplasm | MT | Microtubule |
| DI | Divisions | MTZ | Microtubules (artificially interrupted only in diagram) |
| DK | Developing kinete | | |
| DS | Developing sporozoite | N | Nucleus |
| DW | Double walled organelle (M?) | NH | Nucleus of host cell |
| E | Erythrocyte | NM | Nuclear membranes |
| EL | Lysed erythrocyte | NU | Nucleolus |
| ENH | Enlarged nucleus of host cell | OM | Outer membrane of pellicle |
| ER | Endoplasmic reticulum | P | Polar ring |
| ES | Electron dense part of the thorn | PE | Pellicle |
| FB | Filamentous body | PN | Perinuclear space |
| FI | Filamentous elements | PP | Posterior polar ring |
| FS | Funnel-shaped structure | PT | Protrusions of ray bodies etc. |
| G | Golgi apparatus | R | Rhoptries |
| HC | Host cell cytoplasm | RB | Residual body (remnant) |
| HS | Half spindle | RC | Remnant of cytomere |
| IA | Infected alveoli of the vector tick | RI | Ribosome |
| IIM | Interruption of inner membrane complex | S | Sporozoite |
| | | SCK | Schizont during nuclear division |
| IM | Inner membrane complex of pellicle | SCM | Schizont during merozoite formation |
| IN | Invagination | SD | Salivary gland ductule |
| IV | Intracytoplasmic vacuole into which the developing kinete protrudes | SG | Saliva droplet |
| | | SP | Spindle |
| | | T | Thorn |
| K | Kinete | UA | Normal, uninfected alveoli |
| LB | Laminar body (inclusion) | UM | Unit membrane |
| LH | Limiting membrane of the host cell | V | Vacuole |
| | | VS | Fortifying structure |
| LM | Limiting membrane | WK | Nucleus of host cell |
| LS | Labyrinthine structure | ZY | Zygote |

interpretation of the two different types of schizonts, which can occur side by side. A single schizont may, however, develop through both shapes: during the phase of nuclear divisions it appears as a 'macroschizont', and during (and after) formation of merozoites, as a 'microschizont'. The descriptions of merozoites with significantly different lengths in single species (micromerozoites 1 μm in length, macromerozoites 2–2·3 μm in length; c.f. Levine, 1973) are very probably erroneous, since the fine structure has many similar aspects not only in a single species but also in all others, no matter how the merozoites were formed. Merozoites are pear-shaped with a length of about 1–2 μm and a diameter of 0·6 μm at their nuclear region. Larger merozoites, or others with different shapes, never occur. Each is surrounded by a cell membrane, under which at several places two inner membranes are seen, lying very close to each other (Fig. 3d). Especially at the apical pole, these two inner membranes form a polar ring system, to which subpellicular microtubules are anchored. As already seen in the anlage, three or four rhoptries are present in the differentiated merozoite together with few micronemes (Fig. 3d; R). The nucleus with its relatively homogeneous karyoplasm is very prominent beside a double-walled structure probably representing a mitochondrion. A micropore is not observed at this stage, although it is present in schizonts and in merozoites that have penetrated into erythrocytes (Büttner, 1967a; Schein et al., 1977b). A conoid or similar organelle never occurs at the apical pole.

The merozoites become free and penetrate apparently actively into the erythrocytes beginning from 8 days after infection in T. annulata and 13 days in T. parva (4–7 days after schizogony commenced). The main pathogenic effects on the vertebrate host are found during the phase of intralymphocytic schizogony and lead often to death. At present no specific remedy exists for practical treatment of such clinical cases. Recently, however, two drugs have been shown to be suitable agents for the treatment of such schizonts, Menoctone* (McHardy et al., 1976) and Halofuginone (Schein and Voigt, 1979). Electron microscopic studies (Mehlhorn et al., 1981) of the mode of action of Halofuginone showed that primarily it destroys parasitized lymphocytes, so that the schizonts are released into the extracellular environment (intracellular schizonts are less affected and show only some swelling of the perinuclear spaces). Degeneration of released schizonts then occurs extracellularly, apparently due to the fact that they are bound by a single cell membrane only, whereas merozoites have a typical three-layered pellicle that enables them to survive extracellularly and finally to penetrate the erythrocytes. The destruction of parasitized lymphocytes seems to be relatively sudden, involving all cells simultaneously, since within 24–30 hours after

* Parvaquone (993 C).

treatment intracellular parasites are no longer seen. Intact (i.e. unparasitized) lymphocytes were not affected by Halofuginone.

In order to clarify further the mode of action of Halofuginone, it may be necessary to study the effects on dividing cells in general, since these may also be susceptible to the drug.

In untreated infections with spontaneous recovery, theilerian schizonts disappear from the lymphocytes and only varying numbers of merozoites remain in the erythrocytes.

### B. DEVELOPMENT INSIDE ERYTHROCYTES

The merozoites formed inside the host's lymphocytes are found in erythrocytes beginning 8 days after infection with *T. annulata* or 13 days with *T. parva*. Up to 90% of erythrocytes may be infected with *T. annulata*. In general, two different forms are seen (Figs 1a–d and 4a): slender, comma-shaped stages and spherical or ovoid stages.

The comma-shaped stages measure about 1–1·5 μm in length in *T. parva*, whereas in *T. annulata* or *T. mutans* they may reach up to 2·5 μm (Fig. 4a). Their occurrence among the intraerythrocytic stages varies and seems to be species-specific. Thus they reach 80% (20% are ovoid)* in *T. parva*, whereas this ratio is reversed in *T. annulata* and *T. hirci*. In *T. ovis* and *T. mutans* about 50% of each type of parasite are observed (see Boch and Supperer, 1983). The comma-shaped stages are bordered by a single cell membrane under which remnants of the former inner pellicular complex are often retained. The parasites seem to be situated immediately within the protoplasm of the erythrocyte, although in places an enclosing membrane is visible. This erythrocytic membrane, which is invaginated at the beginning of penetration, disintegrates soon after, since the parasite is surrounded by a single membrane throughout the rest of its life span. One pole of the comma-shaped stage is completely filled by the drop-shaped nucleus. At the other pole the parasite is provided with a typical micropore forming an invagination of 0·22 μm in depth (Fig. 4a, c; MP). In the region of micropores, remnants of the inner pellicular complex are always present and form a cylinder-like structure surrounding the neck of the invagination. A polar ring system and subpellicular microtubules, however, never occur. The comma-shaped forms are characterized by the presence of double-walled vacuoles situated opposite to the nucleus. One of these is dumb-bell-shaped in section and arches over a spherical vacuole measuring 0·3–0·35 μm in diameter. Close to the micropore there are vacuoles (0·3–0·4 μm in diameter) apparently containing erythrocytic material. Close to these vacuoles a number of small electron-dense granules of about 0·1 μm

---

* The proportion increases in older infections.

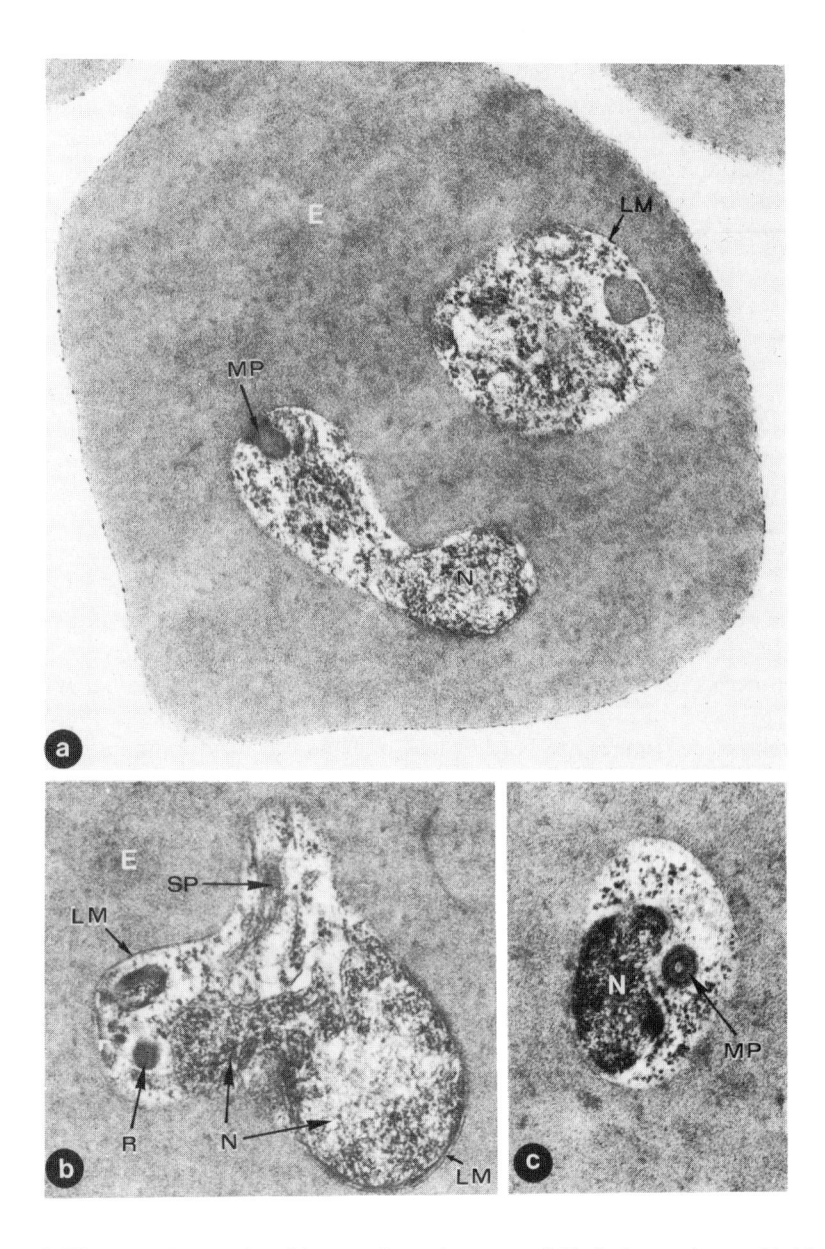

FIG. 4. Electron micrographs of intraerythrocytic stages of *Theileria annulata*. a. Besides the typical kidney-shaped merozoites (lower left) ovoid stages occur (upper right), which are probably gamonts. Magnification ×17 500. b. Binary fission of kidney-shaped stages. Magnification ×21 000. c. Tangential section through micropore, which functions as cytostome. Magnification ×17 500.

diameter always occur. Few lacunae of the endoplasmic reticulum are observed, and no polysaccharide granules or any typical Golgi-apparatus. Only the comma-shaped forms divide by binary fission, passing through a Y-shaped stage (Fig. 4b). True schizonts never occur, although in some species (*T. mutans, T. orientalis*) four merozoites occasionally form a tetrad (the so-called 'Maltese-cross' arrangement). The nuclear membrane is evidently present during division, when extranuclear microtubule-like structures occur. The division of the nucleus is always associated with cellular division so that a schizont-like stage is never present. These divisions apparently often lead to destruction of host cells, so that infections with some *Theileria* species induce an extreme erythrocytopaenia (2 000 000 erythrocytes (mm blood)$^{-3}$; see Friedhoff and Liebisch, 1978), with all the symptoms of anaemia.

The spherical stages are also bounded by a single cell membrane under which no membranous remnant is seen (Fig. 4b; LM). The polar double-walled structures are lacking, but there are mitochondria-like organelles. Vacuoles with erythrocytic material and small, always spherical, electron-dense granules also occur. These have a diameter of about 0·5–0·6 μm and are limited by a membrane enclosing the osmiophilic content. In general, the spherical forms appear more electron-lucent than the comma-shaped ones, which are thought to be their precursors. The spherical forms perhaps represent gamonts from which gamete-like stages are formed in the intestine of ticks which have fed on infected calves. More details of the structure of intra-erythrocytic stages are given elsewhere (Büttner, 1966, 1967a,b; Schein *et al*., 1977).

*Theileria* species (like *Babesia*) never produce pigment as a residue when feeding on erythrocytic cytoplasm, as do all malarial parasites of the genus *Plasmodium*. Apparently they have an alternative metabolism. Two *Theileria* species (*T. velifera* and *T. taurotragi*), however, produce a characteristic alteration of their host cell (Young *et al*., 1978a; Van Vorstenbosch *et al*., 1978). With both species erythrocytic cytoplasm is transformed into crystalline-like material; this appears by electron microscopy as a rectangular veil arranged in a regular pattern with about 25 nm spacing, whatever the shape of the parasite, and reaches a size of 1·8–4·2 × 0·8–2·0 μm. The nature, origin and fate of such crystalline-like material, however, are unknown.

### C. SEXUAL STAGES IN THE INTESTINE OF TICKS AND *in vitro*

In the intestines of ticks that have fed on ruminants infected with *Theileria* species a variety of so-called 'Strahlenkörper' or 'ray-bodies' are seen in smear preparations (Figs 5, 6, 7). The existence of such bodies was first described by Koch (1906) and confirmed by several authors (Dschunkowsky and Luhs, 1909; Gonder, 1910, 1911; Cowdry and Ham, 1932), whereas others ignored their existence or their relationship to the life cycle of *Theileria* (Nuttall and Hindle, 1913; Martin *et al.*, 1964; Hadani *et al.*, 1969). After the rediscovery of these ray-bodies by Schein (1975), our group was able to prove conclusively by electron microscopy (Mehlhorn *et al.*, 1975; Mehlhorn and Schein, 1976) as well as by *in vitro* culture experiments with infected erythrocytes (Mehlhorn *et al.*, 1984) that these stages belong, as apparently sexual forms, to the life cycle of *Theileria* species (which is summarized in Table 4). The development and fate of such stages now appears to be as follows.

(1) Beginning on the second to fourth day after completion of feeding by the ticks (or about 4 days after incubation of infected erythrocytes in a culture medium) ray-bodies are formed by transformation of the ovoidal or

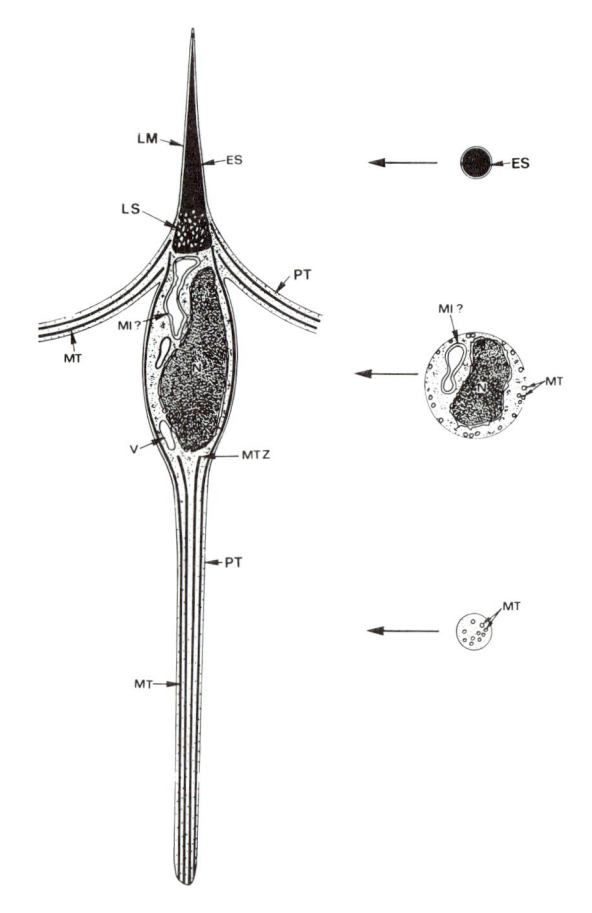

FIG. 5. *Theileria annulata*. Diagrammatic representation of a microgamont before division of the nucleus.

spherical intraerythrocytic stages described above. This process is initiated when these ovoid forms are released from their erythrocytes during lysis inside the intestine (or by manipulation in culture). The ray-bodies (Fig. 7c) are spindle-shaped; they measure about 8–12 μm in length with a diameter in their middle region of about 0·8 μm. These parasites, which are bounded by a unit membrane, have a stiletto-like apex or 'thorn', several flagella-like protrusions (4–5 μm long) and a slender posterior pole (Figs 5, 7b). The stiletto-like structure is electron-dense and measures about 2·2–2·8 μm in length. Its base appears spongy, consisting of coiled, fibrillar elements about 20 nm in diameter. Up to four flagella-like protrusions are found in cross sections taken near the base of the stiletto-like structure. Maximally, ten

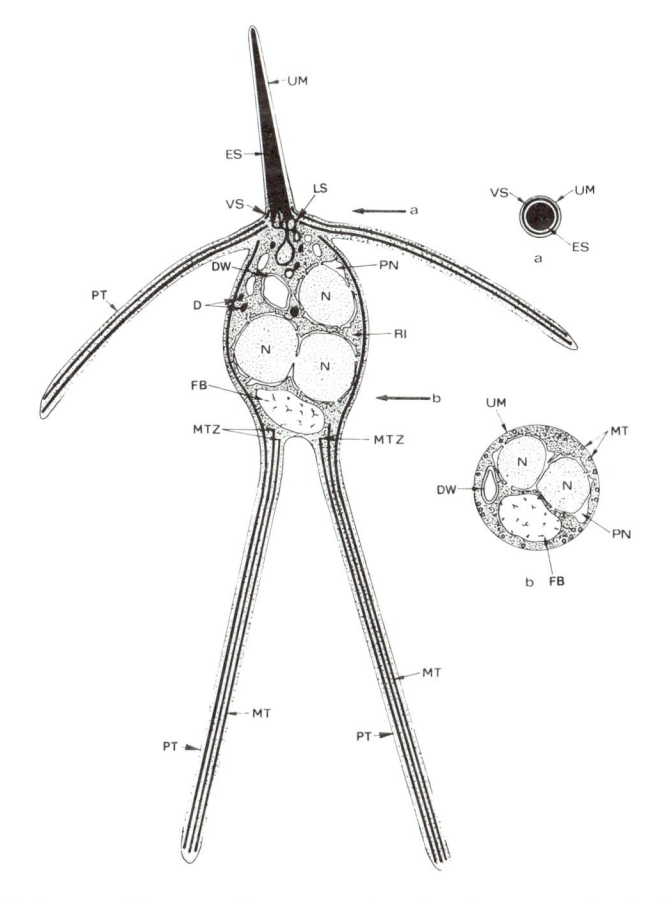

FIG. 6. *Theileria parva*. Diagrammatic representation of a microgamont (ray body). This stage gives rise to uninucleate gametes. a. Cross-section through the thorn; b. cross-section through the nuclear region.

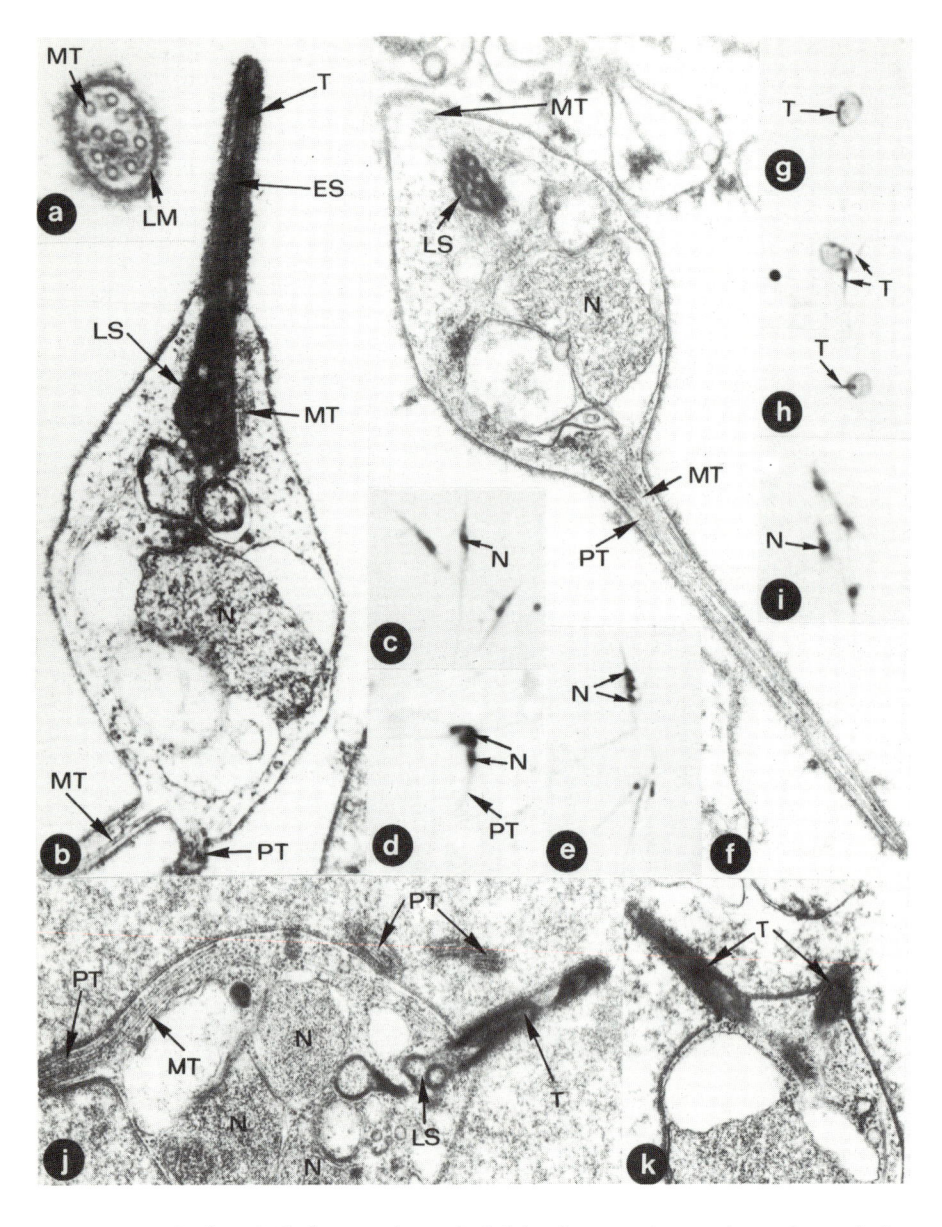

FIG. 7. Ray-bodies of *Theileria* species. a, b, f, j, k: electron micrographs; c, d, e, g, h, i: light micrographs. a. Cross section through a protrusion of a ray-body. Note irregular arrangement of single microtubules. Magnification ×31 500. b, f. Longitudinal sections through uninucleate ray-bodies of *T. annulata*. Magnification ×17 500. c, d, e, g, h, i. Ray-bodies in smear preparations from tick gut and blood culture. Magnification ×1050. j. Longitudinal section of a ray-body of *T. parva* with three nuclei. This stage represents a gamont before division. k. Section through a ray-body of *T. parva* with two thorn-like structures before division. Magnification ×14 000.

TABLE 4

Development of Theileria species in ticks maintained at 28°C

| Theileria species:<br>Vector species:<br>Stages: | T. taurotragi<br>Rhipicephalus appendiculatus<br>Nymph to adult | T. parva<br>Rhipicephalus appendiculatus<br>Nymph to adult | T. annulata<br>Hyalomma anatolicum<br>Nymph to adult | T. mutans<br>Amblyomma variegatum<br>Nymph to adult | T. velifera<br>Amblyomma variegatum<br>Nymph to adult |
|---|---|---|---|---|---|
| Time from feeding to first occurrence of microgametes (days) | 1–2 | 2 | 3 | 5 | 3 |
| Length of microgametes ($\mu$m) | 8 | 10 | 12 | 7 | 7 |
| Diameter of macrogametes ($\mu$m) | 4 | 4–5 | 4 | 4 | 4 |
| Diameter of mature zygotes ($\mu$m) | 12·2 × 9 | 9 | 10 | 6 | 12 |
| Mean time from feeding to moulting (days) | 14 | 19 | 21 | 26 | 26 |
| Earliest transformation of zygotes to kinetes (days) | 12 | 20 | 12 | 30 | 15 |
| Size of kinetes ($\mu$m) | 22·1 × 6·5 | 19·0 × 5·5 | 17·6 × 5·9 | 14 × 4·7 | 18·6 × 7 |
| First detection of kinetes in haemolymph (days) | 16 | 22 | 17 | 34 | 16 |
| First penetration of salivary glands by kinetes (days) | 18 | 25 | 18 | ? | ? |

microtubules are seen within these protrusions, which, however, at their free ends usually contain only two microtubules. Near the base of the stiletto-like structure additional microtubules originate, running beneath the peripheral margin to the posterior cell pole. In cross sections through the middle region of the parasites about 20 microtubules are seen, whereas in the slender posterior part only about 10 appear. The slender posterior part has a diameter 0·25–0·3 μm, nearly double the size of the flagella-like protrusions. The thicker middle region of the parasites is occupied by the nucleus and several double-membraned organelles (Fig. 7f). The ray-bodies are considered to be microgamonts, because beginning from the fifth day after feeding (or after incubation *in vitro*), ray-bodies with up to four nuclei and 'thorns' appear (Figs 6, 7d, e, h, j, k); this indicates that there is division, apparently leading to uninucleated gamete-like stages. In addition to these stages, spherical ones of 4–5 μm in diameter are found, which seem not to form protrusions or which do not divide (Fig. 7g). These are considered to be macrogametes (an interpretation which must be considered tentative, since they may be undeveloped stages incapable of further development).

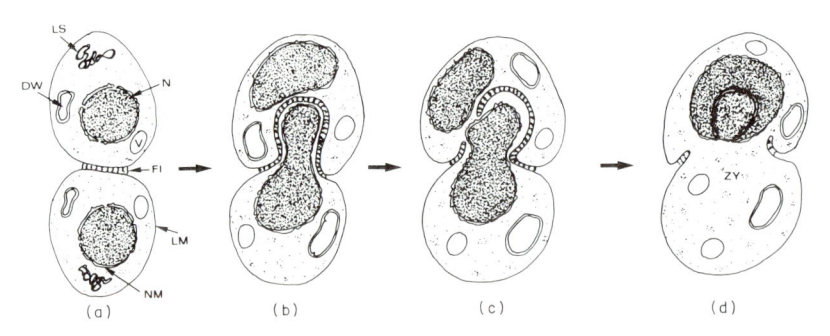

Fig. 8. Diagrammatic representation of the fusion of gametes of *Theileria* species as seen in blood cultures. a. Two gametes in contact. Between their limiting membranes filamentous structures (FI) are formed. b. One gamete penetrates the other. c. Fusion of gamete membranes leading to an open connection. d. This connection becomes larger and the nuclei fuse.

2. Syngamy of gametes (their thorn-like structures having been reduced) occurs 6 days after incubation of intraerythrocytic stages in cultures (Fig. 8a–d). Fusion starts when two gametes come into close contact, by the formation of filamentous structures between their limiting membranes (Figs 8a, 9a; FI). The surface of one of the gametes forms a finger-like protrusion that penetrates into the opposite one (Fig. 8b, d, e). At a point of contact the limiting membranes of the gametes fuse, leading to an open connection (Figs 8c, 9f, g). This connection becomes larger; apparently the nuclei unite through this enlarged opening and fuse. This process is seen up to 13 days

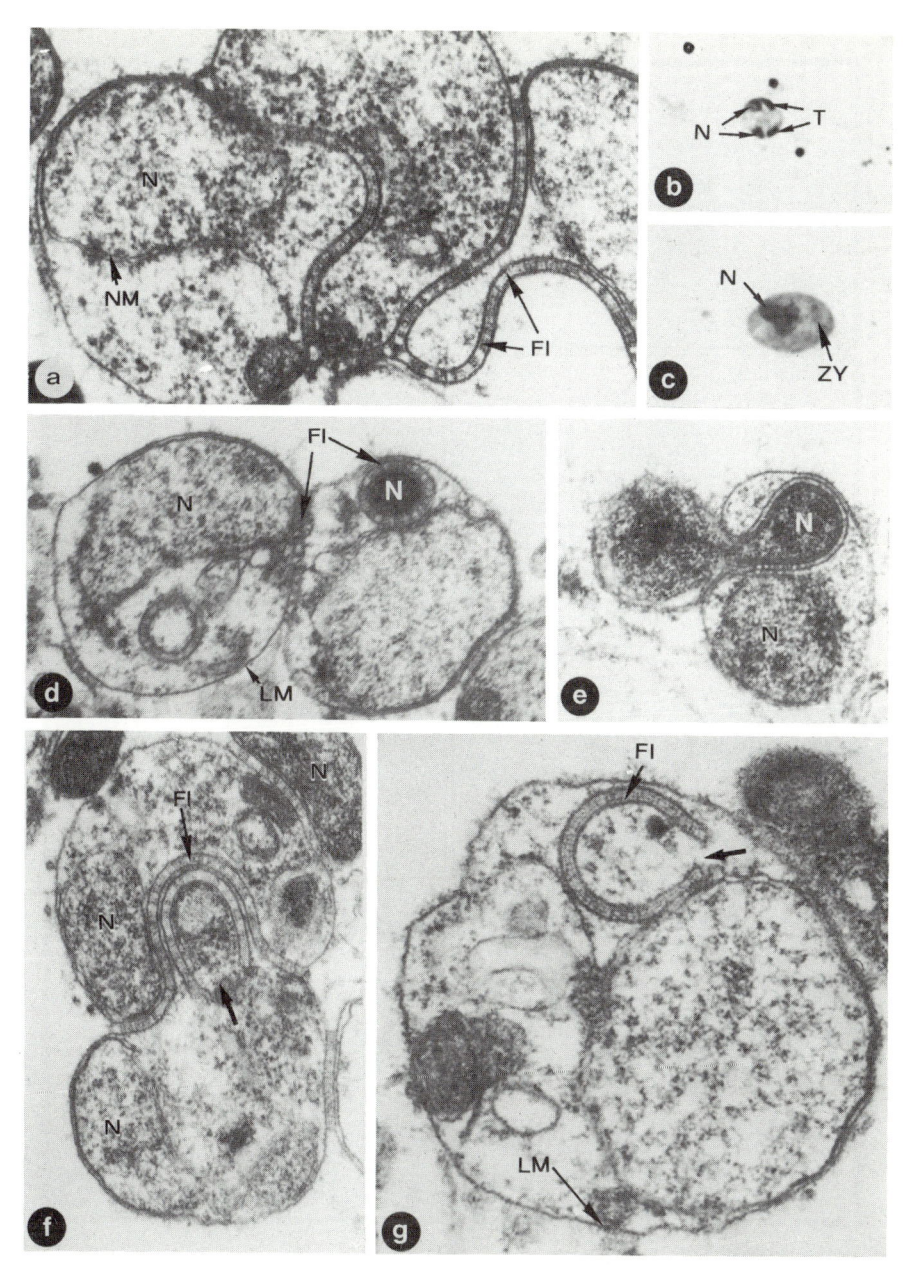

FIG. 9. Micrographs of fusion of gametes in *Theileria* (from blood cultures of *T. parva* and *T. annulata*). a, d, e, f and g, electron micrographs; b and c, light micrographs. a. Gametes in close contact. Magnification ×16 500. b. Gametes during fusion. c. A growing zygote, the nucleus (N) of which is enlarged. Magnification ×990. d. Gametes inside which cross-sectioned finger-like protrusions of other gametes are seen. e. Similar stages as in d, sectioned in another plane. Magnification ×9900. f and g. The cytoplasm of gametes is in direct contact at arrows. Magnification ×11 880.

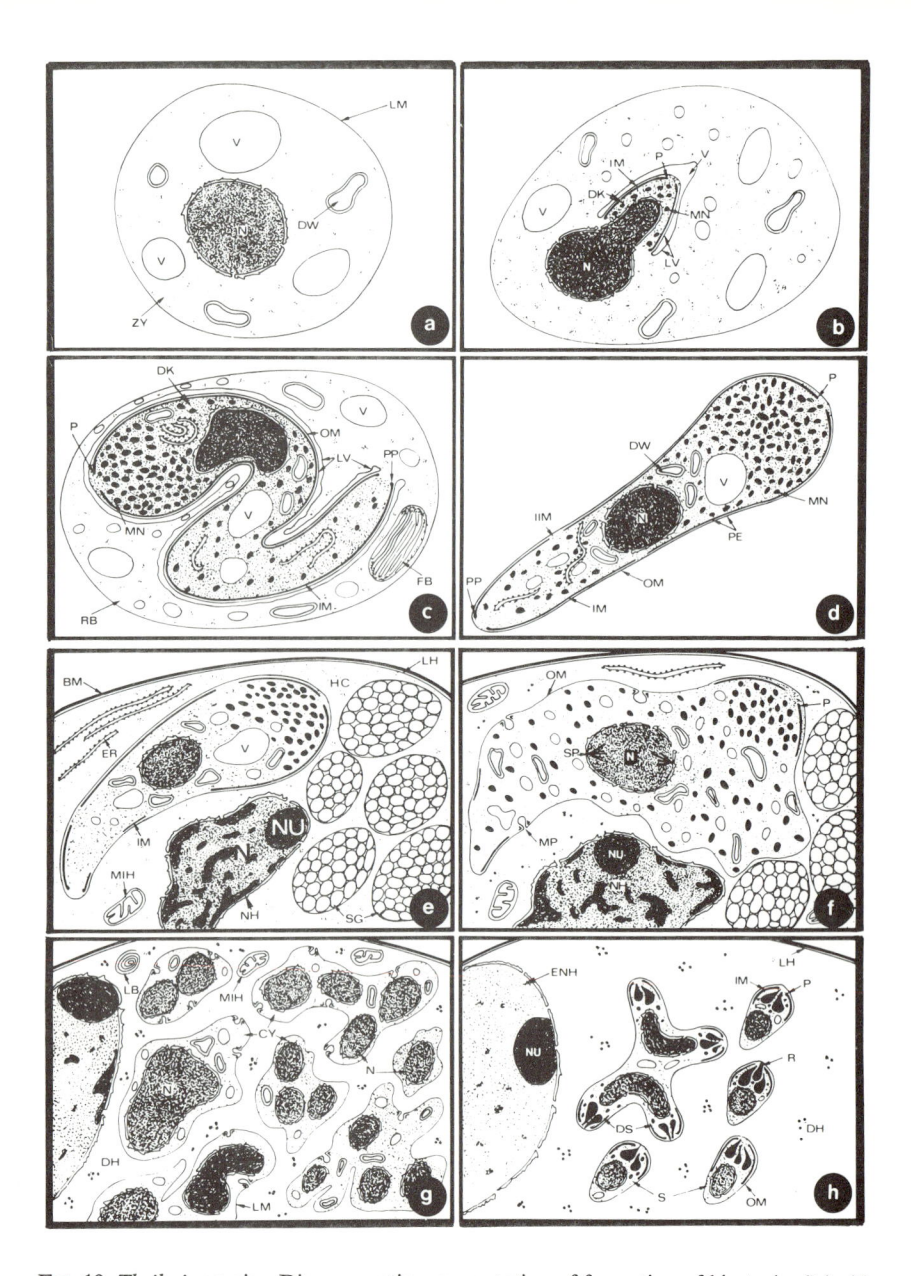

FIG. 10. *Theileria* species. Diagrammatic representation of formation of kinete (a–d) inside intestinal cells of ticks and its fate after having penetrated a cell of the salivary gland (e–h). Note that the developing kinete protrudes into an enlarging vacuole within a zygote (b, c). The mature kinete is directly surrounded by the cytoplasm of its host cell (e, f). Its nucleus elongates, as long strands of nuclear material (g) are formed. Finally the parasite is divided into cytomeres of various sizes, at the surface of which typical sporozoites (h) are formed. The latter are transmitted when the tick next feeds.

after incubation commences; it gives rise to a zygote which can also be observed *in vivo*, in smear preparations of intestines of ticks (Fig. 9b, c).

(3) On days 12–30 after feeding is complete (by the ticks listed in Table 4), transformation of the stationary ovoid or spherical zygote is initiated leading to a club-shaped motile stage (Fig. 10a–d), which—with reference to the so-called ookinete of malaria parasites—is called a kinete (Fig. 13b; for size see Table 4). The mode of transformation is unique in Protozoa and differs from that in the Haemosporidia (see Mehlhorn *et al.*, 1980b). The zygotes are bound by a single cell membrane without micropores (Fig. 10a). Their cytoplasm appears relatively dense and contains some microneme-like organelles, in addition to a larger number of electron-lucent vacuoles 1–2 μm in diameter (Fig. 11). The nucleus of these cells is often situated at the periphery, measures 3 μm in diameter, and has homogeneous nucleoplasm. In other (probably more developed) stages, the cytoplasm and the nucleus form a protrusion into one of the vacuoles adjacent to the nucleus (Figs 10b, 11c). This protrusion is bounded by the vacuolar membrane (Fig. 10b, c), under which appear two newly formed membranes, adjacent to each other. Thus, a typical coccidian pellicle surrounds this protrusion, whose inner complex is interrupted at the apical pole of the protrusion. It forms a thick, electron-dense polar ring (Fig. 12e), which gives rise to ribs. Under the ribs as many as 40 subpellicular microtubules are observed, even in these early stages (Fig. 11c). A conoid or similar organelle is never found at the apical pole; however, numerous micronemes measuring 0·08 × 0·18 μm are seen (Figs 10c, 12a). In this stage of development the subpellicular microtubules stretch from the polar ring to the base of the protrusion. A peculiar vacuole, containing long filaments, is present in the cytoplasm of the residual body in addition to the developing kinete (Fig. 10c). As development progresses the protrusion extends by progressively incorporating cytoplasm of the initially spherical cell (Figs 11b, 12a, e). The membrane of the vacuoles surrounding this stage is now very close to the surface of the developing protrusion (Fig. 10c). Finally only a small zone of cytoplasm surrounds the differentiating new stage. This zone, however, remains connected with the residual body by a small bridge at the posterior pole (Fig. 12a). Always, only a single kinete is formed. Although some electron micrographs suggest the presence of several kinetes, this appearance is due to the large size and tortuous shape of the kinete which may be sectioned in several places (Figs 10c, 11b, 12a, e). The kinete assumes a long drawn-out form before the residual body is ruptured and the motile parasite set free.

(4) The differentiated kinetes are apparently formed more or less simultaneously, since they are always observed in larger numbers from the times indicated in Table 4. (In *T. ovis* this process is completed only after at least 31 days; Mehlhorn *et al.*, 1979.) The kinetes (Fig. 13), the size of which varies

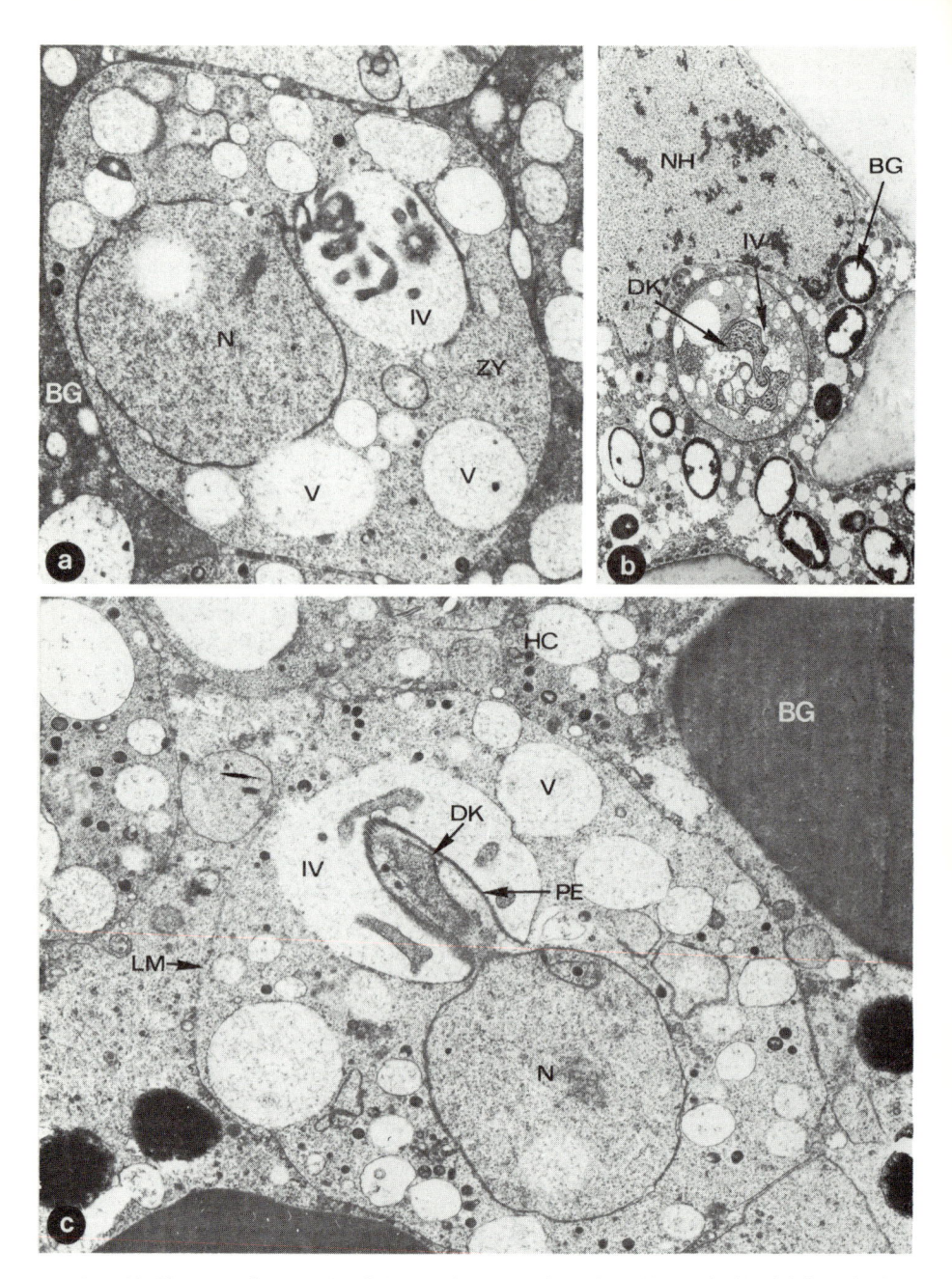

FIG. 11. Electron micrographs of the development of *Theileria* kinetes inside tick intestinal cells. a. Zygote of *T. annulata* with large vacuoles (V). Magnification × 11 200. b. Developing kinete of *T. ovis*. Magnification × 2800. c. *T. annulata*. Early stage of kinete formation. The kinete protrudes into the inner vacuole (IV). During this process a pellicle of 3 membranes (PE) is formed. Magnification × 11 200.

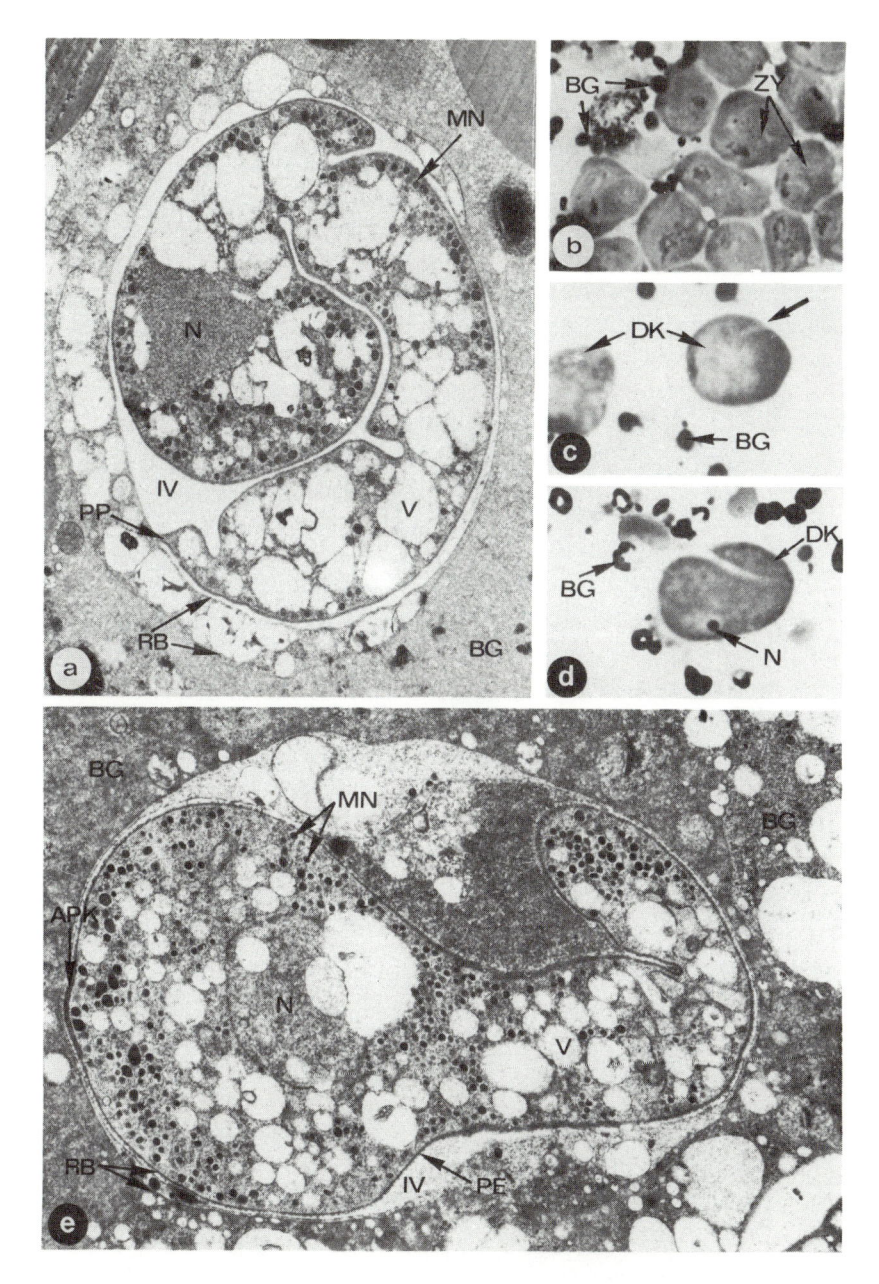

FIG. 12. Micrographs of final differentiation of kinetes from zygotes of *Theileria* species. a, e, Electron micrographs; b–d, light micrographs. a. *T. parva*. This kinete is still connected by its posterior end (PP) with the remnants (RB) of the original spherical zygote. Magnification ×11 200. b. *T. ovis*. Zygotes in tick intestinal cells. Magnification ×1050. c, d. *T. parva*. Kinetes folded (arrow) within remnants of host cells. Magnification ×1050. e. *T. annulata*. Differentiated kinete. Magnification ×7000.

according to the species (see Table 4) and found at first inside the masses of lysed blood within the intestinal cells of the ticks. The differentiated kinetes are surrounded by a typical pellicle, the inner complex of which is absent from only a few places (Fig. 10d; IIM). Typical micropores are not seen. About 40 subpellicular microtubules, anchored at the apical polar ring, stretch for about half the length of the cell. The nucleus, about 3·5 μm in diameter, is often situated centrally or close to the apical pole. It is bounded by a typical nuclear envelope (Fig. 13b). The nucleoplasm often contains dense, granular

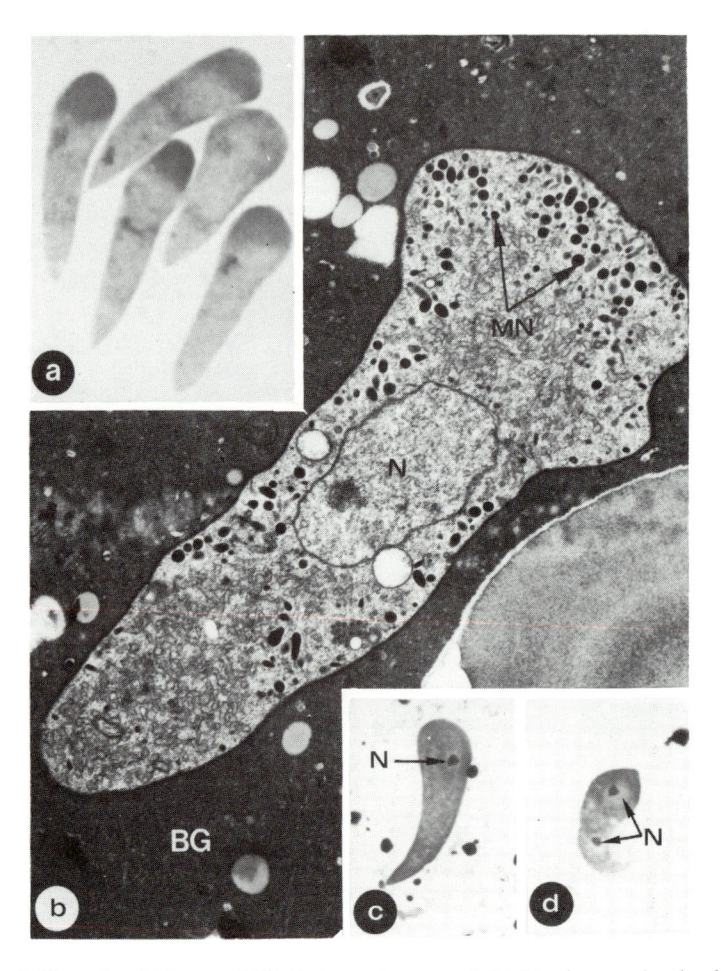

FIG. 13. Differentiated kinetes of *Theileria* species. a, c, d, Light micrographs; b, electron micrograph. a. *T. annulata* kinetes in tick haemolymph. Magnification ×1500. b. *T. velifera* kinete within the digested blood mass inside tick intestinal cell. Magnification ×11 250. c, d. *T. parva*. Division of the kinete nucleus may start even before penetration into salivary gland cells. Magnification ×1350.

inclusions, but appears to be relatively homogeneous. In a few kinetes a spindle apparatus is noted close to the nuclear envelope, but its exact location is impossible to ascertain. Numerous micronemes are present, scattered throughout the entire kinete, with the largest number at the apical pole (Figs 10a, 12e; MN). Typical vacuoles with a maximum size of 1–2 μm are also found in the cytoplasm (Fig. 12a; V). The number of double-walled structures increases during development and, although these organelles have only a few inner protuberances, they resemble mitochondria (Fig. 10a, DW). At the posterior cell pole, lacunae of endoplasmic reticulum become especially evident, often forming a spiral pattern (Fig. 13b). The kinetes of most *Theileria* species are uninucleate when observed inside the tick's intestinal cells (Fig. 13b). In *T. parva* kinetes, however, nuclear division starts so early that kinetes with up to four nuclei (Fig. 13c, d) are often found (Mehlhorn et al., 1978).

The occurrence of kinetes in the haemolymph is often correlated with the moult of the vector ticks (see Table 4); however, it may also occur before the moult as in *T. annulata* or *T. velifera*. Penetration of the kinetes into the cells of the salivary glands is apparently possible only after the moult. This is due to the fact that the salivary glands of ixodid ticks become reduced after each feed (see Fawcett et al., 1981a,b, 1982a) and only begin to develop when moulted tick stages become reattached to a host. Neither penetration nor development was observed in organs other than salivary glands.

D. DEVELOPMENT IN SALIVARY GLANDS OF VECTOR TICKS

Beginning on the first day after attachment of the vector ticks to their hosts, kinetes are to be found inside the cells of the salivarian alveoli (Fig. 14a) of the type III acinus, which contains three glandular cell types (d,e,f). Cells of type d are located close to the acinar duct and contain spotted granules (Fig. 14a; SG), whereas secretory granules of e-cells are uniformly stained dark in electron microscopy. Both cell types are entered by theilerian kinetes and in both development may occur. The kinetes are lodged directly within the cytoplasm of the host cell, not surrounded by a parasitophorous vacuole (Figs 10e, 14a). The typical pellicle of the kinetes, however, is reduced to a single cell membrane in several places (Fig. 10e) owing to the disappearance of the inner complex. Subsequently the shape changes and the parasites become polymorphic, stretching between the secretion granules of the gland cell (see Fig. 10f). The apical polar complex, however, is retained for a longer period; it is still found in parasitic cells that measure 25 μm in diameter. Like the cytoplasm of the parasite, the nucleus increases in size and often has a central accumulation of dense material (Fig. 14a). Parasite growth is probably due to ingestion of host cell cytoplasm through micropores

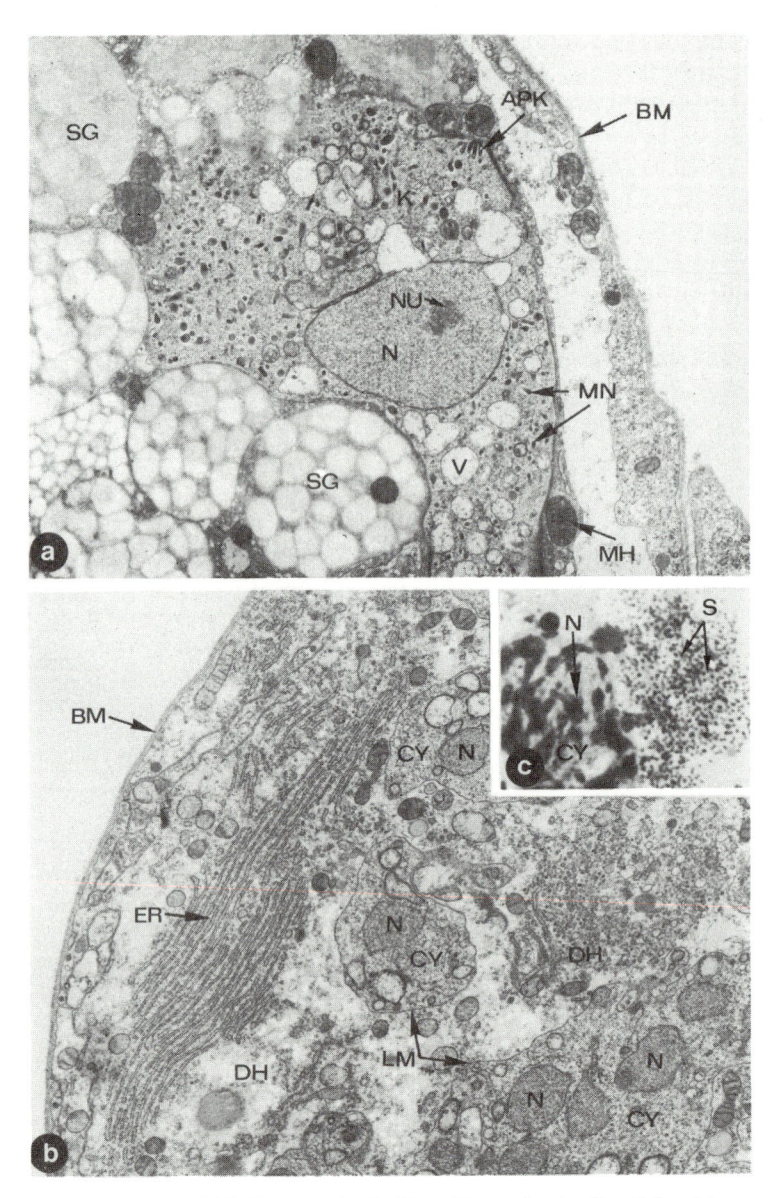

FIG. 14. Development of *Theileria* species inside salivary gland cells of the d-type. a, b, Electron micrographs; c, light micrograph. a. *T. ovis*. The intracellular kinete enlarges and becomes polymorphic; after nuclear division it forms cytomeres. Magnification ×5600. b. *T. parva*. The parasite has filled the host cell with numerous cytomeres (CY); sporozoites are then differentiated at their surface. Magnification ×7000. c. *T. ovis*. Smear of salivary gland alveolus containing sporozoites (S) beside mass of undifferentiated parasitic material (N). Magnification ×1400.

which now appear at the surface of the parasite (Fig. 10f; HP). Concomitant with parasite growth is the disappearance of micronemes, whereas small vacuoles and double-walled mitochondria-like organelles seem to increase in number. Up to three of these organelles are observed in a single section. The nucleus soon starts dividing and nuclei of 5–8 μm in diameter are seen; these are also found in later stages (Figs 10g, 14b). With repeated nuclear divisions, the nuclei become progressively smaller, with thousands being formed (Fig. 14c). The volume of the cytoplasm increases rapidly, but does not lead to the formation of a single giant 'plasmodium'. Numerous invaginations form cytomere-like compartments on the second day after attachment of the new adults and infected salivary gland cells are completely filled with cytomeres that contain relatively few nuclei (maximum 20) (Fig. 14b). When examined by light microscopy, however, the parasite seems to have a compact form, as the cytomeres are very closely packed together (Fig. 14c, left side). This development is a continuous process and is not—as described in the literature (see Fawcett *et al.*, 1982a)—carried out in steps through different types of 'sporoblasts'. The cytoplasm of the host cell is damaged considerably, and only remnants of cytoplasm are seen scattered between the cytomeres (Fig. 14b; DH).

The nucleus of the host cell increases in size immensely and is very often found at the periphery of the alveolus. Cytomere formation is finished 5 days after re-attachment of the ticks. Subsequently, formation of infective sporozoites is observed. It is noted in serial sections that several sporozoite-protrusions (anlagen) develop from a nucleus and its surrounding cytoplasm. Above the poles of the multiply dividing nucleus a protrusion is found, limited by the cell membrane of the cytomere and by the newly formed underlying complex (Fig. 10h). In each protrusion a few micronemes and four rhoptries appear together with a part of the dividing nucleus, as shown diagrammatically in Fig. 10h. The cytoplasm of the cytomeres is completely used up during this process without leaving a residual body, so that after the fifth day following attachment of the ticks the parasitized alveolus is filled with ovoid sporozoites measuring up to 1 μm in diameter (Fig. 15). These infective particles (the fine structure of which is shown in Fig. 15d, e) are now ready for transmission to a vertebrate host when the tick feeds. The results of Fawcett *et al.* (1982a) indicate that the typical apical pole (Fig. 15e; P) is not always completed in the sporozoites of *T. parva*. These structures, which are thought to be essential to the penetration process in other coccidia, are apparently of less importance in theilerian sporozoites, since Fawcett *et al.* (1982b) found that sporozoites of *T. parva* attack at any point on their prospective host cells (i.e. lymphatic cells of cattle) and may enter by either pole. The number of sporozoites in a single salivary gland cell can hardly be estimated. However, comparison of the size of sporozoites with that of the

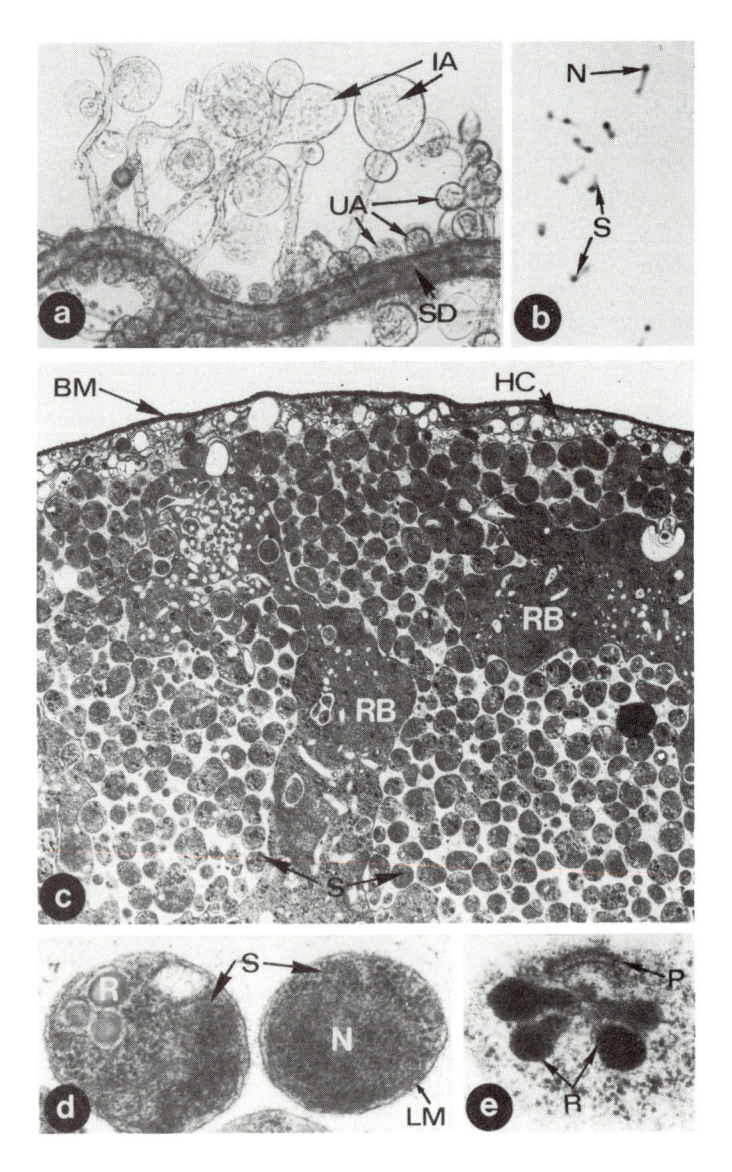

FIG. 15. Final development of *Theileria* species in the cells of the salivary glands. a, b, Light micrographs; c–e, electron micrographs. a. Alveoli of tick salivary gland. Note that parasite-containing alveoli of type III (IA) are much larger than uninfected ones (UA). Magnification ×160. b. Sporozoites of *T. parva* in tick saliva. Magnification ×960. c. A salivary gland cell (type e) filled with sporozoites of *T. parva*. Magnification ×2880. d and e. Higher magnification of sporozoites of *T. ovis*. Magnification ×36 000. Note that in the tangential section (e) a polar ring is present.

swollen cells (Fig. 15a, b) indicates about 50 000 sporozoites per host cell. This amounts to an enormous inoculum from one feeding tick, when many of the alveoli of type III acini are infected as shown in Fig. 15a.

## IV. The Life Cycle of *Babesia* Species

*Babesia* species are common parasites of cattle (and other related ruminants), sheep, goats, horses, pigs, cats, dogs and rodents (Tables 2 and 3). Some *Babesia* species parasitic in cattle (especially *B. divergens*) and rodents (*B. microti*) may even infect human beings and lead to at least as severe symptoms (or death) as in their natural hosts (Anderson *et al.*, 1974; Wernsdorfer, 1983). All species have a range of vector ticks; some are listed in Tables 2 and 3. Apart from the size and shape of intraerythrocytic stages, *Babesia* species differ from *Theileria* species in that *Babesia* are transmitted to the next generation of vector ticks via eggs. This transovarial transmission is apparently the reason why some *Babesia* species have successfully spread throughout the world and are found even in Europe (Tables 2 and 3). The systematic position of at least some of the so-called smaller *Babesia* species of the '*Nuttallia*' type (Table 3) is doubtful, several authors having found reasons to exclude them from the genus *Babesia* and to establish new genera. Some of the reasons are given in Section IV A.

### A. *Babesia* species with intralymphocytic schizonts

The development of *Babesia* species in the vertebrate host, by definition, occurs exclusively within erythrocytes (see Friedhoff, 1981) (Fig. 16). However, in at least two species usually ascribed to this genus—*B. equi* (Schein *et al.*, 1981) and *B. microti* (H. Mehlhorn and E. Schein, unpublished results)—sporozoites invade the lymphocytes of their hosts (and in cell cultures). They are found inside the cytoplasm and develop fissured *Theileria*-like schizonts, often being split up into complex cytomeres between which remain strands of host cell cytoplasm (Fig. 17a, b).

The intralymphocytic schizont cytomeres are bounded by a single membrane with micropores and contain polymorphic nuclei, the chromatin of which is dispersed so that the nucleoplasm has about the same density as the cytoplasm of the schizonts. Small spindle poles are observed in the nuclear envelope. A few cisternae of endoplasmic reticulum and small organelles bounded by two membranes, probably representing mitochondria, lie in the cytoplasm of the schizonts. Precursors of the rhoptries of the subsequently formed merozoites appear as osmiophilic globules in the cytoplasm. A fragmentation process may occur, in the course of which the schizonts appear to be divided into even smaller cytomeres.

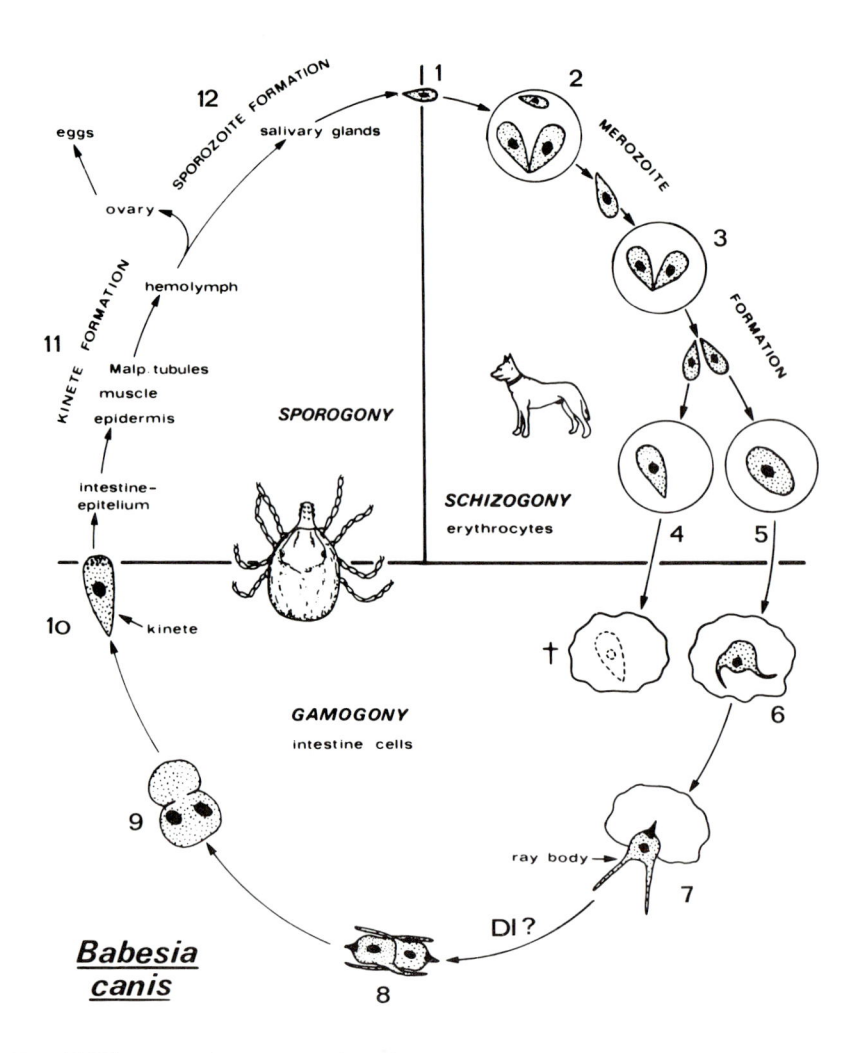

FIG. 16. Diagrammatic representation of the life cycle of *Babesia canis*. **1.** Sporozoite in saliva of feeding tick. **2–3.** Asexual reproduction in erythrocytes of vertebrate host (dog) by binary fission, producing merozoites. **4.** Merozoites are lysed in the gut of the tick. **5–6.** After ingestion, ovoid intraerythrocytic stages form protrusions in the tick's intestine and thus become ray-bodies. **7–8.** Two ray-bodies fuse with each other. **9–10.** Formation of a single motile kinete from a zygote. **11.** After leaving the intestine the kinetes enter various organs and initiate formation of new kinetes (see Fig. 25); (eggs are not penetrated by *B. equi*). **12.** A kinete that has penetrated salivary glands gives rise to thousands of small sporozoites.

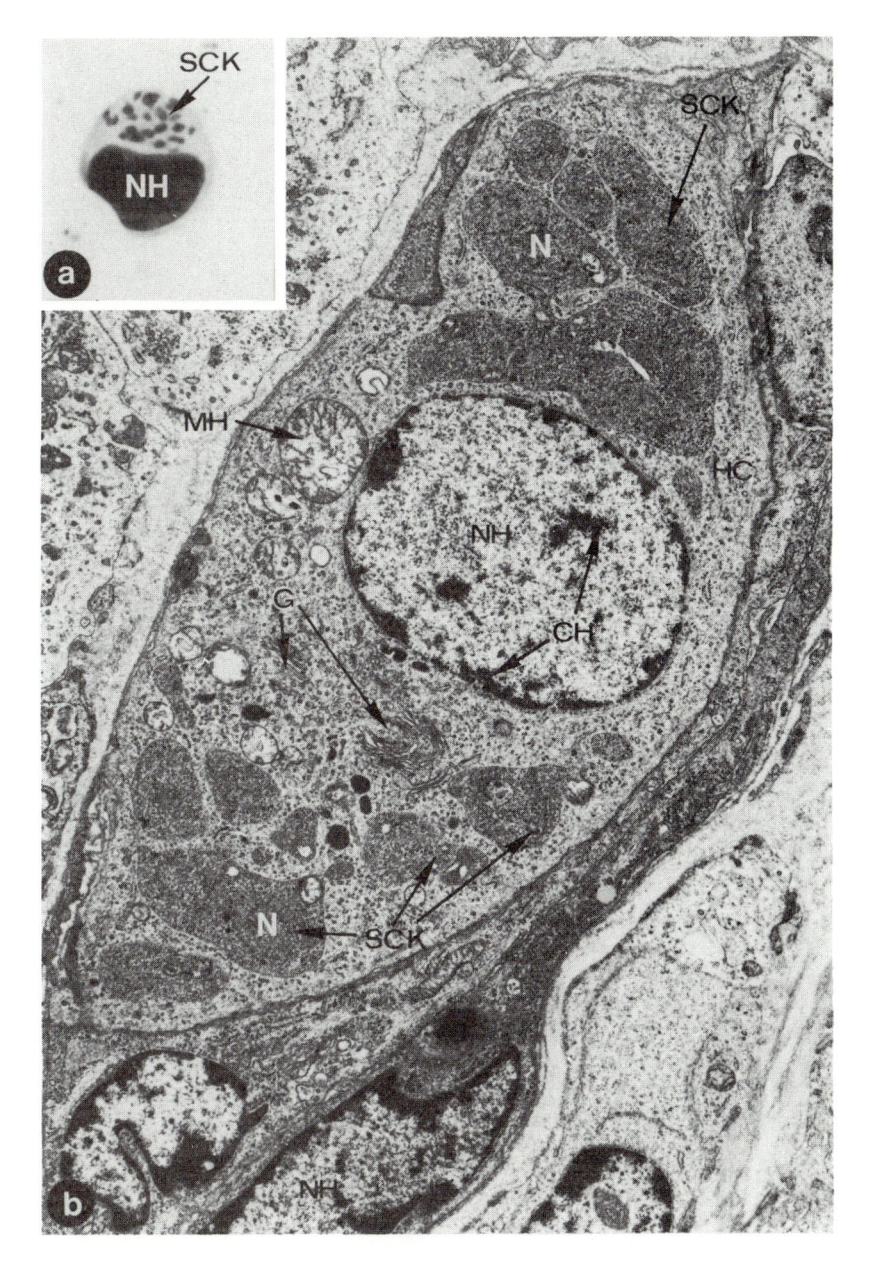

Fig. 17. Schizonts of *Babesia equi* in horse lymphocytes. a. Light micrograph of smear preparation. Magnification ×1600. b. Electron micrograph of section through a lymphocyte containing one or more twisted schizont(s). Magnification ×12 000.

Merozoite differentiation is initiated at several places by the appearance of a double membrane reinforcement beneath the cell membrane. The pellicle protrudes at these points, and an apical complex with rhoptries, micronemes and a polar ring is formed within such merozoite anlagen. Several arise from one cytomere, each incorporating an equal part of the cytoplasm and a portion of the nucleus. During this development the cytoplasm of the schizont progressively merges into the young merozoites until finally all cytoplasm is consumed. Occasionally the last step looks like binary fission of two nearly mature merozoites. On the ninth day after inoculation (*in vitro*) or on days 12–15 after tick attachment (*in vivo*), with *B. equi*, merozoite formation is complete and the mature merozoites occupy most of the host cell. Merozoites are pear-shaped and measure about $2 \cdot 0 \times 1 \cdot 5$ μm. They are bordered by a three-layered pellicle that ends in a polar ring at the anterior and posterior poles. The apical complex consists of rhoptries, micronemes and a ring structure at the very tip. The cytoplasm contains the nucleus and some of the small, mitochondria-like organelles. Finally, the host cell may rupture and release the motile merozoites into the culture medium.

Extracellular merozoites have a fuzzy coat composed of fine fibrils that adhere to the cell membrane of an erythrocyte after contact. Often parasites are found lying in an indentation of the host cell. Subsequently, the merozoites enter the erythrocytes and appear to be constricted during this process. Immediately after invasion, each parasite is found inside a parasitophorous vacuole that disappears later (Fig. 18a).

Because of these findings (Moltmann *et al.*, 1983a), it remains questionable whether *B. equi* and *B. microti* can be included in the genus *Babesia* or whether they should be transferred to *Nuttallia* or *Nicollia*, as suggested by several authors (Trofimov, 1952; Krylov, 1981).

### B. DEVELOPMENT IN ERYTHROCYTES

Location, size and shape of intraerythrocytic stages are characteristic of *Babesia* species and are thus used for diagnostic purposes (Fig. 1e–m; Tables 2 and 3). As described for some species (see Friedhoff and Scholtyseck, 1977; Rudzinska, 1981; Jack and Ward, 1981; Moltmann *et al.*, 1983a), penetration of erythrocytes occurs actively and has the following five phases.

(1) Contact between merozoite and the erythrocyte.
(2) Orientation of the parasite's apical pole towards the erythrocyte, so that the apical organelles, such as rhoptries, come into apposition with the cell surface.
(3) Membrane fusion between merozoite and erythrocyte.
(4) Release of contents of rhoptries.

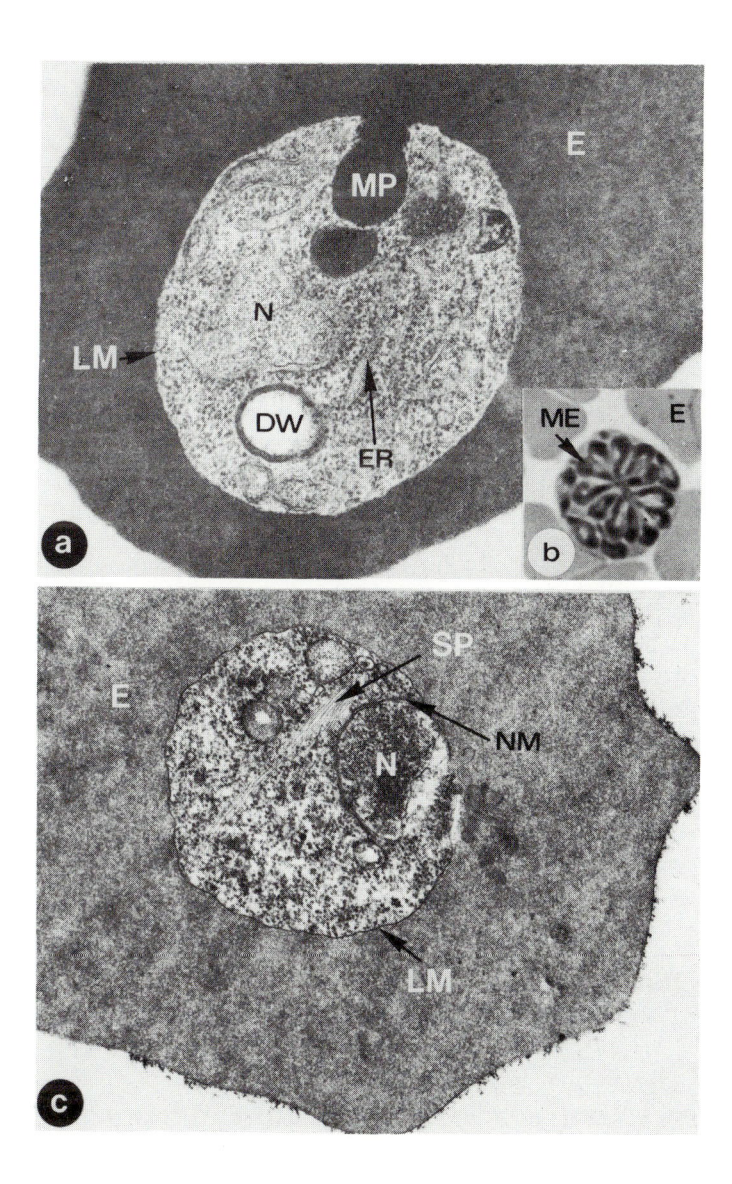

Fig. 18. Micrographs of intraerythrocytic stages of *Babesia* spp. a, c, electron micrographs; b, light micrograph. a. *B. equi*. Magnification ×16 000. b. *B. canis*. Magnification ×1600. c. *B. bigemina*. Magnification ×14 400. Note that the parasites are directly surrounded by the cytoplasm of the erythrocyte, with no intervening parasitophorous vacuole or membrane.

(5) Invagination of the erythrocyte membrane and entry of merozoite. During this final step the limiting membrane of the erythrocyte is not disrupted, so that at first the parasite is closely surrounded by this membrane. During growth the latter disappears and the parasite is released inside the host cell cytoplasm (Fig. 18a, c): this is different from malaria parasites, where parasitic stages always remain enveloped by the limiting membrane of the host cell.

Apart from variations in size and shape—some species having typical pear-shaped merozoites (e.g. *B. motasi, B. bigemina, B. bovis, B. canis, B. divergens, B. caballi*), whereas others are mainly polymorphic (e.g. *B. microti*) —intraerythrocytic merozoites have some common features, which are typical of coccidia *in sensu latu*. They are covered by a more or less complete pellicle consisting of three membranes, have no conoid but possess an apical and a posterior polar ring, rhoptries, micronemes, subpellicular microtubules and a membrane-bound nucleus during division (Fig. 18c; N). Present knowledge suggests that only *B. equi* merozoites possess a micropore (Fig. 18a; MP), a typical feature of malaria parasites and other coccidia. How the other species take up food is not yet understood, although the results of Vivier and Petitprez (1969) indicate possible endocytosis of larger vacuoles at the surface. Some species (e.g. *B. bovis, B. bigemina, B. caballi*) have, in addition, so-called spheroid bodies close to the nucleus; the function of these is unknown (Scholtyseck *et al.*, 1970; Friedhoff and Scholtyseck, 1977).

Merozoites, having entered erythrocytes, rapidly undergo reproduction. This occurs in most species as binary fission leading to the characteristic appearance of paired parasites inside an erythrocyte (see Fig. 1f, i, m). In some species (e.g. *B. equi*; Fig. 1h) four parasites are often formed at the same time, leading to a tetrad or so-called 'Maltese cross' arrangement, whereas in *B. microti* several parasites develop from a polymorphic, schizont-like parent cell (Rudzinska, 1981).\* In any case, rapid reproduction brings about the destruction of the host cell and thus leads to a considerable decrease in erythrocyte numbers and development of haemoglobinuria in *Babesia*-infected animals (see Boch and Supperer, 1983).

Some of the parasites formed inside an erythrocyte do not develop further until they are taken up by ticks during sucking or transferred to blood cultures *in vitro*. These become gamonts. The ratio of gamonts to reproducing stages can, however, only be approximated due to the small size of the intraerythrocytic stages. In some species, such as *B. canis* (Mehlhorn *et al.*, 1980a) and *B. microti* (Rudzinska *et al.*, 1979), morphological transformation into gamonts starts even inside the erythrocytes.

---

\* *B. microti* may also form tetrads of a 'Maltese cross'.

## C. SEXUAL STAGES IN THE INTESTINE OF TICKS AND *in vitro*

The very first description of probable sexual stages was given by Koch (1906) and Kleine (1906) for *Babesia bigemina* and *B. canis*, when they studied development in the tick gut or in a blood culture. These authors observed bizarre-shaped stages with at least one thorn and several stiff flagella-like protrusions. They referred to these stages by the German term 'Strahlenkörper', which should not be confused with the English term 'ciliary body', but was correctly translated by Riek (1964) as 'spiky-rayed stages' or 'ray-bodies'. Since then the existence of 'ray-bodies' was either ignored or doubted in literature for a long time (see Riek, 1964). After the description of similar stages in *Theileria* species by Schein (1975), however, they were rediscovered in several *Babesia* species inside the intestine of experimentally infected vector ticks. Thus a variety of Strahlenkörper (Fig. 22a–h) was found in *B. bigemina* (Friedhoff and Büscher, 1976; Weber and Friedhoff, 1977), *B. microti* (Rudzinska *et al.*, 1979), *B. canis* (Mehlhorn *et al.*, 1980a) and *B. equi* (H. Mehlhorn and E. Schein, unpublished results). Interpretation of the variety of differently shaped stages, however, remained difficult, since they mostly occur closely packed in tick intestinal cells, thus making their origin and derivation impossible to distinguish.

Recent experiments with *B. canis* (Mehlhorn *et al.*, 1981) and *B. equi* (H. Mehlhorn and E. Schein, unpublished results) using culture techniques for infected erythrocytes, however, have allowed the more accurate establishment of a chronological sequence of stages and processes. Thus, the development of the probable sexual stages proceeds as follows (Fig. 19).

(1) The ovoid or spherical intraerythrocytic stages develop into uninucleated Strahlenkörper (= ray-bodies) when they are released from their host cells, whereas the intraerythrocytic kidney-shaped stages degenerate *in vivo*, or divide a few times in cultures before being finally lysed. The uninucleate ray-bodies are spherical, polymorhic or pyramidal with diameters of about 4–7 μm (Fig. 20a, b). They are characterized by a short thorn-like structure (or arrow head), which measures about 1·0–1·2 μm in length and about 0·6–0·8 μm at its base (Fig. 20a; T). At the base, this thorn-like structure is no longer compact but appears labyrinthine (Fig. 20a; LS). The function of the thorn is unknown. Rudzinska *et al.* (1982) suggest that it may be used as an organelle for penetration of the so-called peritrophic membrane, which coats the intestinal canal of some ticks, but is lacking from many tick species. In the latter, intestinal cells become giant due to endocytosis. The ray-bodies are surrounded by a single cell membrane, which forms several typical micropores (Fig. 20b; MP). Five to seven thin protrusions (= Strahlen) often occur opposite to the thorn-like structure (Fig. 20b; PT). These protrusions are up to 8 μm long, about 0·3 μm in diameter, and microtubules of about 20 nm

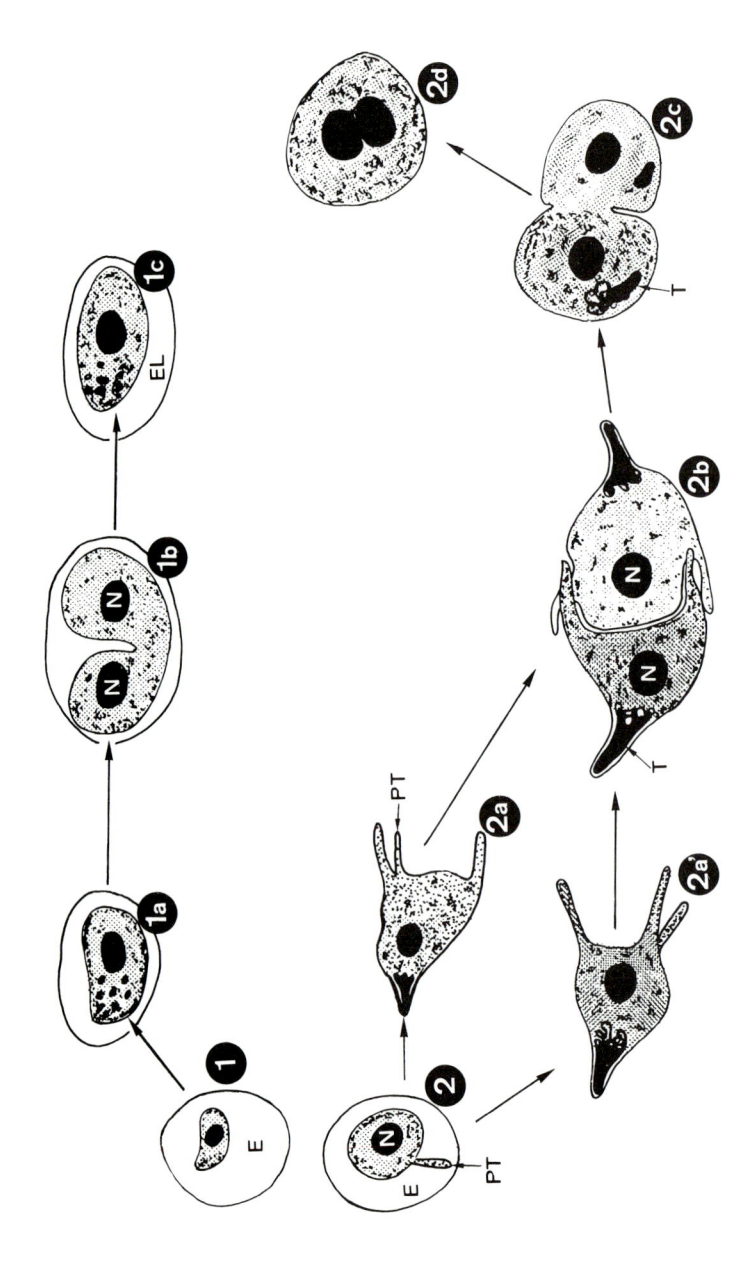

FIG. 19. Diagrammatic representation of stages of *Babesia* species seen successively in blood cultures. **1–2.** Stages seen at the beginning of incubation. **1a–1c.** Formation of merozoite in a degenerating erythrocyte. **2a–2a′.** Strahlenkörper seen both inside and outside erythrocytes about 75 hours after incubation. Two types (dark and light) are present. **2b.** Attachment of two different Strahlenkörper 145 hours after incubation. **2c.** Cytoplasmic connection established in the attached area. **2d.** Fusion of both nuclei. The stages **2b–2d** are interpreted as syngamy of gametes.

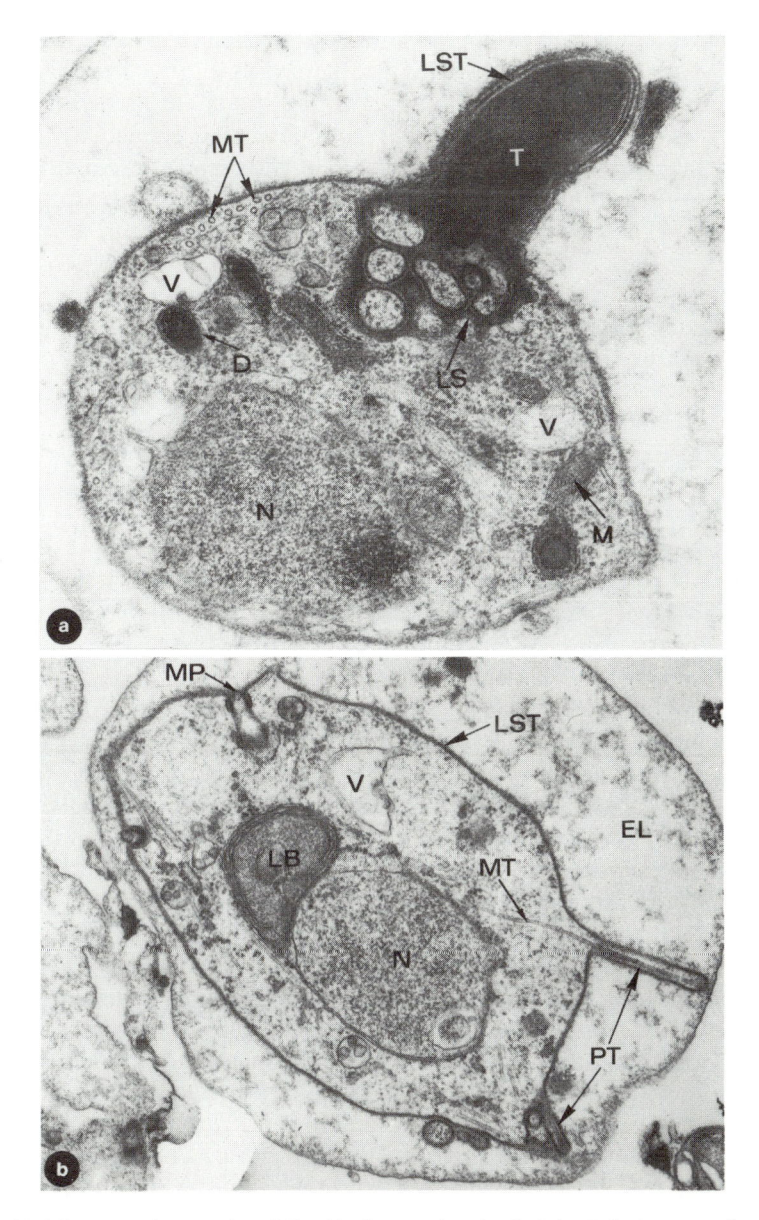

FIG. 20. Electron micrographs of Strahlenkörper (ray-bodies) in *Babesia canis* about 75 hours after incubation in a blood culture. In (b) the parasite is still inside the remnant of the erythrocyte. a. Magnification ×24 000. b. Magnification ×20 400.

thickness are the only structural elements visible in them (Fig. 20b; MT). These microtubules are arranged more or less parallel to the axis of the 'Strahl'. They appear evenly spaced and never form a 9 + 2 configuration or any other pattern. Their number ranges from two up to 12, depending on the thickness of the protrusion (Strahl); each microtubule apparently extends through the entire length of the protrusion. Microtubules are also present in the parasite's cytoplasm, often arranged in bundles in 'corners' of the cell (Fig. 20a; MT). The cytoplasm contains a multilaminated body attached to the single nucleus (Fig. 20b; LB), membrane-bound dense inclusions, small vacuoles, often a bundle of filamentous solid elements, mitochondria-like organelles and cisternae of the rough endoplasmic reticulum. Such stages are seen *in vivo* for about 2 days after feeding or after introduction of infected erythrocytes into cultures. Some indications suggest that these stages may reproduce by binary fission. This, however, is not yet proven.

(2) After incubation in blood cultures for 145 h or about 2–4 days after feeding of ticks, two of the ray-bodies are often found closely attached to each other (Fig. 21a–c). This may occur even inside the remnants of the host cell. Usually the cytoplasm of one ray-body appears denser than that of the other (Fig. 21a). The cell membranes of both these cells fuse to form a single dense line at the site of attachment (Fig. 21a; double arrows). In other pairs it can be seen that the cytoplasm of both ray-bodies is in contact via smaller or larger openings in this attachment zone (Fig. 21b, c). Additional stages occur with two nuclei in one of the joined cells and none in the other and, furthermore, some stages contain two closely attached nuclei (Fig. 22i). The latter are apparently young zygotes, which are visible even by light microscopy in smear preparations (Fig. 22h). From present knowledge, the 'Strahlenkörper' are considered to be anisogametes because of the difference in electron density; however, they appear as isogametes when seen by light microscopy. The various shaped 'Strahlenkörper' seen *in vivo* in the species cited above may be explained by agglomeration of uninuclear ray-bodies, when seen by light microscopy in smears. No cytological finding suggests a meiotic reduction process, which need not necessarily occur during this developmental phase. In *Plasmodium gallinaceum* (Mehlhorn et al., 1980b) no sign of a meiotic process was found at a similar phase of development; further clarification can be expected only after measurements of DNA. (See note on p. 103.)

(3) From a stage that is thought to be a zygote (Fig. 22h, i), a single kinete always develops by a process almost identical with that of theilerian species (see Section III C), the principal features of which are shown in Fig. 23a–d. We were able to verify this for *B. bigemina* (Fig. 24a) and *B. equi*, by making serial sections of intestinal cells of ticks. These kinetes (Fig. 24a, b) measure about 7–8 μm in length and have—apart from slight variations—the same structural elements that characterize the kinetes of theilerian species (Fig.

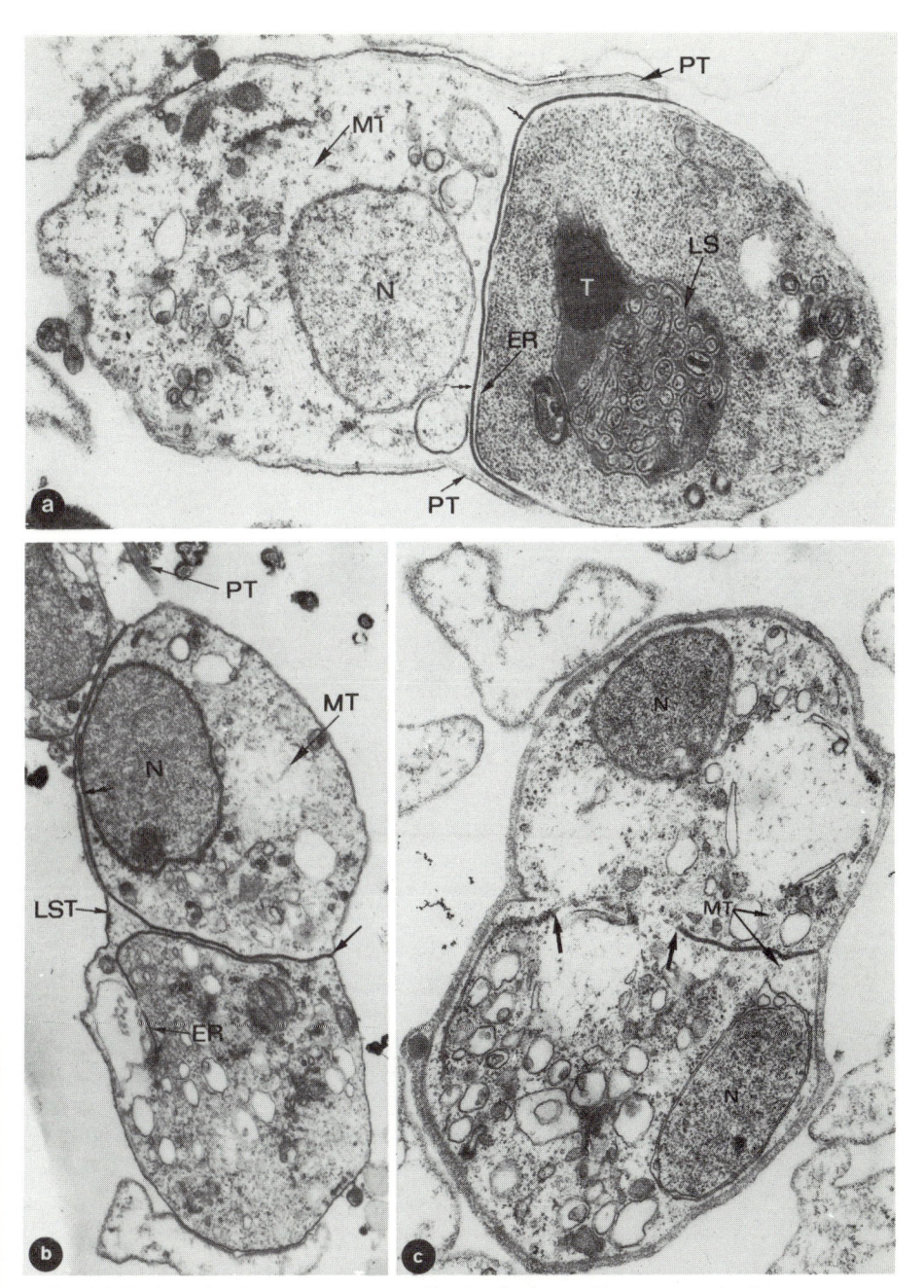

FIG. 21. Fusion of *Babesia canis* gametes after incubation. a. Two gametes in close contact. Note the significant difference in density of both cells. Magnification ×21 600. b. This stage is considered to be the initial fusing stage. Note that the limiting membranes (LST) are already continuous on one side and closely attached at the other (arrow). Magnification ×19 200. c. There is now an opening in the attached membranes (arrows) through which the nuclei (N) come together. Magnification ×19 200.

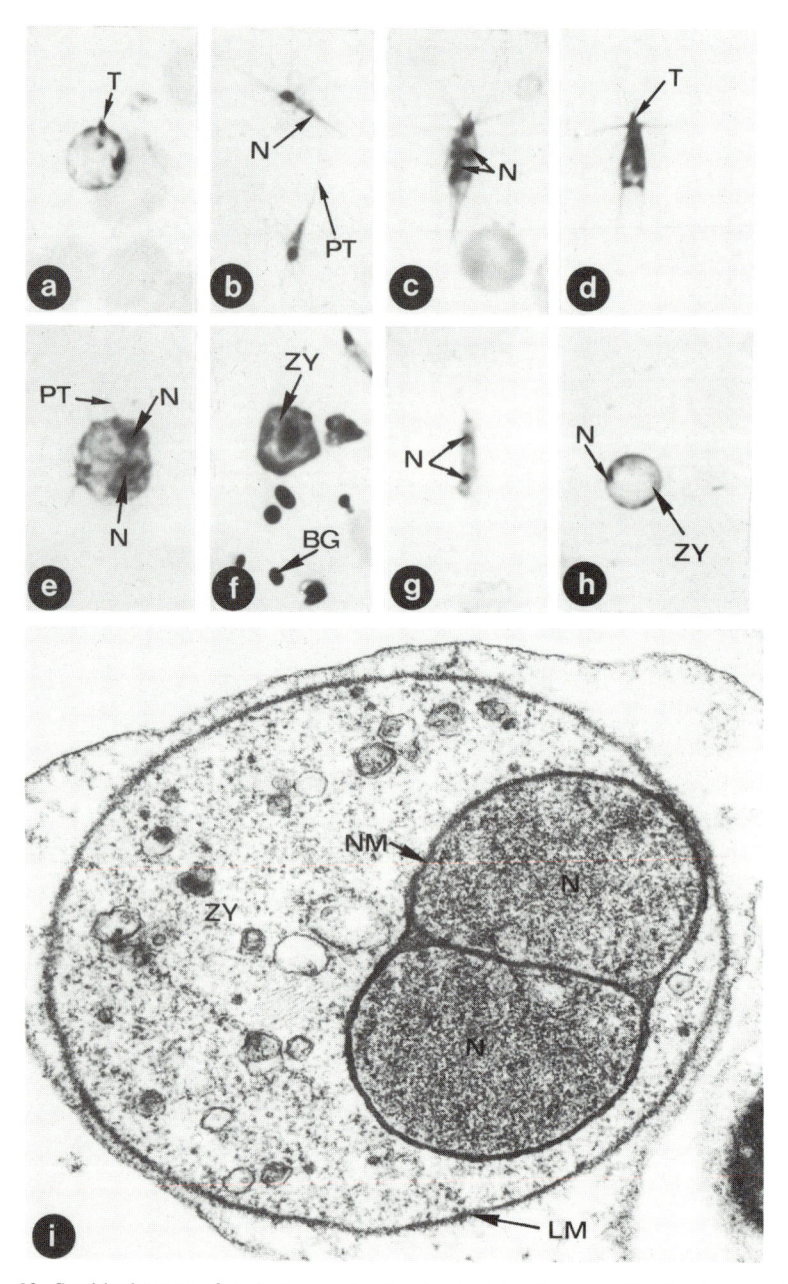

FIG. 22. Strahlenkörper of *Babesia* species *in vivo* and *in vitro*. a–h, light micrographs; i, electron micrograph. a–f. *B. canis*. Magnification ×1125. g–h. *B. bigemina*. Magnification ×1125. i. Section through a zygote, the nuclei of which are fusing. Magnification ×27 000.

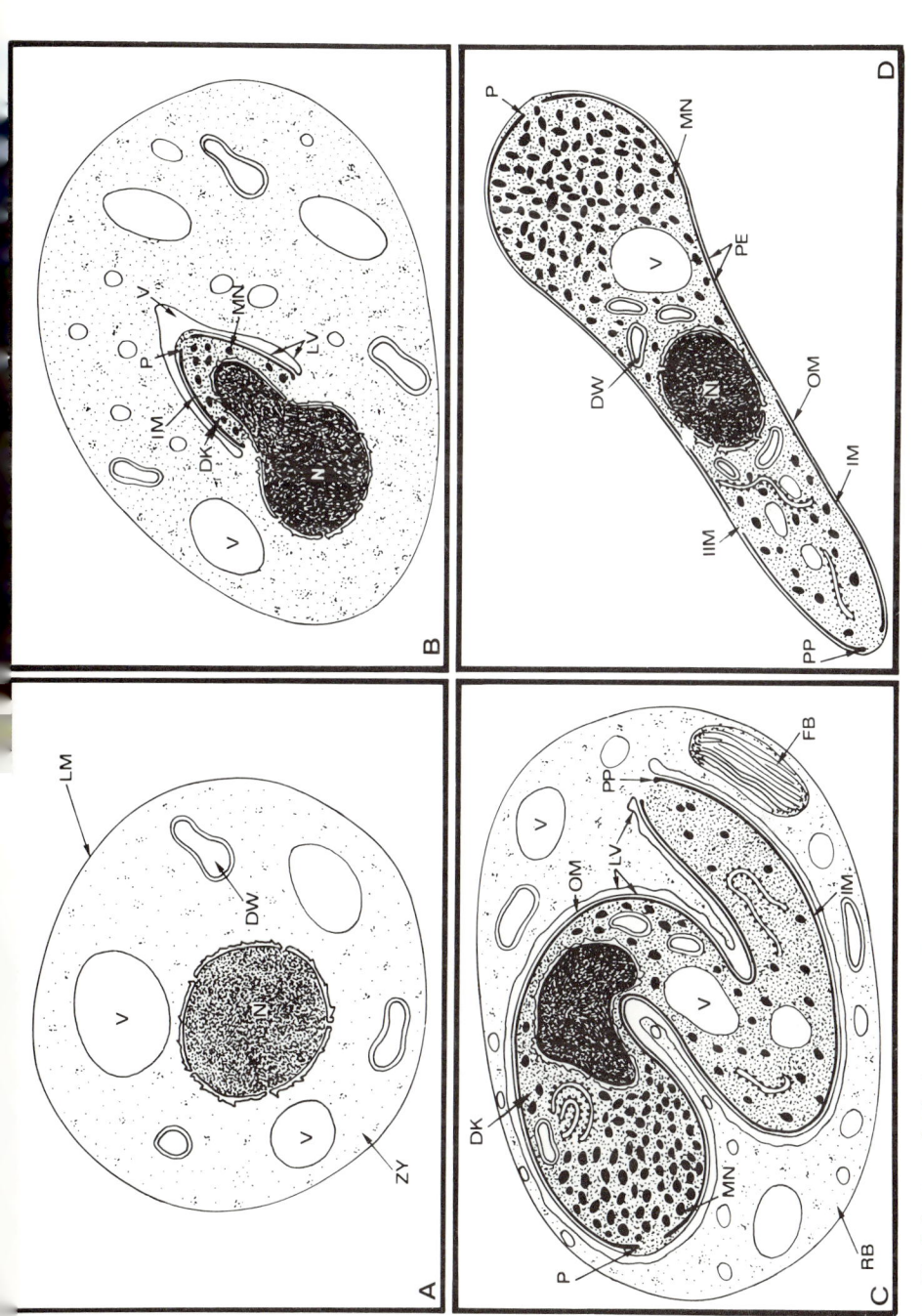

FIG. 23. *Babesia* and *Theileria* species. Diagrammatic representation of the formation of a uninucleate motile kinete (D) from a spherical zygote (A). Note that the developing kinete protrudes into an enlarging inner vacuole (V).

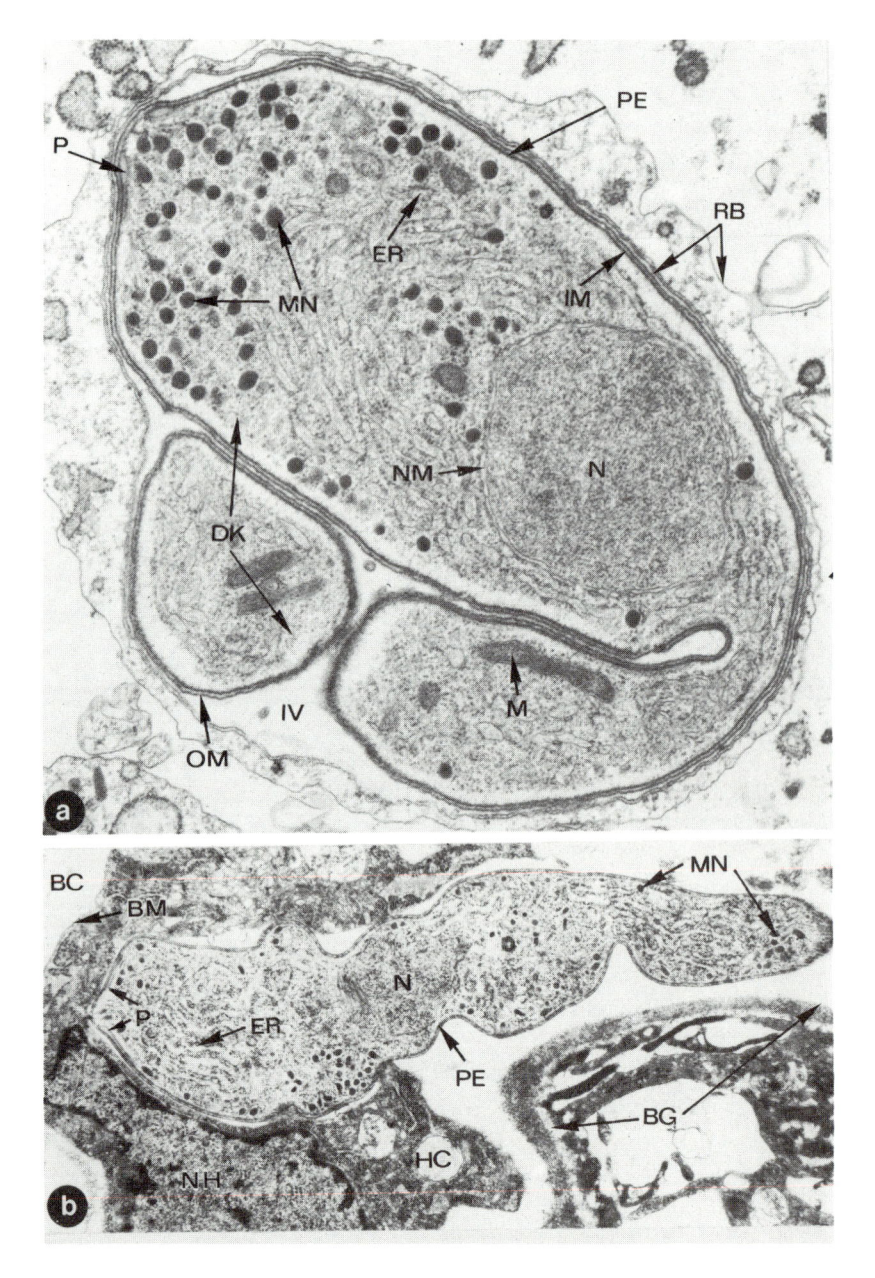

Fig. 24. Electron micrographs of kinetes of *Babesia* species. a. *B. bigemina*. Almost fully differentiated kinete within the original zygote. (Similar appearances seen with *B. equi*.) Magnification ×11 200. b. *B. equi*. Kinete leaving the intestinal cell of the tick. Magnification ×5600.

13b) or the kinetes of *Babesia* species which develop later, after the primary kinetes have left the intestine and penetrated into cells of other organs such as the ovary etc. (see Section IV D). However, for the moment it remains doubtful whether the kinetes derived from zygotes are completely identical with those that are formed later, since we found only a few in the intestine and therefore were, for example, unable to count the subpellicular microtubules, which seem to be specific in number.

### D. DEVELOPMENT IN VARIOUS ORGANS OF TICKS

Kinetes, having left the intestine (Fig. 24b), enter various organs of the vector ticks via the haemolymph. Asexual reproduction (i.e. sporogony) is initiated in the cells of these organs, preference being shown for haemocytes, muscle fibres, malpighian tubule cells and—in female ticks—ovarian cells including the oocytes. Although the details of fine structure of the kinetes derived from this asexual reproduction have been known for some time (Friedhoff and Scholtyseck, 1968, 1969; Potgieter *et al.*, 1976; Potgieter and Els, 1976, 1977; Morzaria *et al.*, 1976, 1978), the complicated process of kinete reproduction has been, until recently, only incompletely understood. Now it seems that the kinete reproduction can be subdivided into three phases (Moltmann *et al.*, 1982a,b), as follows (Fig. 25a–f).

(1) During the first phase, the kinete, which came from the intestine, invades a host cell (often inside the ovary of a female tick) and, without forming a parasitophorous vacuole, lies directly within the host cell cytoplasm (Fig. 27d). The intracellular kinete is transformed into a polymorphic stage, which loses all characteristics of the motile invasive stage and is bounded by only a single cell membrane. During this transformation the nucleus enlarges and becomes lobulate (Figs 25b, 26a).

(2) The polymorphic stage is then subdivided into several single-membrane bound, uninucleate cytomeres that are separated within the host cell cytoplasm (Figs 25c, d, 26b; CY). This process shows all transitions between an inner division of the cytoplasm by endoplasmic reticulum and invagination of the cell boundary, and peripheral formation of cytomeres by cytoplasmic protrusions. Microtubules or centrioles are apparently not involved in the transport of nuclear material into the cytomeres. This second developmental stage has been observed by light microscopy and called a 'fission body' or a 'primary schizont' (Friedhoff, 1969; Potgieter and Els, 1977). The first term now proves to be more appropriate, since this subdivision occurs after gamogony and is thus part of the sporogonic reproduction.

(3) The cytomeres then differentiate into new kinetes (Figs 25e, f; 27a–c). Beneath the membrane of a small intracytoplasmic vacuole, probably derived from the endoplasmic reticulum, a new apical complex is formed (Fig. 25e).

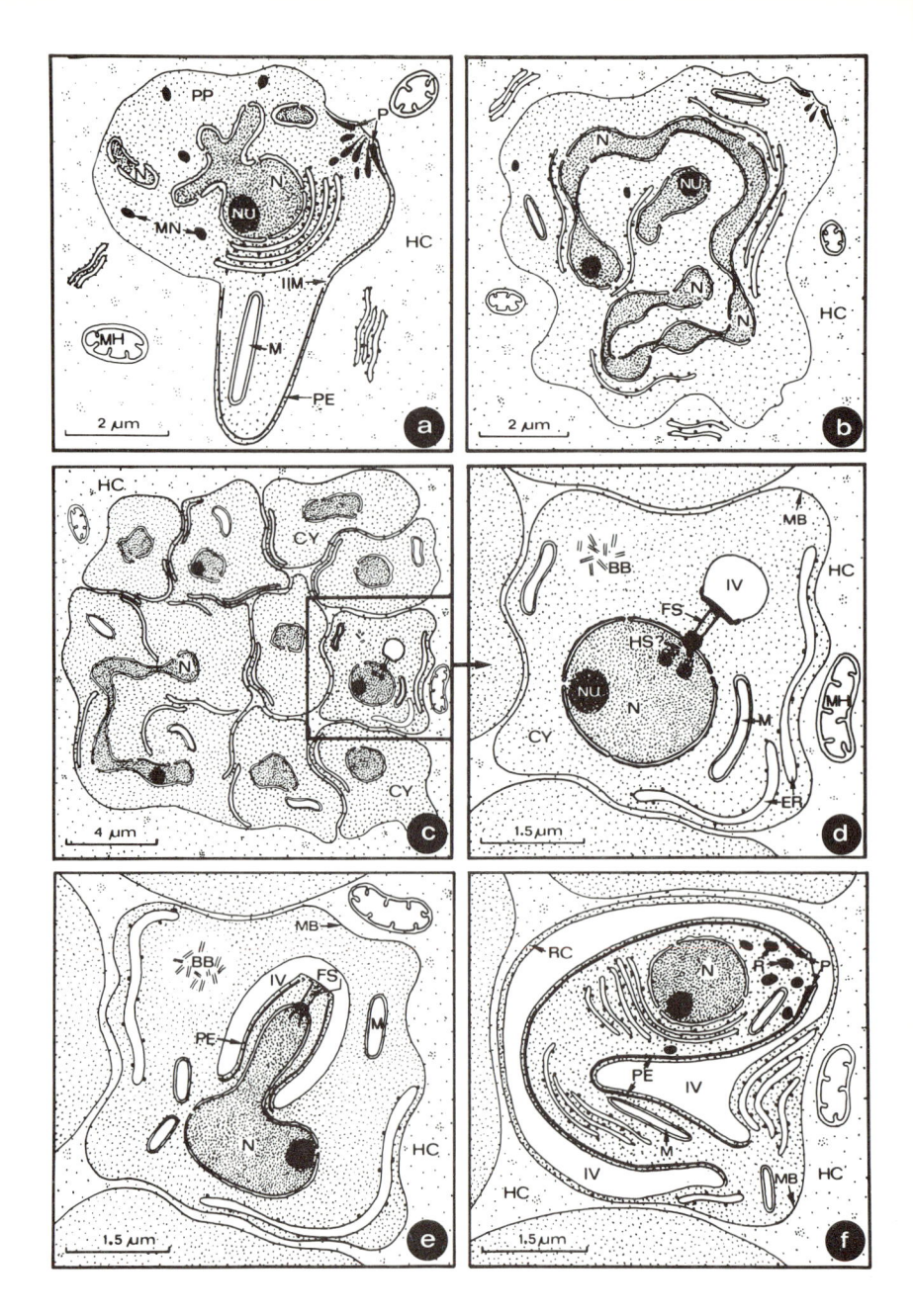

FIG. 25. Diagrammatic representation of formation of kinetes of *Babesia* species (not *B. equi*) in various organs of the tick. At first the kinete enlarges and its nucleus branches (a, b). Then the parasitic cell is divided into several uninucleate cytomeres (c, d), inside each of which a single kinete is formed by protrusion into an inner vacuole (compare the formation of kinetes from a zygote of *Babesia*, Fig. 23, or *Theileria*, Fig. 10a–d).

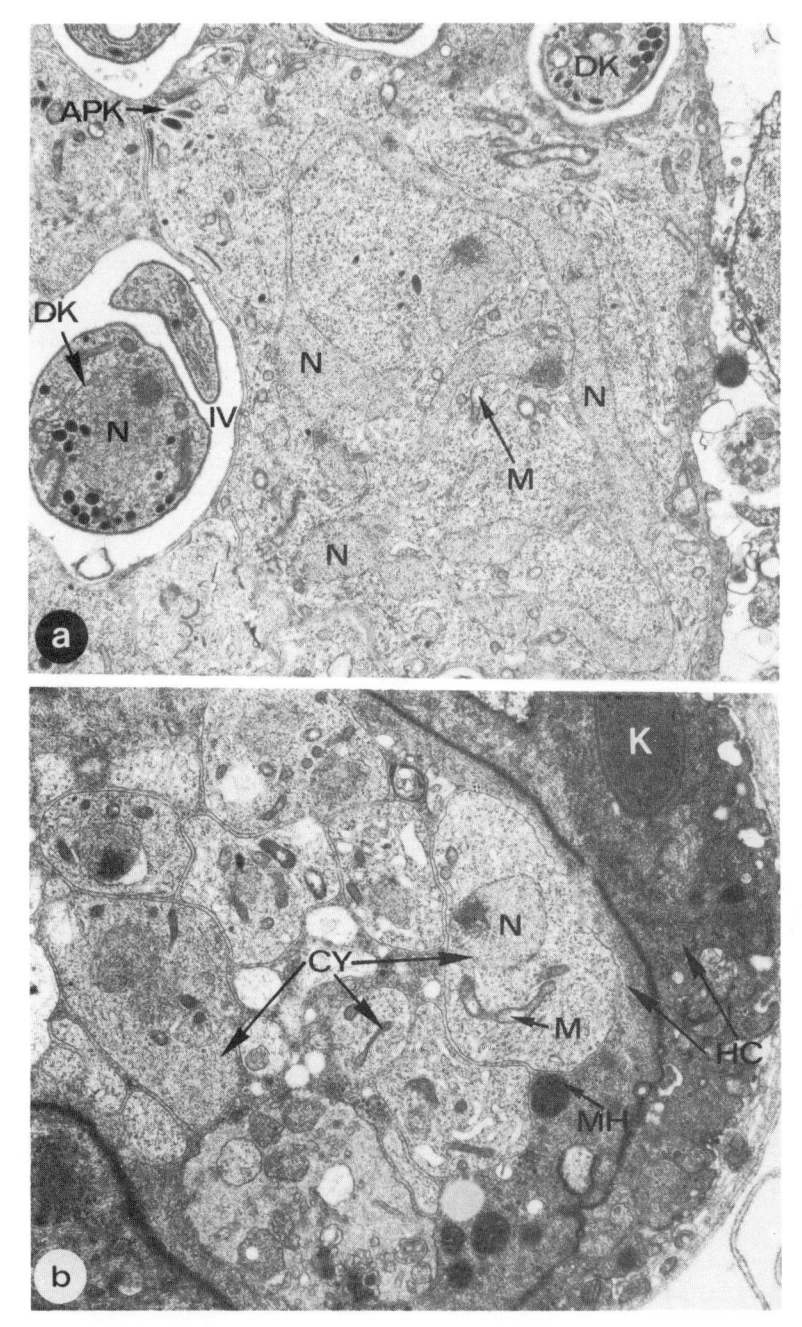

FIG. 26. Electron micrographs of the reproduction of *Babesia ovis* kinetes in the tick ovary. a. Polymorphic growing stage with interconnecting strands of nucleoplasm (N) and remnants of the apical complex (APK). Magnification ×18 000. b. Section through a parasite which has divided into several cytomeres (CY). Magnification ×17 100.

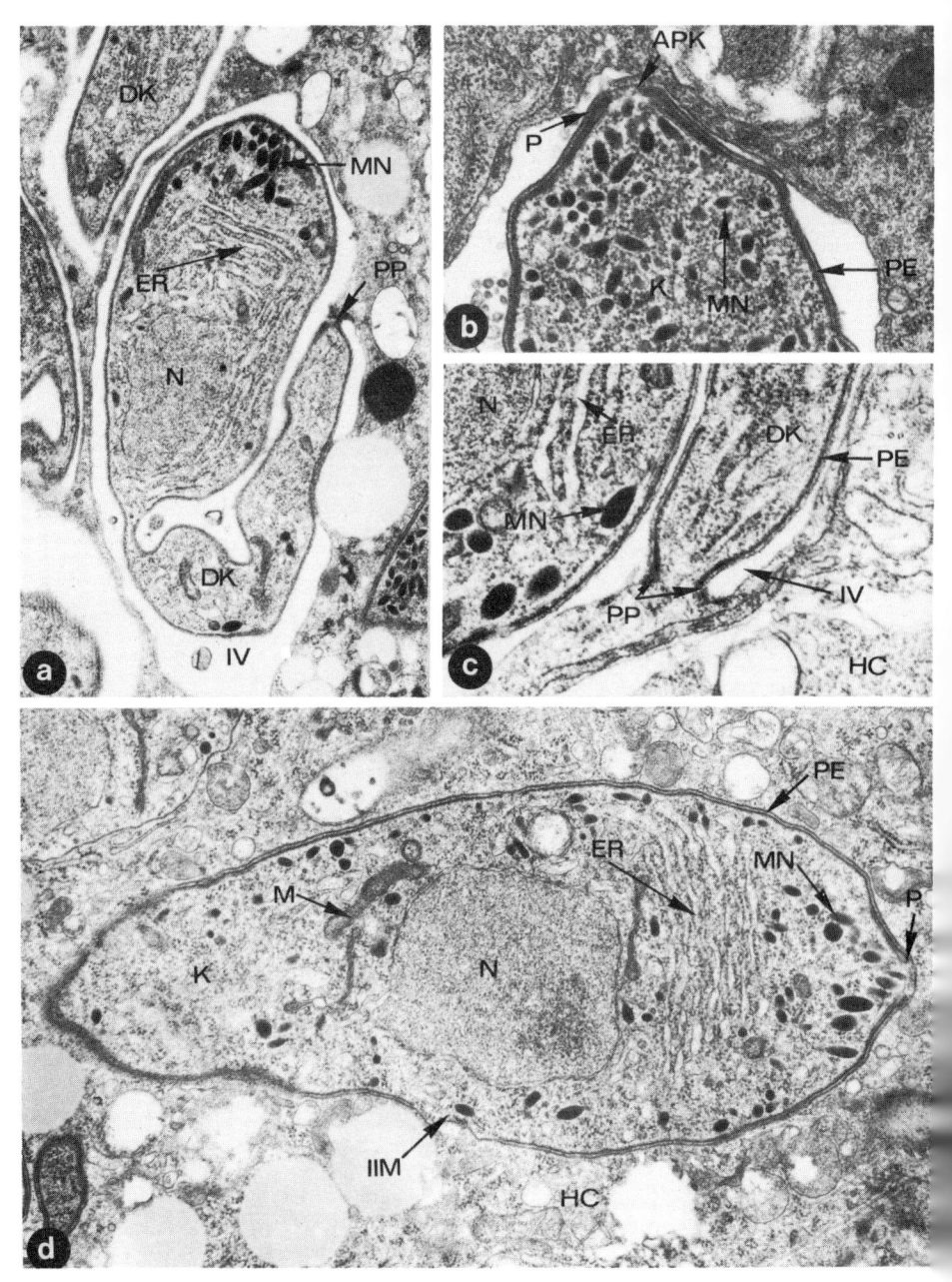

FIG. 27. Electron micrographs of the formation of *Babesia ovis* kinetes (from cytomeres) inside tick ovary cells. a. Almost fully differentiated kinete. Magnification ×12 000. b. Apical pole of a developing kinete. Magnification ×30 000. c. Posterior pole of a developing kinete still connected with the remnants of its cytomere (PP). Magnification ×30 000. d. Longitudinal section through a kinete, entering or leaving a host cell. Magnification ×15 000.

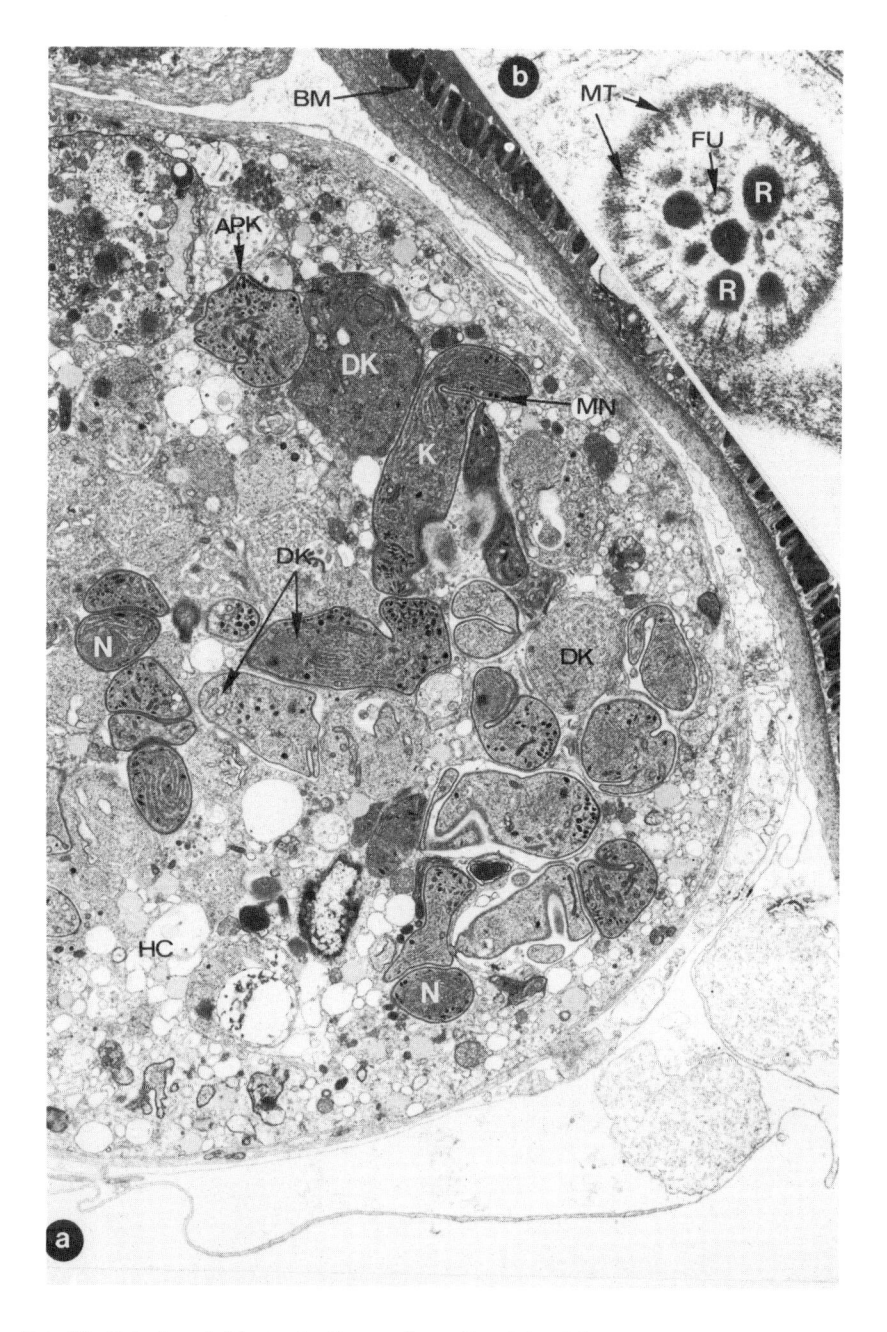

FIG. 28. *Babesia ovis* kinetes in the ovarian cells of the tick. a. General view of heavily infected ovary showing peripheral cytoplasmic degeneration. Various developing kinetes lie directly inside the host cell cytoplasm. Magnification ×7500. b. Cross section through the apical pole of a kinete showing 30 subpellicular microtubules (MT). Magnification ×49 000.

TABLE 5

*Development of* Babesia *species in ticks maintained at* 28°C

| Babesia species | B. bigemina | B. bovis | B. canis | B. caballi | B. ovis | B. equi |
|---|---|---|---|---|---|---|
| Vector species | Boophilus spp. | Boophilus spp. | Dermacentor reticulatus Haemaphysalis leachi Rhipicephalus sanguineus | Dermacentor nitens | R. bursa | Hyalomma spp. |
| Stages | adults, nymphs | adults, larvae | adults | adults | adults | nymphs, adults |
| Time from end of feeding to first occurrence of gametes | 7–19 hours | ? | 2–3 days | ? | ? | 1 day |
| Time from end of feeding to first kinetes in haemolymph | 3 days | 4 days | 6 days | 6 days | 5 days | 2 days |
| Size of kinetes in haemolymph (µm) | $11.0 \times 2.5$ | $15.8 \times 3.0$ | $15.0 \times 2.5$ | $10.0 \times 3.0$ | $9.5 \times 2.0$ | $9.0 \times 2.4$ |
| First detection of kinetes in eggs | 4 days | 5 days | 2 days | ? | 6 days | do not enter eggs |
| Mean time from attachment to occurrence of sporozoites in salivary glands | 9 days (nymphs), 16 days (adults) | 2–3 days (larvae) | 2–3 days (adults) | ? | 3–5 days (adults) | 3–5 days (adults) |
| Size of sporozoites (µm) | $2.5 \times 1.2$ | $1.5 \times 1.0$ | $2.5 \times 1.5$ | $2.5 \times 1.5$ | $2.8 \times 1.2$ | $3.0 \times 1.2$ |

This then protrudes progressively into the lumen of the expanding vacuole. An inner pellicular layer then develops at the base of the protrusion and the nucleus of the cytomere is incorporated into the newly formed kinete. During this process a peculiar funnel-like structure, at the apex of the protrusion, seems to be important. This structure was erroneously interpreted as a tubular structure, centriole, or anterior polar ring (see Weber, 1980). More likely, however, this structure is involved in the protrusion process, since it appears only in the initial phase of kinete formation and disappears later. Nearly all of the cytoplasm is used for the differentiation of the kinete, which finally separates from the remnants of the surrounding cytomere and lies folded within the vacuole. The kinete is released by the rupture of the vacuole and finally numerous kinetes are observed in host cells (Fig. 28). The kinetes may leave these cells and reproduce in other host cells. The last step of this reproduction, i.e. the ultimate formation of the kinete, is identical to that in *Theileria* species (see Section III C) or to that of the transformation of a *Babesia* zygote into a motile kinete (see Section IV C). With respect to the position of this repeated asexual reproduction in the life cycle, the kinetes are sporokinetes and may no longer be termed merozoites or vermicules as in the older literature. These kinetes of *Babesia*, which have a typical fine structure, differ only slightly from those of *Theileria*. Two main differences may be noted. *Babesia* kinetes are mostly smaller (Tables 4, 5) than those of *Theileria* species, which may reach 22 μm in length. The other difference, which may be related to their smaller size, is that *Babesia* kinetes have only about 30 subpellicular microtubules (*B. ovis*, 30; *B. major*, 27 or 28; *B. bovis*, 32; *B. bigemina*, 28) whereas *Theileria* species have about 40 (Moltmann et al., 1982a; Morzaria et al., 1978; Potgieter et al., 1976; Potgieter and Els, 1977; Mehlhorn et al., 1975, 1978). Comparative studies, however, should clarify whether the numbers of subpellicular microtubules are specific for one species or only for one generation of kinetes within one species. Further studies are needed to investigate whether those kinetes, which penetrate organs other than salivary glands, are identical with those that enter cells of the tick salivary glands, and there start a cytologically completely different asexual reproduction, i.e. the formation of the infectious sporozoites described below.

### E. DEVELOPMENT IN SALIVARY GLANDS OF VECTOR TICKS

The first parasitic stages seen in the salivary glands of a vector tick, on the second day after its attachment to its host, are kinetes (Fig. 29a); in *B. equi* they are already there during the moult. During the next 3 days all parasites are found to be more or less in the same developmental stage (indicating synchronous development), until on the fifth day after attachment the infected salivary cells are filled with differentiated sporozoites (Fig. 30a, b).

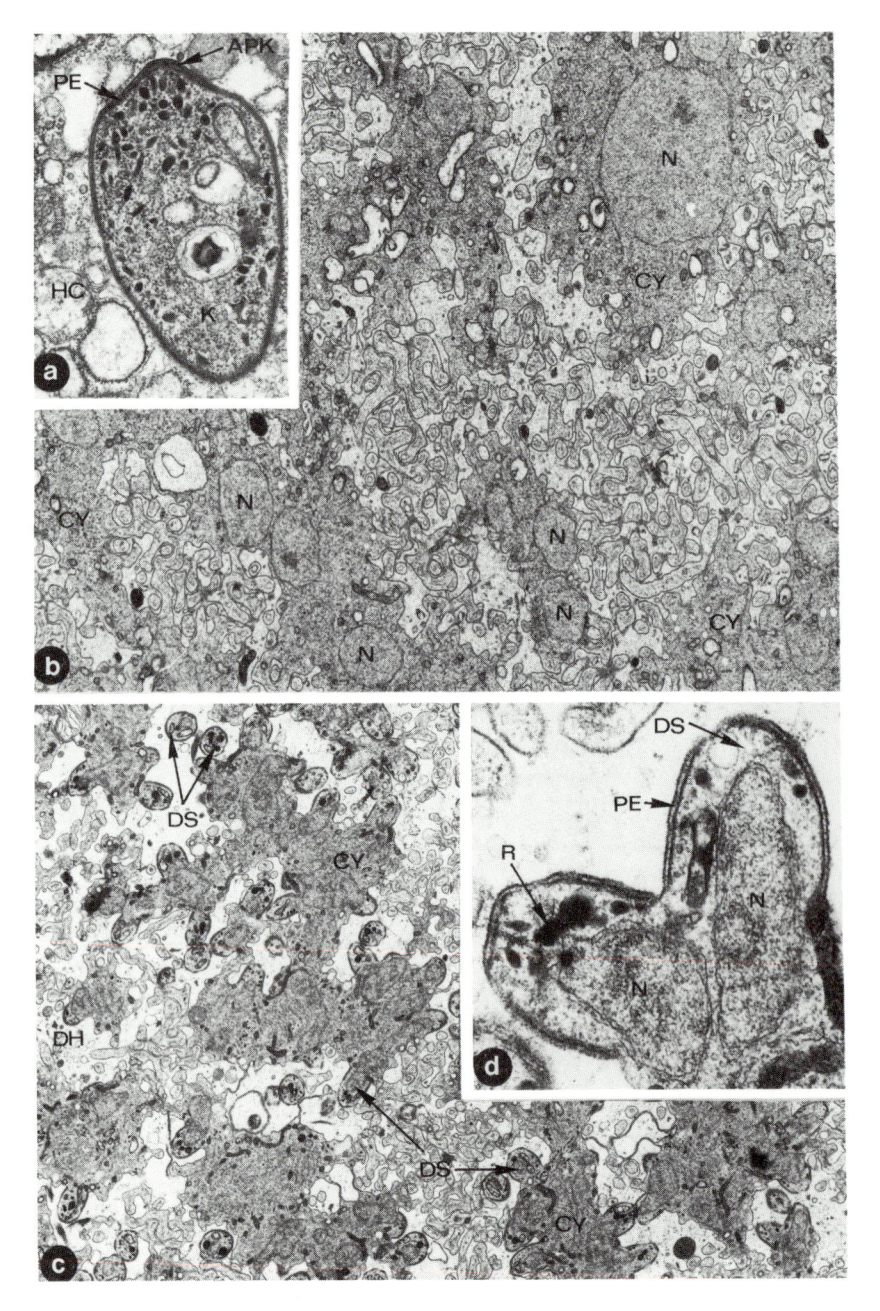

FIG. 29. Electron micrographs showing the development of *Babesia* species (not *B. canis*) in the salivary gland cells of the tick. a. *B. ovis*. A recently penetrated kinete. Magnification ×10 400. b. *B. bigemina*. The penetrated kinete is transformed into a large twisted and infolded mass containing large nuclei. Magnification ×6 500. c, d. *B. equi*. At the periphery of the parasitic body sporozoites are formed. Magnification ×2600 (c) and 26 000 (d).

In most *Babesia* species (including *B. equi*, but excluding *B. canis*), this formation of sporozoites proceeds as follows (see Moltmann et al., 1982a,b, 1983a, b).

Having penetrated salivary gland cells, the kinetes lose their characteristic features and become polymorphic sporonts. The sporonts apparently stimulate their host cells and nuclei to hypertrophy, which in some species (e.g. *B. equi*; not *B. canis*) leads to a significant enlargement of the infected alveoli (diameters up to 300 μm). The host cell cytoplasm appears as a homogeneous matrix within which the mitochondria persist for some time (Fig. 3). The sporonts of *Babesia* are polymorphic, bound by a single membrane, and measure at first about 17 μm in diameter. At the periphery of the sporonts finger-like protrusions are formed, which are twisted among themselves. The nucleus of each sporont is lobed and spindle poles appear at its membrane, from which microtubules radiate into the nucleoplasm. Numerous elongate organelles, presumably representing mitochondria, appear in clusters in the cytoplasm. Some dark globules scattered in the cytoplasm are probably remnants of micronemes.

The sporonts become extensively divided into polymorphic portions of cytoplasm, interconnected by twisted cytoplasmic bridges (Fig. 29b). The cytoplasmic portions contain several nuclei and bundles of mitochondria; occasionally very large nuclei (diameters up to 16 μm) with homogeneous electron-lucent nucleoplasm occur (Fig. 29b; N).

In all species, sporozoite development begins only when the infected tick attaches to a vertebrate host. Development is then mostly completed within 5 days; this rapidity is necessary since the sporozoites must be transmitted during the feeding period, after which the gland cells regularly become reduced. During the first 3 days, no significant ultrastructural change is observed within the sporonts. Beginning often on the fourth day after attachment of the tick*, simultaneous formation of sporozoites is initiated at the periphery of each cytoplasmic portion where a nucleus lies close to the cell membrane (Fig. 29c). Here an osmiophilic layer is seen beneath the cell membrane. Between this layer and the nuclear envelope two small osmiophilic globules occur, probably representing precursors of rhoptries. A spindle pole is attached to the nuclear envelope, with microtubules extending into the nucleoplasm. During the next developmental step, several protrusions of the sporont cell membrane are seen (Fig. 29c,d; DS). These protrusions are already bounded by a pellicle and contain a newly formed apical complex with rhoptries and a polar ring to which microtubules are attached (Fig. 29d). A mitochondrion and a portion of the nucleoplasm are incorporated into each of the protrusions. Single microtubules extend from the spindle pole at the

* Salivary glands of tick larvae infected with *B. bovis* or *B. divergens* contain infective sporozoites 2 days after attachment.

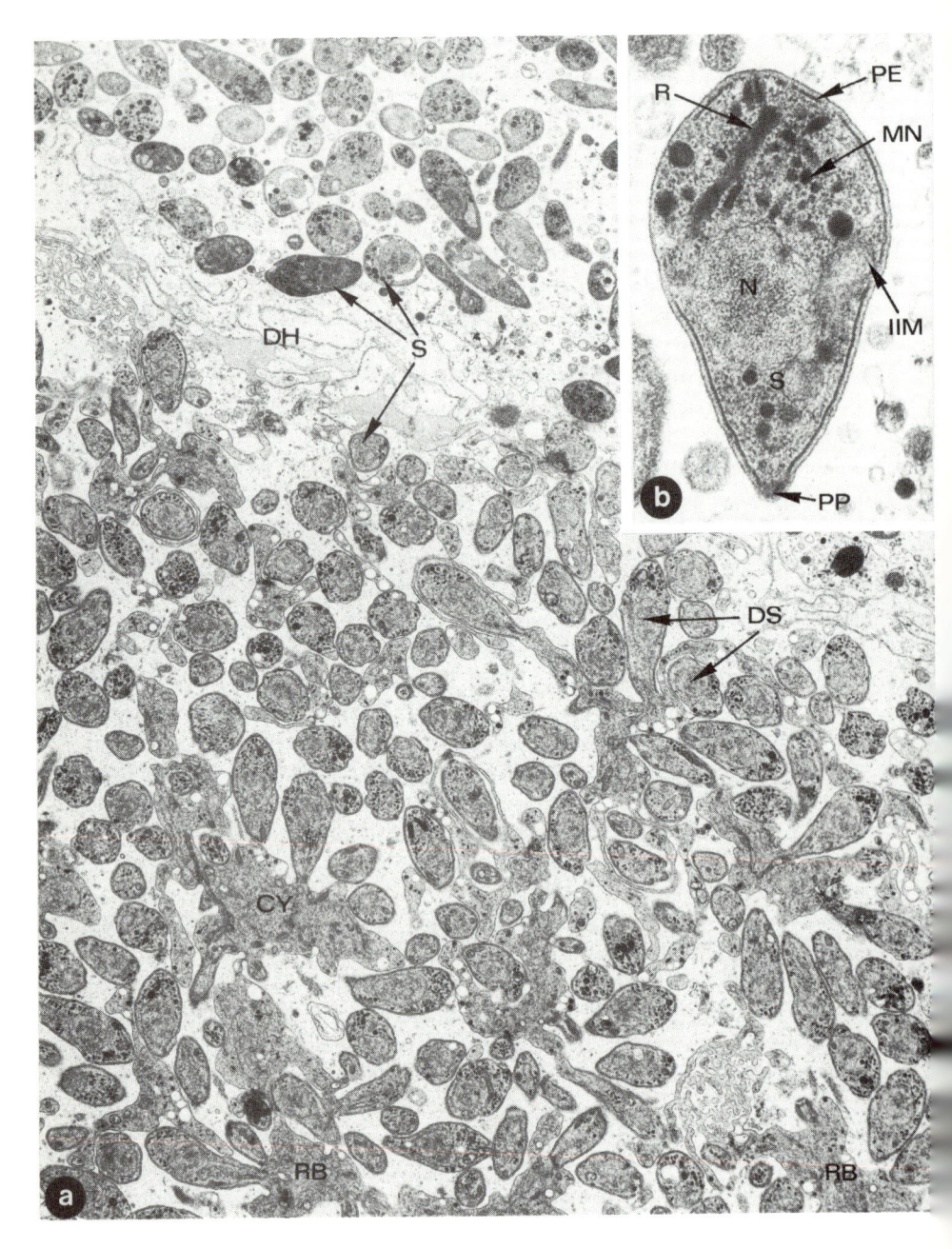

FIG. 30. Electron micrographs of differentiated sporozoites in *Babesia* species. a. Magnification ×7500. b. Magnification ×33 750.

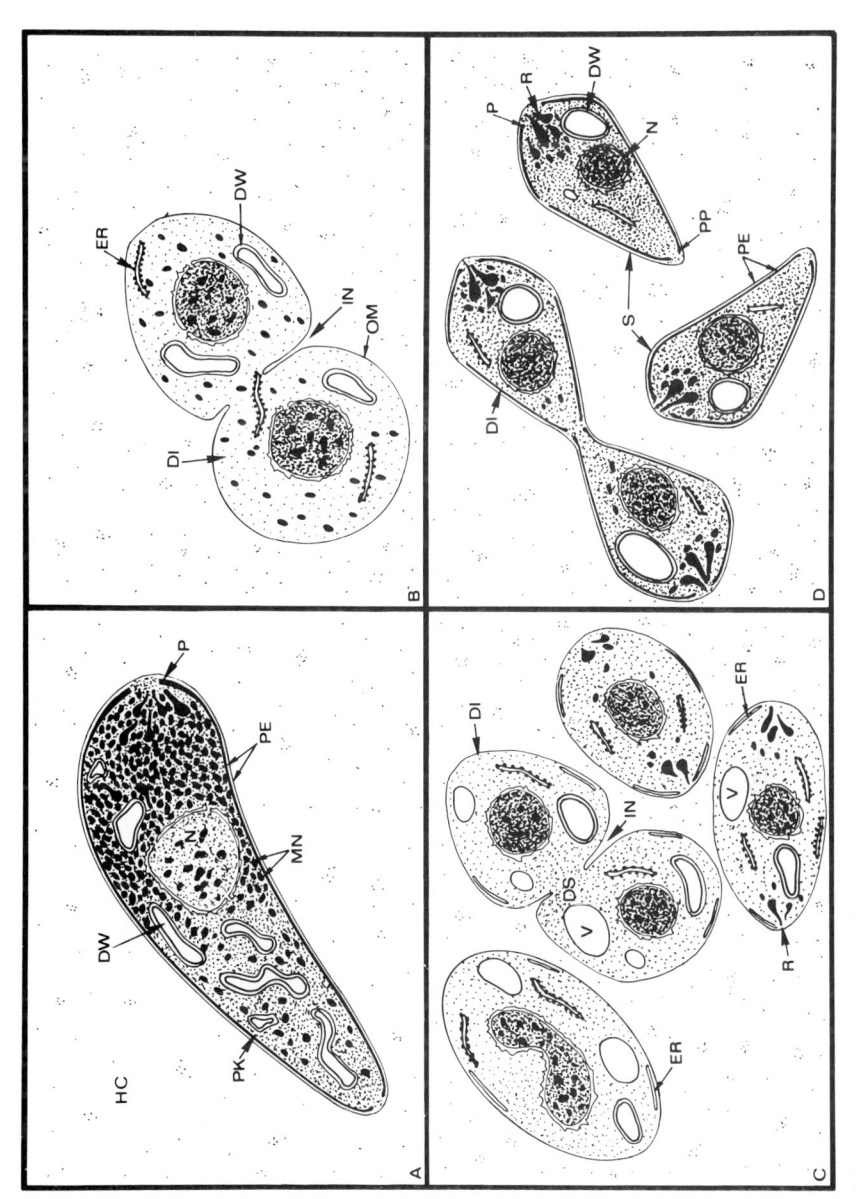

FIG. 31. Diagrammatic representation of the development of sporozoites of *Babesia canis*. In this species the sporozoites are formed by repeated binary fission (A–D), during which the daughter cells develop the shape, size and structure of infective sporozoites.

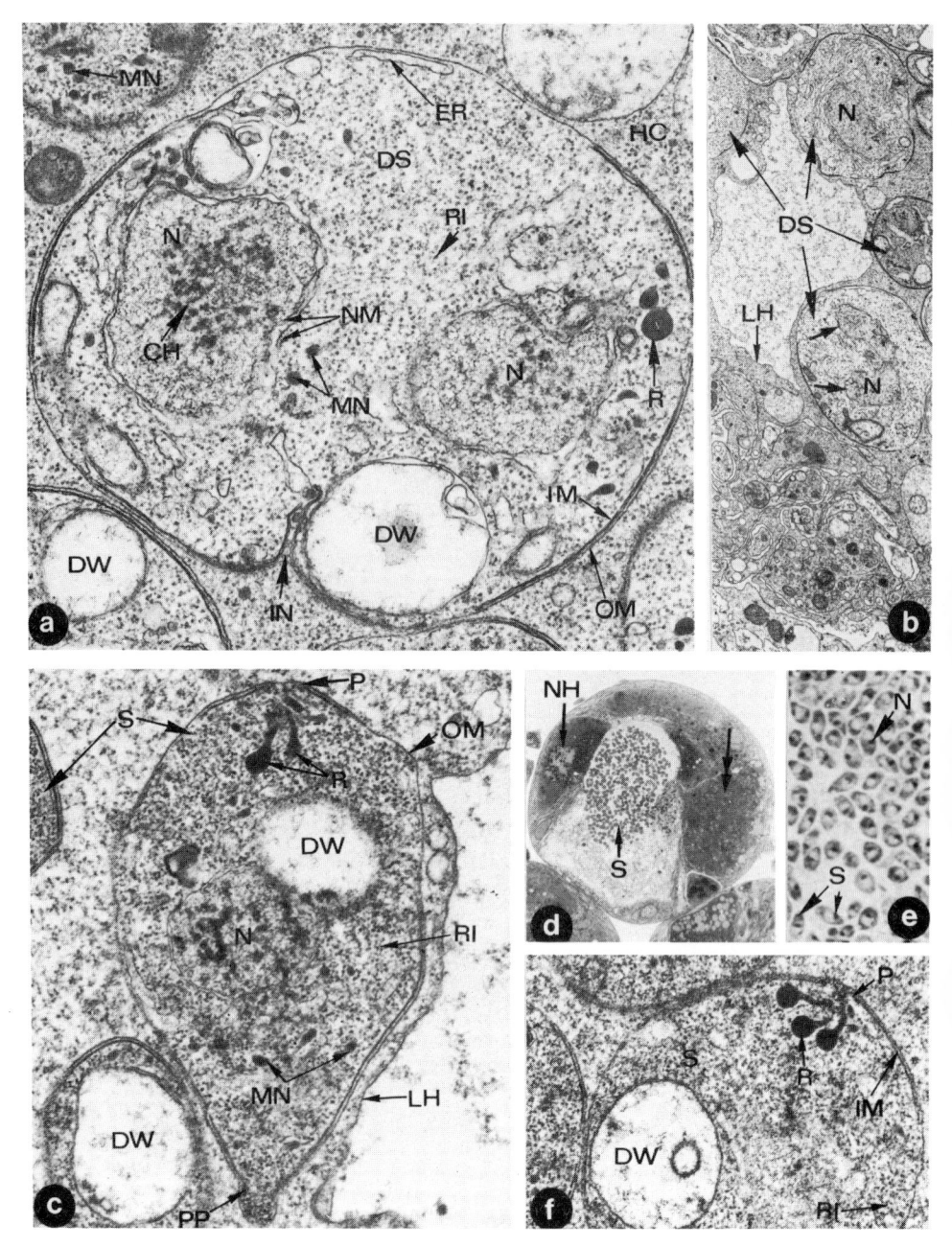

FIG. 32. Formation of *Babesia canis* sporozoites in tick salivary gland cells. a, b, c, f, electron micrographs; d, e, light micrographs. a, b. Cytomeres with two nuclei during division by infolding (IN) of parasite plasmalemma. Magnification ×27 000 (a) and 7500 (b). c, f. Differentiated sporozoites. Magnification ×26 250. d. Section through an alveolus of a tick containing infective sporozoites (S) and, nearby, those still differentiating (double arrow). Magnification ×300. e. Higher magnification of infective sporozoites (S) in an alveolus. Magnification ×11 250.

nuclear envelope to the tip of the protrusion. The development of micronemes is observed during this phase. Small, drop-like, osmiophilic globules without a limiting membrane radiate from a central matrix. The cytoplasm of the sporonts gradually extends into the growing sporozoites, which are finally pinched off at the posterior pole (Fig. 30a, b). The twisted cytoplasmic bridges persist during this phase but disappear later. Sometimes the last remnants of sporonts become spherical and continue to form sporozoites at their periphery. On the fifth day after attachment of the tick, thousands of sporozoites lie in each large host cell (Fig. 30b; S), whereas the other cells of the alveolus have completely degenerated.

This process of sporozoite formation is different from that in *Babesia canis* (see Schein *et al.*, 1979). There the sporozoites are differentiated after constant binary fissions (Figs 31a–d, 32). This process may perhaps not be unique among *Babesia* species, as is indicated by the findings of Potgieter and Els (1976, 1977), who also described binary fissions, but suggested that initially 'schizonts' occur. Thus comparative studies with further *Babesia* species are needed.

The fine structure of sporozoites is described for several *Babesia* species (see Friedhoff, 1981). Apart from differences in size—there are species like *B. microti* and *B. bovis* with sporozoites about 1·5–2·1 μm long, whereas they measure about 2·5 μm in *B. canis* and *B. bigemina*, 3–3·4 μm in *B. equi* and are even somewhat longer in some other species—the sporozoites are structured according to a common plan. They are typically pyriform with a broad apical and small pointed posterior pole (Figs 30b, 32c). They are bounded by a pellicle composed of an outer cell membrane and an inner osmiophilic layer consisting of two membranes, interrupted at some places, and ending in polar rings at the anterior and posterior poles. About 30 subpellicular microtubules are seen in *B. equi* sporozoites; other species also contain subpellicular microtubules but their exact number is not known. Some micronemes and up to five to seven rhoptries belong to the apical complex, which apparently enables the parasite to enter the erythrocytes of the vertebrate host when transmitted by the tick's saliva. The sporozoites always contain a typical nucleus and a mitochondrion. Furthermore, a so-called spherical body is seen in some species (see Friedhoff, 1981) which, however, is lacking from others (such as *B. canis*, *B. microti* and *B. equi*; Figs 30b, 32c); its function and taxonomic significance are not yet understood. The sporozoites, of which about 5000–10 000 are produced within a single alveolus, are finally transmitted to the vertebrate hosts (Tables 2, 3) and initiate the development described in Sections IV A and IV B.

## V. Conclusions

The piroplasms have—as has been shown recently— a typical sporozoan life cycle comprising the following three phases (Figs 2, 16). (1) *Schizogony*, an asexual reproduction phase in the vertebrate host. (2) *Gamogony* with formation and fusion of gametes inside the intestinal cells of ixodid ticks (which suck blood only three times during their life). (3) *Sporogony*, an asexual reproduction in the salivary gland of the tick leading to the infectious, saliva-transmitted sporozoites (in most *Babesia* species further reproductions belonging to this phase proceed in various other organs).

The establishment of such a three-phase life cycle depended on the following findings. (a) The fine structure of stages involved in the life cycle is very similar to that of the Sporozoa (Apicomplexa), especially the Haemosporida (malaria parasites), which are transmitted by blood-sucking mosquitoes. (b) The fact that the so-called 'ray-bodies', described in 1906 but neglected until recently, belong (as gamonts and gametes) to the life cycle of piroplasms, as was proven by *in vitro* experiments with infected erythrocytes and studies of their further development over a time period as long as that in the intestine of ticks.

Comparative biological and morphological studies showed that the economically important piroplasms comprise three groups: (1) *Babesia* species *sensu strictu*; (2) *Babesia equi*, *B. microti* (and others?); and (3) *Theileria* species.

The *Babesia* species *sensu strictu* differ from *Theileria* in the following ways.

(1) *Babesia* species are transovarially transmitted to the next generation of ticks, whereas with *Theileria* a tick stage (larve, nymph or adult) is parasite-free after feeding and parasites are not transmitted to the progeny of ticks.

(2) *Babesia* species develop exclusively inside erythrocytes of their vertebrate hosts, whereas *Theileria* reproduces mainly in lymphocytes.

(3) Intraerythrocytic stages of *Babesia* are at least double the size of *Theileria* (Fig. 1).

(4) Gamonts and the fusion of gametes differ in the two groups (Figs 8, 19).

(5) Reproduction of kinetes of *Babesia* occurs in cells of various organs including oocytes (Fig. 25), whereas in *Theileria* species kinetes penetrate directly into the salivary gland cells.

(6) Kinetes of *Babesia* species are significantly smaller than those of *Theileria* (Figs 13b, 27d), and have only about 30 instead of 40 subpellicular microtubules.

(7) Sporozoites of *Babesia* are at least double the size of those of *Theileria* and possess a more complex pellicle (Figs 15b–e, 30b, 32c).

(8) Chemotherapy of infected vertebrate hosts with diamidine compounds (e.g. Berenil) affects only *Babesia* species; this indicates that the metabolism of the members of the two genera is probably different.

The 'intermediate babesian species' *B. equi* and *B. microti* differ from *Babesia* species *sensu strictu* by their additional reproduction in lymphocytes (Fig. 17). Furthermore, at least for *B. equi*, evidence suggests that no transovarial transmission occurs (because there is no reproduction of kinetes in oocytes).

The fine structure of erythrocytic stages of *B. equi* is more similar to other *Babesia* species than to *Theileria*, although they possess a typical micropore which is also present in erythrocytic stages of *Theileria*. The gametes of *B. equi* are also more of the babesian type. However, further studies are needed in order to clarify the question, whether these differences justify the exclusion of *B. equi* (*B. microti* and others?) from the genus *Babesia* and the establishment of a new genus.

REFERENCES

Anderson, A. E., Cassaday, P. B. and Healy, G. R. (1974). Babesiosis in man. *American Journal of Clinical Pathology* **62**, 50–77.
Babès, V. (1869). Die Aetiologie der seuchenhaften Hämoglobinurie des Rindes. *Archiv für Pathologie und Anatomie* **115**, 81–108.
Babès, V. (1888). Sur l'hémoglobinurie bactérienne des boeufs. *Comptes Rendus de l'Académie des Sciences (Paris)* **107**, 693–694.
Boch, J. and Supperer, R. (1983). "Veterinärmedizinische Parasitologie." Parey, Berlin.
Bool, P. H., Goedbloed, E. and Keidel, H. J. W. (1961). The bovine babesia species in the Netherlands: *Babesia divergens* and *Babesia major*. *Tijdschrift voor diergeneeskunde* **86**, 28–37.
Büttner, D. W. (1966). Über die Feinstruktur der erythrocytären Formen von *Theileria mutans*. *Zeitschrift für Tropenmedizin und Parasitologie* **17**, 397–406.
Büttner, D. W. (1967a). Die Feinstruktur der Merozoiten von *Theileria parva*. *Zeitschrift für Tropenmedizin und Parasitologie* **18**, 224–244.
Büttner, D. W. (1967b). Elektronenmikroskopische Studien der Vermehrung von *Theileria parva* im Rind. *Zeitschrift für Tropenmedizin und Parasitologie* **18**, 245–268.
Christophers, S. R. (1907). Preliminary note on the development of the *Piroplasma canis* in the tick. *British Veterinary Journal* **1**, 76–78.
Cowdry, E. V. and Ham, A. W. (1932). Studies on East Coast Fever. I. The life cycle of the parasites in ticks. *Parasitology* **24**, 1–43.
Crawley, H. (1916). Note on the stage of *Piroplasma bigeminum* which occurs in the cattle tick, *Margaropus annulatus*. *Journal of Parasitology* **2**, 87–92.

Davis, L. J. (1929). On the piroplasm of the Sudanese wild cat (*Felis ocreata*). *Transactions of the Royal Society of Tropical Medicine and Hygiene* **22**, 523–535.

Dennig, H. K. (1969). *Babesia herpailuri*, eine neue Babesienart der Feliden. *Progress in Protozoology*, 3rd International Congress on Protozoology, Leningrad, 2–10 July 1969, p. 256.

Dennig, H. K. and Brocklesby, D. W. (1972). *Babesia pantherae* sp. nov. a piroplasm of the leopard (*Panthera pardus*). *Parasitology* **4**, 525–532.

Dennis, E. W. (1932). The life cycle of *Babesia bigemina* (Smith and Kilborne) of Texas cattle-fever in the tick *Margaropus annulatus* (Say) with notes on the embryology of *Margaropus*. *University of California Publications in Zoology* **36**, 263–298.

Donnelly, J. and Peirce, M. A. (1975). Experiments on the transmission of *Babesia divergens* to cattle by the tick *Ixodes ricinus*. *International Journal of Parasitology* **5**, 363–367.

Dschunkowsky, E. and Luhs, I. (1909). Entwicklungsformen von Piroplasmen in Zecken. *Proceedings, 9th International Veterinary Congress, Den Haag*, 51–61.

Dyakonow, L. P. and Godzhaev, A. N. (1971). Development of *Theileria annulata* in *Hyalomma anatolicum*. *Veterinariya* **48**, 61–65.

Enigk, K. (1943). Die Überträger der Pferdepiroplasmose, ihre Verbreitung und Biologie. *Archiv für Tierheilkunde* **78**, 209–240.

Enigk, K. (1944). Weitere Untersuchungen zur Überträgerfrage der Pferdepiroplasmose. *Archiv für Tierheilkunde* **79**, 58–80.

Fawcett, D. W., Doxsey, S. and Büscher, G. (1981a). Salivary gland of the tick vector of East Coast Fever. I. Ultrastructure of the type III acinus. *Tissue and Cell* **13**, 209–230.

Fawcett, D. W., Doxsey, S. and Büscher, G. (1981b). Salivary gland of the tick vector of East Coast Fever. II. Cellular basis for fluid secretion in the type III acinus. *Tissue and Cell* **13**, 231–251.

Fawcett, D. W., Doxsey, S. and Büscher, G. (1982a). Salivary gland of the tick vector of East Coast Fever. III. The ultrastructure of sporogony in *Theileria parva*. *Tissue and Cell* **14**, 183–206.

Fawcett, D. W., Doxsey, S., Stagg, D. A. and Young, A. S. (1982b). The entry of sporozoites of *Theileria parva* into bovine lymphocytes *in vitro*. Electron microscopic observations. *European Journal of Cell Biology* **27**, 10–21.

Friedhoff, K. T. (1969). Lichtmikroskopische Untersuchungen über die Entwicklung von *Babesia ovis* in *Rhipicephalus bursa*. *Zeitschrift für Parasitenkunde* **32**, 191–219.

Friedhoff, K. T. (1981). Morphologic aspects of *Babesia* in the tick. *In* "Babesiosis" (M. Ristic and J. P. Kreier, eds), pp. 143–169. Academic Press, New York.

Friedhoff, K. T. and Büscher, G. (1976). Rediscovery of Koch's Strahlenkörper of *Babesia bigemina*. *Zeitschrift für Parasitenkunde* **50**, 345–347.

Friedhoff, K. T. and Liebisch, A. (1978). Piroplasmeninfektionen der Haustiere. *Tierärztliche Praxis* **6**, 125–139.

Friedhoff, K. T. and Scholtyseck, E. (1968). Feinstruktur von *Babesia ovis* in *Rhipicephalus bursa*. Transformation sphäroider Formen zu Vermiculaformen. *Zeitschrift für Parasitenkunde* **30**, 347–359.

Friedhoff, K. T. and Scholtyseck, E. (1969). Feinstrukturen der Merozoiten von *Babesia bigemina* im Ovar von *Boophilus microplus* und *Boophilus decoloratus*. *Zeitschrift für Parasitenkunde* **32**, 266–283.

Friedhoff, K. T. and Scholtyseck, E. (1977). Fine structural identification of erythrocytic stages of *Babesia bigemina*, *B. divergens*, and *B. bovis*. *Protistologica* **13**, 195–204.

Friedhoff, K. T. and Smith, R. D. (1981). Transmission of *Babesia* by ticks. *In* "Babesiosis" (M. Ristic and J. P. Kreier, eds), pp. 267–321. Academic Press, New York.

Friedhoff, K. T., Scholtyseck, E. and Weber, G. (1972). Die Feinstruktur der differenzierten Merozoiten von *Babesia ovis* in den Speicheldrüsen weiblicher Zecken. *Zeitschrift für Parasitenkunde* **38**, 132–140.

Gonder, R. (1910). Die Entwicklung von *Theileria parva*, dem Erreger des Küstenfiebers der Rinder in Afrika. Teil I. *Archiv für Protistenkunde* **21**, 143–164.

Gonder, R. (1911). Die Entwicklung von *Theileria parva*, dem Erreger des Küstenfiebers der Rinder in Afrika. Teil II. *Archiv für Protistenkunde* **22**, 170–178.

Groves, M. G. and Dennis, G. L. (1972). *Babesia gibsoni*: field and laboratory studies of canine infections. *Experimental Parasitology* **31**, 153–159.

Hadani, A., Tsur, I., Pipano, E. and Senft, Z. (1963). Studies on the transmission of *Theileria annulata* by ticks (Ixodoides, Ixodidae). I. *Hyalomma excavatum*. *Journal of Protozoology* **10**, Supplement, 35.

Hadani, A., Pipano, E. and Dinür, Y. (1969). Studies on the transmission of *Theileria annulata* by ticks. *Journal of Protozoology* **16**, Supplement, 37.

Holbrook, A. A., Johnson, A. J. and Madden, P. A. (1968). Equine piroplasmosis: intraerythrocytic development of *Babesia caballi* (Nuttall) and *Babesia equi* (Laveran). *American Journal of Veterinary Research* **29**, 297–303.

Jack, R. M. and Ward, P. A. (1981). Mechanisms of entry of *Plasmodia* and *Babesia* into red cells. *In* "Babesiosis" (M. Ristic and J. P. Kreier, eds), pp. 445–457. Academic Press, New York.

Joyner, L. P., Davies, S. F. M. and Kendall, S. B. (1963). The experimental transmission of *Babesia divergens* by *Ixodes ricinus*. *Experimental Parasitology* **14**, 367–373.

Kleine, F. K. (1906). Kultivierungsversuch der Hundepiroplasmen. *Zeitschrift für Hygiene und Infektionskrankheiten* **54**, 11–16.

Knuth, P. and Du Toit, P. J. (1921). Die Piroplasmose des Schweines. Die durch *Piroplasma trautmanni* (nov. spec.) verursachte Piroplasmose. *In* "Handbuch der Tropenkrankheiten" (Hrsgb. C. Mense, ed.), 2nd edition, Vol. 6 ("Tropenkrankheiten der Haustiere"), pp. 409–411. Fischer, Jena.

Koch, R. (1898). "Reiseberichte über Rinderpest, Bubonenpest in Indien und Afrika, Tsetse oder Surrakrankheit, Texasfieber, tropische Malaria, Schwarzwasserfieber." Springer, Berlin.

Koch, R. (1906). Beiträge zur Entwicklungsgeschichte der Piroplasmen. *Zeitschrift für Hygiene und Infektionskrankheiten* **54**, 1–9.

Krylov, M. V. (1981). ["Piroplasms".] Nauka, Leningrad. [In Russian.]

Levine, N. D. (1973). "Protozoan Parasites of Domestic Animals and of Man," 2nd edition. Burgess Publishing Company, Minneapolis.

Martin, H. and Brocklesby, D. W. (1960). A new parasite of eland. *Veterinary Record* **72**, 331–332.

Martin, H. M., Barnett, S. F. and Vidler, B. O. (1964). Cyclic development and longevity of *Theileria parva* in the tick *Rhipicephalus appendiculatus*. *Experimental Parasitology* **15**, 527–555.

McHardy, N., Haigh, A. J. B. and Dolan, T. T. (1976). Chemotherapy of *Theileria parva* infection. *Nature* **171**, 134–135.

Mehlhorn, H. and Schein, E. (1976). Elektronenmikroskopische Untersuchungen an Entwicklungsstadien von *Theileria parva* im Darm der Überträgerzecke *Hyalomma anatolicum* excavatum. *Zeitschrift für Tropenmedizin und Parasitologie* **27**, 182–191.

Mehlhorn, H., Weber, G., Schein, E. and Büscher, G. (1975). Elektronenmikroskopische Untersuchungen an Entwicklungsstadien von *Theileria annulata* im Darm und in der Hämolymphe von *Hyalomma anatolicum excavatum*. *Zeitschrift für Parasitenkunde* **48**, 137–150.

Mehlhorn, H., Schein, E. and Warnecke, M. (1978). Electron microscopic studies on the development of kinetes of *Theileria parva* in the gut of the vector tick *Rhipicephalus appendiculatus*. *Acta Tropica* **35**, 123–136.

Mehlhorn, H., Schein, E. and Warnecke, M. (1979). Electron microscopic studies on *Theileria ovis*: development of kinetes in the gut of the vector tick, *Rhipicephalus evertsi evertsi*, and their transformation within cells of the salivary glands. *Journal of Protozoology* **26**, 377–385.

Mehlhorn, H., Schein, E. and Voigt, W. P. (1980a). Light and electron microscopic study on developmental stages of *Babesia canis* within the gut of the tick *Dermacentor reticulatus*. *Journal of Parasitology* **66**, 220–228.

Mehlhorn, H., Peters W. and Haberkorn, A. (1980b). The formation of kinetes and oocysts in *Plasmodium gallinaceum* and considerations on phylogenetic relationships between Haemosporidia, Piroplasmida and other Coccidia. *Protistologica* **16**, 135–154.

Mehlhorn, H., Moltmann, U., Schein, E. and Voigt, W. P. (1981). Electron microscopical study on the effects of Halofuginone on *Theileria parva*. *Zeitschrift für Tropenmedizin und Parasitologie* **32**, 231–233.

Mehlhorn, H., Moltmann, U. and Schein, E. (1984). *In-vitro* culture and syngamy of supposed gametes in *Theileria annulata*. *Zentralblatt für Bakteriologie, Hygiene und Parasitologie, Abteilung I* (in press).

Moltmann, U. G., Mehlhorn, H. and Friedhoff, K. T. (1982a). Ultrastructural study of the development of *Babesia ovis* in the ovary of the vector tick *Rhipicephalus bursa*. *Journal of Protozoology* **29**, 30–38.

Moltmann, U. G., Mehlhorn, H. and Friedhoff, K. T. (1982b). Electron microscopic study on the development of *B. ovis* in the salivary glands of the vector tick *Rhipicephalus bursa*. *Acta Tropica* **39**, 29–40.

Moltmann, U. G., Mehlhorn, H., Schein, E., Rehbein, G., Voigt, W. P. and Zweygarth, E. (1983a). Fine structure of *Babesia equi* within lymphocytes and erythrocytes of horses: an *in-vivo* and *in-vitro* study. *Journal of Parasitology* **69**, 111–120.

Moltmann, U. G., Mehlhorn, H., Schein, E., Voigt, W. P. and Friedhoff, K. T. (1983b). Ultrastructural study on the development of *Babesia equi* (Coccidia: Piroplasmia) in the salivary glands of its vector ticks. *Journal of Protozoology* **30**, 218–225.

Morzaria, S. P., Bland, P. and Brocklesby, D. W. (1976). Ultrastructure of *Babesia major* in the tick *Haemaphysalis punctata*. *Research in Veterinary Science* **21**, 1–11.

Morzaria, S. P., Bland, P. and Brocklesby, D. W. (1978). Ultrastructure of *Babesia major* vermicules from the tick *Haemaphysalis punctata* as demonstrated by negative staining. *Zeitschrift für Parasitenkunde* **55**, 119–125.

Muratov, E. A. and Kheisin, E. M. (1959). [The development of *Piroplasma bigeminum* in the tick *Boophilus calcaratus*.] *Zoologicheskii Zhurnal* **38**, 970–985. [In Russian.]

Nuttall, G. M. F. and Hindle, E. (1913). Conditions influencing the transmission of east coast fever. *Parasitology* **6**, 321–332.

Potgieter, F. T. and Els, H. J. (1976). Light and electron microscopic observations on the development of small merozoites of *Babesia bovis* in *Boophilus microplus* larvae. *Onderstepoort Journal of Veterinary Research* **43**, 123–128.

Potgieter, F. T. and Els, H. J. (1977). Light and electron microscopic observations on the development of *Babesia bigemina* in larvae, nymphae and non-replete females of *Boophilus decoloratus*. *Onderstepoort Journal of Veterinary Research* **44**, 213-232.

Potgieter, F. T., Els, H. J. and van Vuuren, A. S. (1976). The fine structure of merozoites of *Babesia bovis* in the gut epithelium of *Boophilus microplus*. *Onderstepoort Journal of Veterinary Research* **43**, 1-10.

Purnell, R. E. and Joyner, L. P. (1968). The development of *Theileria parva* in the salivary glands of the tick, *Rhipicephalus appendiculatus*. *Parasitology* **58**, 725-732.

Regendanz, P. (1936). Über den Entwicklungsgang von *Babesia bigemina* in der Zecke *Boophilus microplus*. *Zentralblatt für Bakteriologie, Parasitenkunde, Infektionskrankheiten und Hygiene, Abteilung I*, **137**, 423-428.

Regendanz, P. and Reichenow, E. (1933). Die Entwicklung von *Babesia canis* in *Dermacentor reticulatus*. *Archiv für Protistenkunde* **79**, 50-71.

Reichenow, E. (1937). Über die Entwicklung von *Theileria parva*, dem Erreger des Küstenfiebers der Rinder in *Rhipicephalus appendiculatus*. *Zentralblatt für Bakteriologie, Parasitenkunde, Infektionskrankheiten und Hygiene, Abteilung I* **140**, 223-226.

Reichenow, E. (1938). Über die Entwicklung des Erregers des Küstenfiebers der Rinder und die Pathogenese dieser Krankheit. *Acta Conversa Malariae Tertianae et Tropica Morbis*, Part 1, 681-686.

Reichenow, E. (1940). Der Entwicklungsgang des Küstenfiebererregers im Rinde und in der übertragenden Zecke. *Archiv für Protistenkunde* **94**, 1-56.

Riek, R. F. (1964). The life cycle of *Babesia bigemina* in the tick vector *Boophilus microplus*. *Australian Journal of Agricultural Research* **15**, 802-821.

Riek, R. F. (1966). The life cycle of *Babesia argentina* (Lignières, 1903) in the tick vector *Boophilus microplus* (Canestrini). *Australian Journal of Agricultural Research* **17**, 247-254.

Riek, R. F. (1968). Babesioses. *In* "Infectious Blood Diseases of Man and Animals" (D. Weinmann and M. Ristic, eds), Vol. 2, pp. 219-268. Academic Press, New York.

Rosenbusch, F. (1927). Estudios sobre la tristeza; evolucion del *Piroplasma bigemina* en la garrapata (*Boophilus microplus* Can. Lah.). *Revista de la Universidad de Buenos Aires* **5**, 863-867.

Rudzinska, M. A. (1981). Morphologic aspects of host cell parasite relationships in babesiosis. *In* "Babesiosis" (M. Ristic and J. P. Kreier, eds), pp. 87-141. Academic Press, New York.

Rudzinska, M. A., Spielman, A., Riek, R. F., Lewengrub, S. J. and Piesman, I. (1979). Intraerythrocytic "gametocytes" of *Babesia microti* and their maturation in ticks. *Canadian Journal of Zoology* **57**, 424-434.

Rudzinska, M. A., Spielman, A., Lewengrub, S., Piesman, J. and Karakashian, S. (1982). Penetration of the peritrophic membrane of the tick by *Babesia microti*. *Cell and Tissue Research* **221**, 471-481.

Sarwar, S. M. (1935). A hitherto undescribed Piroplasma of goats (*Piroplasma taylori*). *Indian Journal of Veterinary Science* **5**, 171-176.

Schein, E. (1975). On the life cycle of *Theileria annulata* in the midgut and hemolymph of *Hyalomma anatolicum excavatum*. *Zeitschrift für Parasitenkunde* **47**, 165-167.

Schein, E. and Friedhoff, K. T. (1978). Lichtmikroskopische Untersuchungen über die Entwicklung von *Theileria annulata* in *Hyalomma anatolicum excavatum*. II. Die Entwicklung in Hämolymphe und Speicheldrüsen. *Zeitschrift für Parasitenkunde* **56**, 287-303.

Schein, E. and Voigt, W. P. (1979). Chemotherapy of bovine theileriosis with Halofuginone. *Acta Tropica* **36**, 391–394.

Schein, E., Büscher, G. and Friedhoff, K. T. (1975). Lichtmikroskopische Untersuchungen über die Entwicklung von *Theileria annulata* (Dschunkowsky und Luhs, 1904) in *Hyalomma anatolicum excavatum* (Koch, 1844). I. Die Entwicklung im Darm vollgesogener Nymphen. *Zeitschrift für Parasitenkunde* **48**, 123–136.

Schein, E., Warnecke, M. and Kirmse, P. (1977a). Development of *Theileria parva* in the gut of *Rhipicephalus appendiculatus*. *Parasitology* **75**, 309–316.

Schein, E., Mehlhorn, H. and Warnecke, M. (1977b). Zur Feinstruktur der erythrocytären Stadien von *Theileria annulata*. *Zeitschrift für Tropenmedizin und Parasitologie* **28**, 349–360.

Schein, E., Mehlhorn, H. and Warnecke, M. (1978). Light and electron microscopic studies on the schizogony of 4 *Theileria* species of cattle (*T. parva parva*, *T.p. lawrencei*, *T. mutans* and *T. annulata*). *Protistologica* **14**, 337–348.

Schein, E., Mehlhorn, H. and Voigt, W. P. (1979). Electron microscopical studies on the development of *Babesia canis* (Sporozoa) in the salivary glands of the vector tick *Dermacentor reticulatus*. *Acta Tropica* **36**, 229–241.

Schein, E., Rehbein, G., Voigt, W. P. and Zweygarth, E. (1981). *Babesia equi* (Laveran 1901). 1. Development in horses and in lymphocyte culture. *Zeitschrift für Tropenmedizin und Parasitologie* **32**, 223–227.

Scholtyseck, E., Mehlhorn, H. and Friedhoff, K. T. (1970). The fine structure of the conoid of sporozoa and related organisms. *Zeitschrift für Parasitenkunde* **34**, 68–94.

Sergent, E., Donatien, A., Parrot, L. and Lestoquard, F. (1936). Cycle évolutif de *Theileria dispar* du boeuf chez la tique *Hyalomma mauritanicum*. *Archives de l'Institut Pasteur d'Algérie* **14**, 259–294.

Shortt, H. E. (1973). *Babesia canis*: the life cycle and laboratory maintenance in its arthropod and mammalian hosts. *International Journal for Parasitology* **3**, 119–148.

Smith, T. and Kilborne, F. H. (1893). Investigations into the nature, causation and prevention of Texas or southern cattle-fever. *Bulletin of Bureau of Animal Industry, U.S. Department of Agriculture, Washington* **1**, 177–304.

Starcovici, C. (1893). Bemerkung über den durch Babès entdeckten Blutparasiten und die durch denselben hervorgebrachten Krankheiten, die seuchenhafte Hämoglobinurie des Rindes (Babès), das Texasfieber (Smith) und der Carceag der Schafe (Babès). *Zentralblatt für Bakteriologie, Parasitenkunde, Infektionskrankheiten und Hygiene* **14**, 1–8.

Theiler, A. (1904). East coast fever. *Transvaal Agriculture Journal* **2**, 421–438.

Theiler, A. (1906). *Piroplasma mutans* (n. spec.) of South African cattle. *Journal of Comparative Pathology and Therapy* **19**, 292–438.

Trofimov, I. T. (1952). Patologičeskaja anatomija i patogenez nuttallioza lošadej. *Uchenye Zapiski Kazanskogo Veterinarnogo Instituta* **58**, 45–96.

Uilenberg, G., Rombach, M. C., Perié, N. M. and Zwart, D. (1980). Blood parasites of sheep in the Netherlands. II. *Babesia motasi* (Sporozoa, Babesiidae). *Veterinary Quarterly* **2**, 3–14.

Van Vorstenbosch, C. J., Uilenberg, G. and Dijk, I. E. (1978). Erythrocytic forms of *Theileria velifera*. *Research in Veterinary Science* **24**, 214–221.

Vivier, E. and Petitprez, A. (1969). Observations ultrastructurales sur l'hématozoaire *Anthemosoma garnhami*. *Protistologica* **5**, 363–379.

Warnecke, M., Schein, E., Voigt, W. P. and Uilenberg, G. (1979). On the life cycle of *Theileria velifera* (Uilenberg, 1964) in the gut and the haemolymph of the tick vector *Amblyomma variegatum* (Fabricius, 1794). *Zeitschrift für Tropenmedizin und Parasitologie* **30**, 318–322.

Warnecke, M., Schein, E., Voigt, W. P., Uilenberg, G. and Young, A. S. (1980). Development of *Theileria mutans* (Theiler, 1906) in the gut and the haemolymph of the tick *Amblyomma variegatum* (Fabricius, 1794). *Zeitschrift für Parasitenkunde* **62**, 119–125.

Weber, G. (1980). Ultrastrukturen und Cytochemie der Pellikula und des Apikalkomplexes der Kineten von *Babesia bigemina* und *B. ovis* in Hämolymphe und Ovar von Zecken. *Journal of Protozoology* **27**, 59–71.

Weber, G. and Friedhoff, K. T. (1977). Preliminary observations on the ultrastructure of supposed sexual stages of *Babesia bigemina* (Piroplasmea). *Zeitschrift für Parasitenkunde* **53**, 83–92.

Weber, G. and Walter, G. (1980). *Babesia microti* (Apicomplexa: Piroplasmida): electron microscope detection in salivary glands of the tick vector *Ixodes ricinus* (Ixodoides; Ixodidae). *Zeitschrift für Parasitenkunde* **64**, 113–115.

Wernsdorfer, G. (1983). Differentialdiagnose von Malaria und Babesiose beim Menschen. *Proceedings of the Joint Meeting of the German speaking Societies of Tropical Medicine, Garmisch-Partenkirchen*, 31.

Young, A. S., Grootenhuis, I. G., Smith, K., Flowers, M. J., Dolan, T. T. and Brocklesby, D. W. (1978a). Structures associated with *Theileria* parasites in eland erythrocytes. *Annals of Tropical Medicine and Parasitology* **72**, 443–454.

Young, A. S., Purnell, R. E., Payne, R. C., Brown, C. G. D. and Kanhai, G. K. (1978b). Studies on the transmission and course of infection of a Kenyan strain of *Theileria mutans*. *Parasitology* **76**, 99–115.

Young, A. S., Grootenhuis, J. G., Leitch, B. L. and Schein, E. (1980). The development of *Theileria* = *Cytauxzoon taurotragi* (Martin and Brocklesby, 1960) from eland in its tick vector *Rhipicephalus appendiculatus*. *Parasitology* **81**, 129–144.


*Note added in proof*

The fusion of 'Strahlenkörper' of *Babesia microti in vivo* (see p. 78) has been further confirmed by Rudzinska *et al.* (1983).

*Reference*

Rudzinska, M. A., Spielman, A., Lewengrub, S., Trager, W. and Piesman, J. (1983). Sexuality in piroplasms as revealed by electron microscopy in *Babesia microti*. *Proceedings of the National Academy of Sciences of the USA* **80**, 2966–2970.

# Metabolism of *Entamoeba histolytica* Schaudinn, 1903

RICHARD E. REEVES

*Departments of Biochemistry and of Tropical Medicine and Medical Parasitology, Louisiana State University Medical Center, New Orleans, Louisiana, USA*

| | | |
|---|---|---|
| I. | Introduction | 106 |
| II. | Carbohydrate Metabolism | 108 |
| | A. Glucose to End Products in Cells Grown Axenically | 108 |
| | B. Other Activities in Cells Grown Axenically | 113 |
| | C. Activities Not Present in Cells Grown Axenically | 115 |
| | D. Activities Reported Only from Cells Grown with Bacteria | 116 |
| | E. Carbohydrate Hydrolases | 117 |
| III. | Other Metabolic Capabilities | 118 |
| | A. Protein Metabolism | 118 |
| | B. Lipid Metabolism | 119 |
| | C. Nucleic Acid Metabolism | 120 |
| | D. Electron Transport | 121 |
| | E. Substrate Transport | 122 |
| IV. | Effect of Associate Cells | 122 |
| | A. Enzyme Activities | 122 |
| | B. Pathogenicity | 123 |
| V. | Growth Requirements | 124 |
| | A. Vitamins and Cofactors | 124 |
| | B. Substrates | 125 |
| VI. | Regulation | 126 |
| VII. | Evolutionary Considerations | 127 |
| VIII. | Speculation | 128 |
| IX. | Conclusions | 131 |
| | Acknowledgement | 132 |
| | Appendix A. Quantitation | 132 |
| | Appendix B. Free Amino Acids in Axenic Growth Media | 133 |
| | References | 133 |

## I. Introduction

A recent report by the World Health Organization (1981) and a review by Albach and Booden (1978) consider many aspects of the metabolism of *Entamoeba histolytica*.

Because it is directly implicated in human disease, research on *E. histolytica* has been relatively well supported. However, studies on the organism have been very unbalanced. The statement in the World Health Organization (1981, p. 29) report, 'knowledge of the biology of *Entamoeba histolytica* has recently been well developed . . .', greatly overstates recent progress. A defined culture medium is as remote as it was a decade ago. The understanding of nuclear division is inadequate at the microscopic level and completely lacking at the molecular level. The role of virus-like cellular inclusions is not clear in amoebae axenically grown, much less so *in vivo*. Knowledge of the metabolism of carbohydrates is reasonably complete, but the areas of protein, lipid and nucleic acid metabolism are virtually untouched. Two features which may have delayed progress are firstly, the unabashed focus on the feeble pathogenic properties of axenic amoebae, and secondly the literature on amoebae is exceptional in having the major fraction of its recent research papers* published without external critical review, an inadequate substitute for publication in the primary literature.

The challenge presented by *E. histolytica* is that of an unfamiliar eukaryotic organism having several steps in its glycolytic pathway catalysed by newly recognized enzyme activities. This being so, caution is clearly indicated in extrapolating to amoebae a conventionality with respect to more-investigated organisms. This theme will be repeated.

The two stages in the life cycle of *E. histolytica*, cyst and trophozoite, play the clearly defined roles of transmission and explosive growth. To date only the trophozoite is amenable to many types of study. It is readily grown in rich, undefined axenic culture systems of a complexity that defeat most nutritional studies.

---

* Despite efforts of the organizers to disseminate published papers from the interesting and stimulating seminars on amoebiasis, I am uncertain that the following publications are widely accessible: *Archivos de Investigacion Medica* (Mexico), Vol. 2 (1971), supplement 1, pp. 155–452; Vol. 3, (1972), supplement 2, pp. 229–458; Vol. 4 (1973), supplement 1, pp. 1–262; Vol. 5 (1974), supplement 2, pp. 265–564; Vol. 9 (1978), supplement 1, pp. 85–415; Vol. 11 (1980), supplement 1, pp. 1–328; Vol. 13 (1982), supplement 3, pp. 1–326; and *Proceedings of the International Conference on Amebiasis* (1976) (Sepulveda, B. and Diamond, L. S. eds), pp. 1–892. Instituto Mexicano del Seguro Social, Mexico, D.F.

Work on the intermediary metabolism of *E. histolytica* began with Kun and Bradin (1953) at a time when amoebae were often grown in a biphasic soup with an unidentified multibacterial flora. Such growth conditions were not reproducible. Better-controlled conditions were provided by monobacterial culture systems (Shaffer and Frye, 1948; Reeves *et al.*, 1957; Reeves and Ward, 1965). These systems employed a clear monophasic fluid medium with an inhibited bacterial associate. Of greater importance were the axenic cultures developed by Diamond (1968) and Diamond *et al.* (1978). But, a lack of accord that characterized studies of the earlier era has not vanished. Contributing to this are features of the axenic media not widely recognized: (1) their high osmolarity (0·42 and 0·36 osmol for TP-S-1 and TYI-S-33 media, respectively); (2) their components, without added sugar, contribute sufficient glucose to allow the survival of serial transplants of small numbers of amoebae; (3) growth rates and cell harvests vary widely with different lots of media components; (4) some of the added vitamins and growth factors are of doubtful utility (see Section V A); and (5) it has always been a difficult task to transfer amoebae into axenic culture.

This review deals almost entirely with activities of typical *E. histolytica* originally isolated from human infections. Strain designations will seldom be emphasized since significant differences have not been encountered among those studied; indeed, the related species, *E. invadens*, *E. terrapinae*, *E. moshkovski*, and the *E. histolytica*-like strains possess some of the unusual metabolic characteristics of the typical *E. histolytica* strains. Throughout this review where 'amoeba' is used without qualification it refers to typical *E. histolytica*. We still pay a high toll for the taxonomic error of incorrectly classifying the *Entamoeba histolytica*-like organisms in the species *E. histolytica* (for a similar opinion see Elsdon-Dew, 1976). Workers new to the field, those most needed, will tend to work on the strains readily adaptable to a variety of cultural conditions. This often means *E. histolytica*-like organisms (see Gicguaud, 1979).

In Sections II and III evidence will be presented that amoebae grown axenically have a complete, but circumscribed, metabolic capability; and that, when living in contact with other cells, they have a faculty for expanding their metabolic competence. The latter idea has long been widely entertained. If true, it is not surprising that divergent reports of amoebic metabolic activities are not yet resolved. Section IV addresses possible modes of transfer of activities from associated cells to the amoeba and evidence that pathogenicity is thereby influenced. Later sections consider growth factors and nutrients, metabolic regulation, and speculative views on how the metabolism of amoebae may shed light on primitive metabolic pathways.

## II. Carbohydrate Metabolism

### A. Glucose to End Products in Cells Grown Axenically

A report stating that amoebae do not use free glucose (Hallman et al., 1954) was quickly corrected by Becker and Geiman (1955). Labelled [$^{14}$C]-glucose catabolized by amoebae is more than 95% accounted for by label in glycogen, carbon dioxide, ethanol and acetate (Montalvo et al., 1971). Only traces of label are found in protein, lipid and nucleic acids.

Glucose catabolism begins with transport across the cell membrane by facilitated diffusion. The $K_m$ value for glucose transport is low and its maximum rate of transport is about the same as the estimated rate of glycolytic flux in rapidly growing cells, suggesting that transport may be a rate-limiting step in glycolysis (Serrano and Reeves, 1974, 1975).

Enzyme activities that catalyse glucose fermentation are listed in Table 1 with the EC numbers assigned to them by the Enzyme Nomenclature Committee (1979). For each of the metabolic intermediates Table 1 lists the name, an identifying code, and a reference to a paper reporting the presence of each enzyme in amoebae. The metabolic intermediates are not novel, but several of the steps are catalysed by recently discovered enzyme activities involving unusual cosubstrates. The steps are shown, schematically, in Fig. 1a. The enzymes catalysing steps 3, 7, 10, 11 and 14 through to 19 are not known to occur widely in other eukaryotic organisms. A brief description of the unusual steps in this glycolytic pathway follows (numbered according to Table 1 and Fig. 1a; references appear in the table).

*Step. 3.* The phosphofructokinase of *E. histolytica* employs inorganic pyrophosphate (PP$_i$) rather than a nucleoside triphosphate as phosphate donor. This enzyme has since been found in *Propionibacterium shermanii* (O'Brien et al., 1975), and in various other prokaryotes and eukaryotes (Reeves et al., 1982; Van Schaftingen et al., 1982). In their initial report Reeves et al. (1974a) stated that an adenosine triphosphate (ATP)-dependent phosphofructokinase activity was present along with the major PP$_i$-dependent enzyme. A great deal of further work has convinced me that the activity we considered to be ATP-dependent phosphofructokinase was due to the functioning of the amoebic glycogen-formation pathway by which PP$_i$ is formed (see Section II B). Pyrophosphate analogue inhibitors for this enzyme were studied by Eubank and Reeves (1982).

*Step 7.* The phosphoglycerate kinase is selective for guanine nucleotides rather than adenine nucleotides.

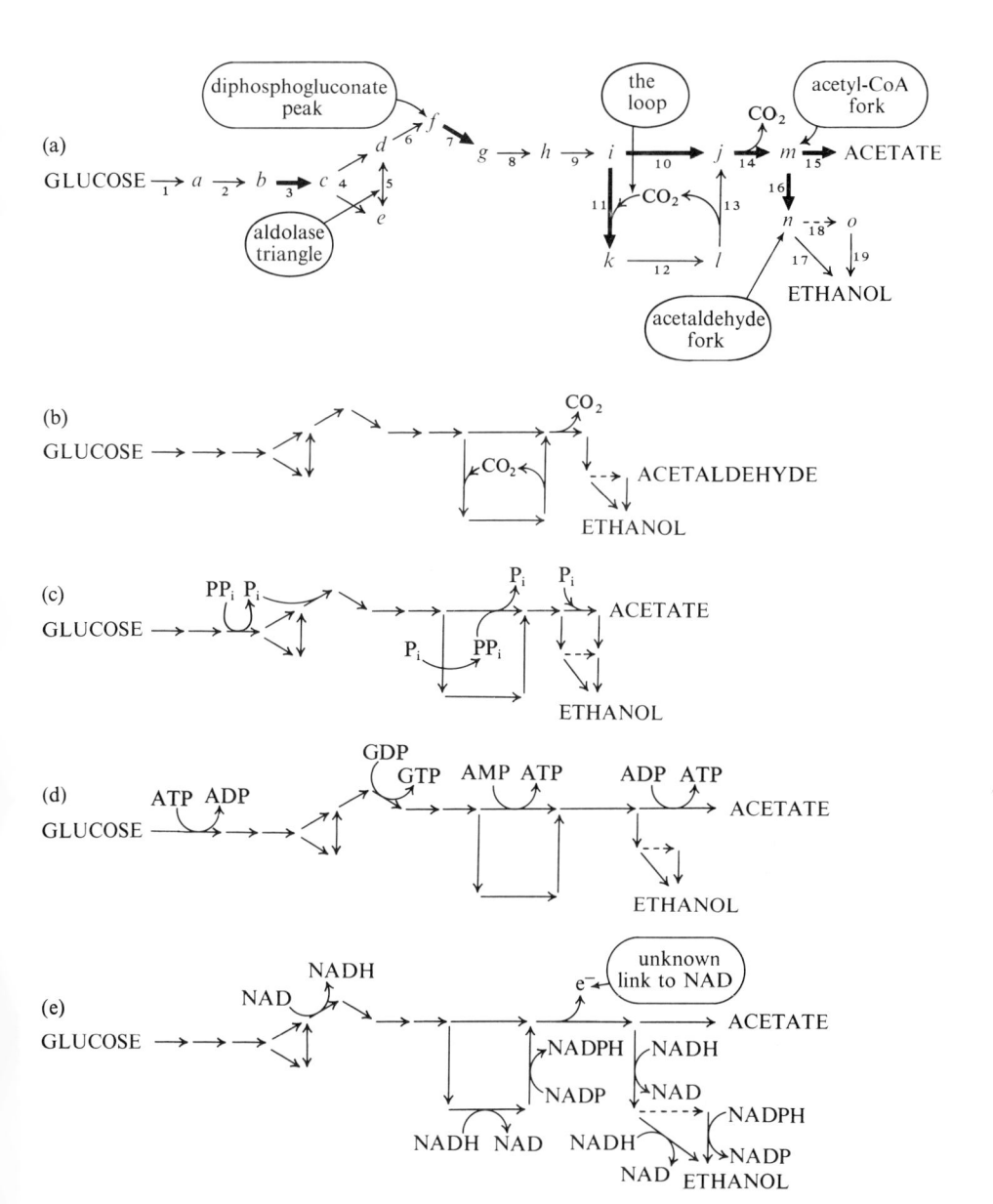

FIG. 1. The glycolytic pathway for *Entamoeba histolytica* grown axenically. a. Under the partially aerobic conditions of growth which prevail in axenic culture medium. Unusual steps are shown with bold arrows. Prominent features of the pathway are indicated. Enzymes and intermediates are identified as in Table 1. The broken arrow, step 18, represents a non-enzymic reaction. b. The anaerobic pathway. c. The linkage of orthophosphate ($P_i$) and inorganic pyrophosphate ($PP_i$). d. The linkage with adenine and guanine nucleotides. e. The linkage with the pyridine nucleotides (NAD and NADP).

TABLE 1

Steps in the glycolytic pathway of Entamoeba histolytica

| Step | Enzyme (EC no.) | Product (glycolytic direction) Code | Name | Reference (to presence in amoeba) |
|---|---|---|---|---|
| 1 | Glucokinase (2.7.1.2) | a | Glucose 6-phosphate | Reeves et al. (1967) |
| 2 | Glucosephosphate isomerase (5.3.1.9) | b | Fructose 6-phosphate | Montalvo and Reeves (1968) |
| 3 | Phosphofructokinase (PP$_i$) (2.7.1.90) | c | Fructose 1,6-bisphosphate | Reeves et al. (1974a, 1976) |
| 4 | Aldolase (4.1.2.13) | {d, e | Glyceraldehyde 3-phosphate Dehydroxyacetone phosphate | Kalra et al. (1969); Susskind et al. (1982) |
| 5 | Triosephosphate isomerase (5.3.1.1) | d | Glyceraldehyde 3-phosphate | Reeves (1974) |
| 6 | Glyceraldehydephosphate dehydrogenase (1.2.1.12) | f | 1,3-Diphosphoglycerate | Reeves (1974) |
| 7 | Phosphoglycerate kinase (GTP) (2.7.2.10) | g | 3-Phosphoglycerate | Reeves and South (1974) |
| 8 | Phosphoglycerate mutase (2.7.5.3 or 5.4.2.1) | h | 2-Phosphoglycerate | Reeves (1974) |
| 9 | Enolase (4.2.1.19) | i | Phosphoenolpyruvate | Reeves (1974) |
| 10 | Pyruvate, phosphate dikinase (2.7.9.1) | j | Pyruvate | Reeves (1968); Reeves et al. (1968) |
| 11 | PEP carboxyphosphotransferase (4.1.1.38) | k | Oxaloacetate | Reeves (1970) |
| 12 | Malate dehydrogenase (1.1.1.37) | l | L-Malate | Reeves and Bischoff (1968) |
| 13 | Malate dehydrogenase decarboxylating (1.1.1.40) | j | Pyruvate | Reeves and Bischoff (1968) |
| 14 | Pyruvate synthase (1.2.7.1) | m | Acetyl-CoA + CO$_2$ | Reeves et al. (1977) |
| 15 | Acetyl-CoA synthetase (ADP-forming) (6.2.1.13) | — | Acetate | Reeves et al. (1977) |

| | | | | |
|---|---|---|---|---|
| 16 | Acetaldehyde dehydrogenase (acylating) (1.2.1.10) | n | Enzyme-bound thiohemiacetal | Lo and Reeves (1978) |
| 17 | Alcohol dehydrogenase (1.1.1.1) | – | Ethanol | Lo and Reeves (1978) |
| 18 | Non-enzymic hydrolysis of thiohemiacetal | o | Acetaldehyde | |
| 19 | Alcohol dehydrogenase (1.1.1.2) | – | Ethanol | Reeves et al. (1971a); Lo and Chang (1982) |

*Steps 10 and 11.* These steps come into play in a branched pathway (loop) between phosphoenolpyruvate (PEP) and pyruvate. One enzyme is pyruvate-phosphate dikinase, the other is PEP carboxyphosphotransferase, the latter discovered in *P. shermanii* by Sui and Wood (1962). The two other loop enzymes are the familiar nicotinamide adenine dinucleotide (NAD)-dependent malate dehydrogenase and malate dehydrogenase (decarboxylating); the latter converts L-malate into pyruvate plus carbon dioxide employing NADP as electron carrier.

*Step 14.* Pyruvate synthase is known in bacteria, in *E. invadens* (Lindmark, 1976) and in *Tritrichomonas foetus* (Lindmark and Muller, 1973). It catalyses a reaction between pyruvate and CoA to form acetyl-CoA and carbon dioxide, yielding electrons which may be accepted by ferredoxin or flavin, but not by NAD or NADP. Pyruvate synthase replaces the activity of pyruvate decarboxylase or pyruvate dehydrogenase, both of which *E. histolytica* lacks. The pyruvate oxidase reported by Takeuchi *et al.* (1975) will be considered in Section II C.

*Step 15.* The enzyme that catalyses the formation of acetate and ATP from acetyl-CoA resembles a previously known acetate thiokinase of bacterial origin (EC 1.2.1.1), except that the latter functions with adenosine monophosphate (AMP) and $PP_i$ and the amoebic enzyme with adenosine diphosphate (ADP) and $P_i$.

*Step 16.* Acetaldehyde dehydrogenase (acylating) is an NAD-linked enzyme activity which reversibly converts the acetyl group of acetyl-CoA into the aldehyde level of oxidation. In amoebae the immediate product of acetyl-CoA and NADH is apparently an enzyme-bound acetaldehyde thiohemiacetal which slowly equilibrates with free acetaldehyde, probably by a non-enzymic step (step 18).

*Step 17.* Formally, this step is the common NAD-linked alcohol dehydrogenase, but in amoebae it seems to be located on the same protein which bears the acetaldehyde dehydrogenase activity. Steps 16 and 17 are thus believed to be catalysed by a bifunctional enzyme protein, the second activity utilizing as substrate the thiohemiacetal product of the first.

*Step 18.* This step is believed to be relatively unimportant in alcohol production by amoebae, and to proceed non-enzymically.

*Step 19.* This step is catalysed by an abundant NADP-linked amoebic alcohol dehydrogenase. This enzyme has the property of reducing acetaldehyde, its normal substrate, and of reacting strongly with certain other aldehydes and ketones, particularly with acetone, a non-physiological substrate. Only the enzyme from a bacterial source described by Hoshino and Udagawa (1960) seems to have this last property to such a pronounced degree.

Fig. 1 shows (a) the pathway for partially aerobic growth as in axenic culture, and (b) the anaerobic pathway. In the presence of a bacterial associate containing hydrogenase, pathway (b) may be modified to produce a small amount of acetate and hydrogen gas.

### B. OTHER ACTIVITIES IN CELLS GROWN AXENICALLY

Three pathways diverge from the amoebal glycolytic pathway: the glycogen cycle, the galactose link to glycolysis, and hexose–pentose interconversion.

### 1. *The glycogen cycle*

The enzymes of amoebic glycogen formation and glycogenolysis were studied by Blytt and Reeves (1976) and Takeuchi *et al.* (1977a,b). They appear to be those normally found in eukaryotic organisms except for a lack of reported regulatory features. Phosphoglucomutase and glucose-1-phosphate uridylyltransferase are cytoplasmic enzymes, but glycogen phosphorylase occurs loosely bound to particulate glycogen and is slowly liberated in homogenates by the action of amylase.

Since amoebae utilize $PP_i$, glycogen cycling in this organism proceeds from glucose 1-phosphate to glycogen and back to glucose 1-phosphate without the expenditure of a high-energy phosphate bond. The overall effect of one cycle is to convert one molecule of orthophosphate and nucleoside triphosphate into a nucleoside diphosphate and $PP_i$. Glycogen cycling may be the principal source of the elevated intracellular $PP_i$ concentrations found in the organism (Reeves *et al.*, 1974a).

### 2. *Galactose link to glycolysis*

D-Galactose is transported and actively metabolized by amoebae, as first shown by Weinbach and Diamond (1974). I wish to correct statements emanating from this laboratory that the growth of *E. histolytica* requires glucose or a glucose-containing polymer (Reeves, 1972; Serrano and Reeves 1974, 1975). Galactose, when substituted for glucose, will support the growth of amoebae in axenic media. When 1-[$^{14}$C]galactose replaces added glucose in axenic growth medium the label appears in glycogen and the ethanol and acetate end-products. Virtually no label appears in carbon dioxide (L. G. Warren, personal communication, 1981). This suggests that galactose is metabolized via the amoebic version of the Embden-Meyerhof pathway. We have found galactose metabolism to proceed in the following manner (P. A. Lobelle-Rich and R. E. Reeves, unpublished work, 1982). D-Galactose enters the cell by the same transport system that brings in glucose. Once in the cell it is converted into α-galactose 1-phosphate by an ATP-dependent galactokinase (EC 2.7.1.6). The α-galactose 1-phosphate is uridylated to

UDPgalactose by one of two UTP-utilizing uridylyltransferases present in *E. histolytica* (Lobelle-Rich and Reeves, 1982, 1983a). The UDPgalactose formed is converted into UDPglucose by UDPglucose 4-epimerase (EC 5.1.3.2), and thence to the glycolytic pathway via glycogen. Having found in glucose-grown cells the enzymes that link galactose to glycolysis, we grew cells for many transfers with galactose substituted for the added glucose of TP-S-1 medium. The cells grew more slowly, but the same enzymes were found. The activity of these enzymes was about 2·5 to 7 μmol min$^{-1}$ (ml of packed cells)$^{-1}$, the same activities we had found in the glucose-grown cells. Some of our cultures had been in glucose-containing medium for 10 years without loss of their ability instantly to metabolize galactose. Growth on galactose caused no induction of additional galactose enzyme activity (R. E. Reeves and P. A. Lobelle-Rich, unpublished work, 1981).

3. *Hexose to pentose interconversion*

Susskind *et al.* (1982) clarified the hexose–pentose interconversion pathway in amoebae. The organism lacks transaldolase, but a pathway similar to that involved in the Calvin cycle was found to operate. This is shown schematically in Fig. 2. The authors demonstrated that *E. histolytica* possesses transketolase, that amoebic aldolase utilizes sedoheptulose bisphosphate as a substrate, and that the reversible amoebic phosphofructokinase catalyses conversion of sedoheptulose bisphosphate into sedoheptulose 7-phosphate. They reconstituted this system *in vitro* using partially purified amoebic enzymes and demonstrated that it functions.

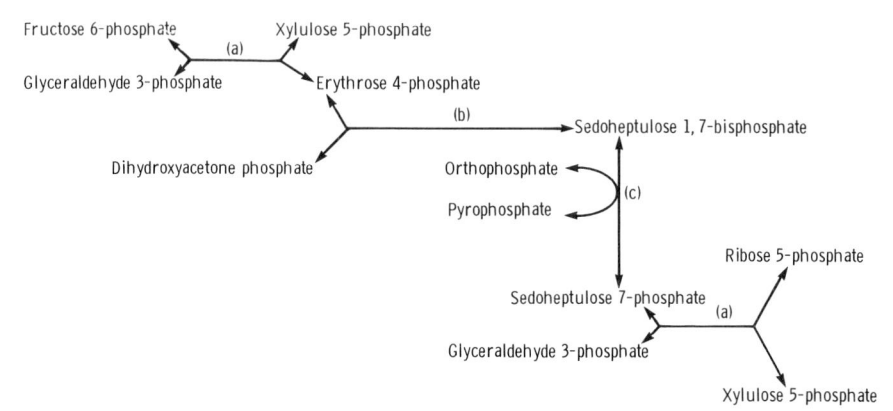

FIG. 2. A schematic pathway for hexose–pentose interconversion in *Entamoeba histolytica*. Amoebal enzymes comprising the enzyme system are: (a) transketolase; (b) aldolase; (c) phosphofructokinase (PP$_i$). (Reproduced, with permission, from *Biochemical Journal* **204**, 195.)

## 4. Nucleotide reductase

It appears that amoebae must possess the capability of reducing nucleotides to 2′-deoxynucleotides for the synthesis of deoxyribonucleic acid (DNA). Indirect evidence for this reduction is the inhibition of amoebal growth by hydroxyurea observed by Austin and Warren (1982).

### C. ACTIVITIES NOT PRESENT IN CELLS GROWN AXENICALLY

We examined the soluble proteins from amoebae grown axenically for a wide variety of enzyme activities. Those activities not found after careful and repeated searches are listed in Table 2. For five of the activities listed there are published reports claiming the positive finding of the activity in axenic amoebae. I have exhaustively reexamined each of these positive claims and I am satisfied that each is invalid. The reexamination included the growth condition, the assay protocol and, save in one instance, the strain of amoeba used by the original authors. Since not all of the rebuttals are in print each is briefly outlined below.

Lactate dehydrogenase was claimed to be present in a published seminar report by Lee *et al.* (1971). The protocol used was not included in the publication, but was kindly supplied by one of the authors (L. Landa, personal communication, 1972). The protocol was that of Nielands (1955) plus 8 mM sodium cyanide. This unfortunate modification of Nielands' NAD-containing assay completely invalidates the assay since cyanide chemically reacts with NAD giving a product with absorbancy similar to that of reduced NAD. In our hands, without added cyanide, there was no evidence of lactate dehydrogenase.

TABLE 2

*Enzyme activities not found in axenic* E. histolytica *(EC number in parentheses)*

1. Lactate dehydrogenase (1.1.1.27)
2. 6-Phosphofructokinase (ATP) (2.7.1.11)
3. Hydrogenase (1.18.3.1)
4. Phosphate acetyltransferase (2.3.1.8)
5. Glucose-6-phosphate dehydrogenase (1.1.1.43)
6. Phosphogluconate dehydratase (4.2.1.12)
7. Phospho-2-keto-3-deoxy-gluconate aldolase (4.1.2.14)
8. Glycerol-3-phosphate dehydrogenase (1.1.1.8 or 1.1.1.94)
9. Pyruvate oxidase (acylating) (1.2.3.6)
10. Transaldolase (2.2.1.2)
11. Pyruvate kinase (2.7.1.40)
12. Pyruvate dehydrogenase (1.2.2.2)
13. Pyruvate decarboxylase (4.1.1)
14. UDPglucose–hexose-1-phosphate uridylyltransferase (2.7.7.12)

6-Phosphofructokinase (ATP). A rebuttal of the claim of finding this activity (Reeves et al., 1974a) is in Section II A of this review.

Hydrogenase. Band and Cerrito (1979) did not demonstrate the activity of hydrogenase in the HK9 strain of amoeba, but cited in support of its presence their inability to grow the amoebae under an atmosphere containing 10% hydrogen. The argument is that if a functioning hydrogenase were present all pyridine nucleotides would be completely reduced, thus causing metabolism to cease. In this laboratory the axenic NIH:200 strain of amoeba was successfully grown under atmospheres containing up to 40% hydrogen (J. Hauswirth, unpublished work, 1979). It seems probable that something other than hydrogenase adversely affected growth in the experiments of Band and Cerrito (1979). The absence of hydrogen production by axenic amoebae was effectively proven by Tanowitz et al. (1975).

Glycerol-3-phosphate dehydrogenase was claimed by Takeuchi et al. (1980), but Reeves and Lobelle-Rich (1983) showed that the positive finding was due to another enzyme acting on an impurity in their enzyme substrate.

Pyruvate oxidase (acetylating). The activity reported by Takeuchi et al. (1975) and identified by them as a new enzyme (EC 1.2.3.6) may be that of pyruvate synthase (EC 1.2.7.1) (Reeves et al., 1977). Further reinvestigation of this problem failed to reveal any amoebic enzyme capable of catalysing the transfer of electrons from pyruvate directly to oxygen without the intervention of an electron carrier. Indeed, the standard enzyme assay employed by the prior authors contained added flavin, an electron carrier functioning with pyruvate synthase.

Three enzymes in Table 2 are those involving the Entner-Doudoroff pathway. They will be discussed in Section II D, below. References to the absence, from axenic amoebae, of some of the activities are transaldolase (Susskind et al., 1982), pyruvate kinase (Reeves, 1968, 1972), pyruvate dehydrogenase, pyruvate decarboxylase and phosphate acetyltransferase (Reeves et al., 1977), UDPglucose-hexose-1-phosphate uridylyltransferase (Lobelle-Rich and Reeves, 1983a).

The absence of cytochromes and of a functioning tricarboxylic acid cycle has been reported by Weinbach et al. (1976b).

### D. ACTIVITIES REPORTED ONLY FROM CELLS GROWN WITH BACTERIA

Phagocytotic and pinocytotic protozoa contain undigested foodstuff, an array of more or less degraded products, and those destined for excretion. When grown with bacterial associates, amoebae contain a variety of exogenous enzymes. Examples are the first seven of the activities mentioned in Table 2. Evidence is mounting that some of these enzymes may function, for a time, in amoebae before being inactivated. The simplest premise is that when an activity not present in axenic amoebae appears in amoebae grown with bacteria it is of exogenous origin. The three enzymes considered in the next paragraph challenge this premise.

1. *An enigma, the Entner-Doudoroff pathway*

Two of the activities listed in Table 2 comprise the pathway Entner and Doudoroff (1952) discovered for the fermentation of gluconate, and are connected to glucose fermentation by glucose-6-phosphate dehydrogenase. Hilker and White (1959) found evidence in amoebae for the functioning of three enzymes. Two of the three enzymes were reported by Bragg and Reeves (1962) from an *E. histolytica*-like organism. Employing a Shaffer-Frye culture, Entner (1958) found amoebae grown with bacteria to have a labelling pattern from [1-$^{14}$C]glucose which indicated the functioning of the Entner-Doudoroff pathway. The enzymes of the Entner-Doudoroff pathway were not found in the bacteria (Entner, 1958; Bragg and Reeves, 1962); nor have they been found in amoebae grown axenically (R. E. Reeves, unpublished work, 1975 and 1978).

2. *Other activities*

An unusual triosephosphate dehydrogenase was reported by Kun *et al.* (1956). This enzyme, strongly activated by cysteine, produced hydrogen sulphide. This activity has not since been reported from any source, nor have I been able to duplicate the original finding. Lactate dehydrogenase in cells grown with bacteria was reported by Entner and Anderson (1954) and Mignani and Bickel (1961). An ATP-linked phosphofructokinase in cells grown with bacteria was reported by Hilker and White (1959) and the functioning of hydrogenase in such cells was demonstrated by Montalvo *et al.* (1971). This activity has been repeatedly observed in anaerobically prepared homogenates of bacteria-grown cells (L. G. Warren and R. E. Reeves, unpublished work, 1971 to 1979). The presence of succinate dehydrogenase in bacteria-grown cells was reported by Entner and Anderson (1954) and Kalra *et al.* (1968). Phosphate transacetylase was reported by Reeves *et al.* (1977).

E. CARBOHYDRATE HYDROLASES

Workers agree that amoebae contain amylase (Hallman and DeLameter, 1953; Baernstein *et al.*, 1954; Hilker *et al.*, 1957; Mukhtar and Mohan Rao, 1972; Serrano *et al.*, 1977). Blytt and Reeves (1976) found the amylase to be associated with particulate glycogen. Jarumilinta and Maegraith (1960) found hyaluronidase in amoeba grown with bacteria but the bacterial flora of their cultures did not have this enzyme. The absence of sialic acids from the amoebal membrane was noted by Feria-Velasco *et al.* (1973).

## III. OTHER METABOLIC CAPABILITIES

So little is published in the areas covered by this section that for some of the subtopics it will be more instructive to begin with the glaring vacuities and follow with such positive information as is available. It will be necessary to draw on published abstracts and unpublished work to avoid leaving some important points completely without comment.

### A. PROTEIN METABOLISM

Although serological studies by Noya Gonzalez et al. (1980) have shown that amoebae contain membrane proteins directly derived from the growth medium there is no estimate of the quantitative extent of such incorporation. To cast some light on this problem the HK9 strain was grown in TP-S-1 medium containing 15 tritiated amino acids (0·8 μCi ml$^{-1}$). The washed cells were sonically disrupted and divided by centrifugation into pellet and soluble fractions. Protein in each fraction was precipitated with trichloroacetic acid. Protein from the soluble fraction contained 31 000 c.p.m. mg$^{-1}$, and from the pellet, 18 000 c.p.m. mg$^{-1}$ (R. E. Reeves, unpublished work, 1982). Barring a remarkable difference in amino acid content one would conclude that the protein of the pellet fraction (membrane) includes at least 40% exogenous protein; more if, as seems likely, the soluble fraction also contains exogenous protein. Our findings seem to be contradicted by those of Mendoza et al. (1982), who found greater incorporation of label into the pellet fraction of homogenates of cells grown with a single tritiated amino acid, L-leucine.

Although total amoebic protein contains a full complement of the common amino acids (Becker and Geiman, 1954; Lopez-Revilla and Navarro, 1978), there is no published evidence that amoebic protein synthesis utilizes all of them. Some of the assayed amino acids may have been derived from protein from exogenous sources. Carter et al. (1967), employing an *in vitro* system of amoebal ribosomes and enzymes, found lysine, phenylalanine and proline to be incorporated into acid-insoluble polymers. Entner and Grollman (1973) obtained incorporation of leucine into amoebic protein. Only these four amino acids have been shown to be employed by the protein synthesis apparatus, three with the expected codon. Acceptor tRNA species and a number of isoacceptors were reported by Hernandez-Velarde et al. (1980) for twelve L-amino acids, including the four mentioned above. To extend information on amino acid utilization, protein arising from axenic growth in medium containing labelled amino acids was hydrolysed and subjected to split-stream amino acid assay (for method see Coulson and Hernandez, 1968); 16 amino acids were found to be incorporated. Those not tested were asparagine, glutamine and tryptophane. The incorporation of L-tyrosine is uncertain. Not surprisingly, it appears that amoebae are able to incorporate most, or all, of the common L-amino acids into proteins (R. A. Coulson and R. E. Reeves, unpublished work, 1982).

An abstract by Meza et al. (1982) and a seminar by Sabanero and Meza (1982) report a welcome investigation of one *E. histolytica* protein (actin) by a variety of techniques.

Working with bacteria-grown amoebae, Mohan Rao and Dutta (1966) found significant transamination between α-oxoglutarate and eight amino acids. Less activity was seen with pyruvate as the nitrogen acccceptor and still less with oxaloacetate. Indirect evidence indicated that pyridoxal phosphate was a cofactor. In cells grown axenically, Lee et al. (1971) observed amino transfer between glutamic and oxaloacetic acids. We have confirmed this, and found transamination between aspartate and pyruvate, but not between alanine and α-oxoglutarate (P. Lobelle-Rich, unpublished work, 1982).

Early work on proteinases in bacteria-grown amoebae is reviewed by Von Brandt (1973). Neal (1960) concluded that the evidence does not demonstrate that high proteolytic activity is required for tissue invasion. Proteinases studied in amoebae grown axenically included an acid proteinase and a thiol-activated neutral proteinase (McLaughlin and Faubert, 1977), a collagenase (Munoz et al., 1982), and a cytotoxic agent in axenic cells which may be a proteinase (Lushbaugh and Pittman, 1982).

Weinbach et al. (1976a) and Reeves et al. (1980) reported non-haeme iron and acid-labile sulphide in amoebae. The latter workers isolated a low molecular-weight ferredoxin containing about 1 μmol of each per mg of protein. Guthrie and Reeves (1977) stated that filtration through Sephadex G-100 reveals the presence of at least two iron–sulphur proteins, one of high and another of low molecular weight.

We have attempted to introduce into protein isotopically labelled carbon from [$^{14}$C]glucose. Growing cultures will tolerate a high burden of $^{14}$C-labelled amino acids, but [$^{14}$C]glucose appears to limit growth, possibly because it gains entry into the adenine nucleotides as shown by Susskind et al. (1980). Growth for 40 hours in TP-S-1 medium containing [U-$^{14}$C]glucose (0·6 μCi ml$^{-1}$) gave 3-fold multiplication. The purified protein contained 14 150 c.p.m. mg$^{-1}$; 95% of the counts were in aspartate (Asp + Asn) and 5% in alanine. No radioactivity was detected in any other amino acid (R. E. Reeves and R. A. Coulson, unpublished work, 1983).

B. LIPID METABOLISM

No recent study on amoebic lipid metabolism has come to my attention. Almost all the lipid in cultivated cells was found to be derived from the medium (Sawyer et al., 1967). Much could be learned by studies on the nutritional value of the high- and low-density lipoproteins. Sawyer *et al.* (1967) investigated the lipids in amoebae grown in the modified Shaffer-Frye culture system. They found 10 to 16 mg of lipid per ml of cells (about 8 to 14% of the dry weight), of which 60% was phospholipid. They determined the fatty acid composition of total lipid and of the fractionated neutral and phospholipids. The growth medium contained all the fatty acids found in the amoebae save a few containing 21 and 24 carbon atoms. When grown in the presence of [U-$^{14}$C]glucose only 10% of the lipid carbon was derived from glucose, most of it appearing in fatty acids. Sawyer *et al.* (1967) concluded that amoebae are able to elongate preformed fatty acids, if not to synthesize them *de novo*. After saponification of the radiolabelled lipid a trace of label was in the unsaponifiable fraction and in the water-soluble fraction. This work should be repeated with amoebae grown axenically since Reeves and Lobelle-Rich (1983) suggest that axenic amoebae cannot synthesize glycerolphosphate.

Other published work on amoebic lipids has concerned cholesterol as a growth requirement and as an agent enhancing pathogenicity (see Sections IV B and V B).

C. NUCLEIC ACID METABOLISM

Results of the utilization of modern methods in research on amoebic nucleic acids have barely reached the printed page. The work by Reeves *et al.* (1971b) and Gelderman *et al.* (1971a,b) did little more than establish a low guanine plus cytosine content of amoebal DNA. Seminar reports by Boonlayangoor *et al.* (1978b) and Heebner and Albach (1982) note sucess in obtaining high molecular-weight RNA species from the organism. Work by the former group on the transport of bases and nucleosides is mentioned in Section III E. Boonlayangoor *et al.* (1980) found that amoebae do not synthesize purines from labelled glycine or formate precursors. P. Lobelle-Rich and R. E. Reeves (unpublished work, 1982) hybridized amoebic DNA, using the Southern blot technique, to a radiolabelled probe prepared from a plasmid containing yeast ribosomal DNA. Their results indicate that amoebal ribosomes are of the eukaryotic variety, as expected. In combination with DNA cleavage by restriction enzymes, the hybridization procedure may offer some promise as a means of differentiating among strains. An adenosine kinase from amoebae was reported by Lobelle-Rich and Reeves (1983b).

By absorbing solutions of Panmede and Trypticase with activated carbon, Reeves and West (1980) were able to arrive at a medium deficient in nucleic acid precursors. With this medium they confirmed the requirement of the organism for purines; AMP was the most effective growth promoter, followed by adenine, and then by adenosine. GMP alone did not support growth but it enhanced the effect of added AMP. In confirmation of the idea that amoebae can synthesize the pyrimidine bases they found no requirement for them. Albach and Booden (1978) mention that labelled orotic acid is incorporated into amoebic RNA. In this laboratory it was found that amoebae contain aspartate transcarbamylase (H. D. Bradshaw, unpublished work, 1982). Thus, the first and last steps in the synthesis of pyrimidines have been identified in amoebae.

### D. ELECTRON TRANSPORT

Under anaerobic conditions the unresolved feature of electron transport is the route by which electrons liberated at the pyruvate oxidation step are transferred to NAD. Under aerobic conditions, the route by which electrons from reduced flavins transfer to oxygen without the production of peroxide is also unresolved. A scheme for electron transport in amoebae grown axenically was presented by Weinbach *et al.* (1976b, 1980). My speculations on the subject are presented in Section VIII.

The amoebic glycolytic system affords two sites where transhydrogenation occurs between the two nicotinamide nucleotide species (NAD and NADP) to maintain a suitable reduced and oxidized nucleotide balance. One of these sites is at the level of ethanol and acetaldehyde oxidation and reduction. There are two alcohol dehydrogenases, one linked to NAD, the other to NADP. These two enzymes are located in the cytoplasmic compartment and they will function as a system to maintain a thermodynamically balanced concentration of the four nucleotide species. The other site for substrate level transhydrogenation is the four-enzyme loop between phosphoenolpyruvate and pyruvate. These enzymes are all physiologically reversible. One complete clockwise cycle, pyruvate to pyruvate, around this loop effects the net change:

$$NAD^+ + NADPH + PP_i + AMP \leftrightarrows NADH + NADP^+ + ATP$$

In the counterclockwise direction one cycle of the loop results in the same equation, read right to left. An interesting feature of the nicotinamide nucleotide transhydrogenase described by Harlow *et al.* (1976) is that the activity they describe reflects the activity of the loop enzymes. Upon adding ATP to their reaction system they found a stimulation of NADH → NADP$^+$ transfer and an inhibition of NADPH → NAD$^+$ transfer. These effects correspond to those expected of the loop enzyme system operating in the counterclockwise mode, as it should with added ATP. It would be interesting to learn if PP$_i$ and AMP, added instead of ATP, would produce the opposite nicotinamide nucleotide transhydrogenations and the formation of ATP.

E. SUBSTRATE TRANSPORT

The carbohydrate transport system described by Serrano and Reeves (1974, 1975) functions with the D-sugars glucose, galactose, xylose, 2-deoxyglucose, 3-O-methylglucose, and probably with mannose. Judged by a lack of inhibition of glucose transport, maltose, sucrose and fructose are not transported by the same system. The high temperature coefficient of glucose transport ($Q_{10} = 5$) is consistent with the inability of amoebae to grow in culture below about 33°C. Only in a minor respect, inhibition by methyl-α-D-glucoside, does the amoebic system differ markedly from the human erythrocyte transport system described by LeFevre (1961).

Weinbach and Diamond (1974) and Takeuchi *et al.* (1975, 1979) reported the stimulation of oxygen uptake in intact amoebae by L-serine. Their investigatory method involved use of the oxygen electrode. Although they did not mention transport across the cell membrane their findings could have been achieved only with a functioning L-serine transport system. This challenging observation led us to attempts to repeat their work. Over the 7 years since their initial claim, I have repeated their experiments on 30 occasions and have been unable to confirm their findings. The cell preparations showed stimulation of oxygen uptake by glucose, but not by serine. I conclude that a membrane transport system for serine has not been demonstrated in amoebae.

Lo and Reeves (1979a) suggested that the uptake of riboflavin was accomplished by pinocytosis rather than a mediated transport process. A similar conclusion for the uptake of niacin was reached by Weik and Reeves (1980) and for the uptake of AMP by Reeves and West (1980).

Boonlayangoor *et al.* (1978a) studied the uptake of bases and nucleosides. They found adenine, adenosine and guanosine to be taken up partly by carrier-mediated transport. Their evidence supported two transport sites; one for adenine–adenosine and another for adenosine–guanosine. This was a carefully conducted study; however, pinocytotic uptake might still satisfy amoebic growth requirements (Reeves and West, 1980).

IV. Effect of Associate Cells

A. ENZYME ACTIVITIES

In early work I failed to find the bacterial enzyme 1-phosphofructokinase (EC 2.7.1.56) in amoebae grown with bacteria. This ATP-dependent enzyme (Reeves *et al.*, 1966) is present in the Shaffer-Frye bacterial associate at high activity. It should have been easily identifiable if picked up by the amoebae.

Later, we found that a different enzyme, bacterial hydrogenase, is picked up and retained by amoebae; and that it remains functional in amoebic cytoplasm for several hours (L. G. Warren and R. E. Reeves, unpublished work, 1971–1979). Amoebae utilize this enzyme to produce hydrogen gas (Montalvo et al., 1971). Another enzyme, phosphate transacetylase (EC 2.3.1.8), not found in axenic amoebae, was regularly present in cells grown in the Shaffer-Frye culture system (Reeves et al., 1977).

Bacterial activities, like hydrogenase, have only a transient existence in amoebic cytoplasm. They need constant renewal by ingestion of more bacteria. After washing the cells free from bacteria, hydrogenase persists in amoebae for several hours at 37°C (Hauswirth and Warren, 1981). This enzyme appears in amoebae soon after feeding on the bacteria, even if the amoebae are subjected to cycloheximide to stop their protein synthesis; thus hydrogenase is marked as a product of bacterial origin (L. G. Warren, unpublished work, 1971–1974).

### B. PATHOGENICITY

A review by Neal (1983) on the subject of antiamoebic compounds helps clarify this problem. The review contains 93 references, many of which are not mentioned in the present review. Wittner and Rosenbaum (1970) found the virulence of amoebae grown axenically to be increased when they were reassociated for 6 to 12 hours with certain strains of bacteria before inoculation into the livers of weanling hamsters. Upon returning these amoebae to axenic culture a loss of virulence took place slowly over a period of 3 to 5 months. This aspect was further investigated by Bos and Hage (1975) and Bos (1976) who employed amoebae associated with crithidia as well as with bacteria. A slow decay of virulence upon returning to axenic growth was again observed. Further recent studies were reported by Raether (1976). One cannot read early work on the infectivity, pathogenicity and cyst-forming capability of amoebae without realizing the profound influence exerted by past association of the amoeba with bacteria or animal tissue cells. Chang's (1945) paper is one of many which describe experiments giving a picture of an organism which alters the expression of its genome upon association with other cells. Whether this is accomplished by the temporary addition of DNA to the genome, by the regulation of DNA expression, or by some other process, is within the province of molecular biologists to determine. New hands with new techniques may be required.

Sharma (1959) found that hypercholesteraemia of the host animal increased susceptibility to pathological damage on challenge by amoebae. This has been verified by Biagi et al. (1962), Singh et al. (1971), Bos and Van de Griend (1977) and by Meerovitch and Ghadrian (1978).

Important studies have been made on the experimental production of pathological damage in the livers of newborn hamsters (Mattern and Keister, 1977). This technique is being employed to study possible cytotoxic agents in axenic amoebae (Mattern et al., 1978b), and the effects of viruses on amoebic pathogenicity (Mattern et al., 1978a, 1979). Eventually, viruses may be found to play a role in amoebal metabolism. Studies on amoebic viruses are reported by Miller and Swartzwelder (1960), Mattern et al. (1972, 1979), Diamond et al. (1972, 1976), Bird and McCaul (1976) and Villegas Gonzalez et al. (1976). It seems that their ecological niche positions amoebae to be virus disseminating agents without rival. With the exposure to viruses that amoebae encounter in their normal habitat, one wonders why reports of virus affecting amoebal pathogenicity are sparse in clinical studies. That they have not been so implicated may be due to the amoebae shedding viruses during encystation.

Wessenberg (1974) speculated upon the effect of host heat stress on the pathogenicity of amoebae.

## V. Growth Requirements

### A. VITAMINS AND COFACTORS

Early work on the nutrition of *E. histolytica* was reviewed by Nakamura (1953). In a medium with an antibiotic-inhibited mixed bacterial flora, Nakamura (1955) and Nakamura and Baker (1956) reported stimulation of amoebic growth by 25 substances, including autoclaved bacterial cells. At one time or another we have tested 23 of the 25 substances in axenic growth and found nine of them to be either stimulatory or essential. That little is known with certainty regarding the vitamin requirements of axenic amoebae is due to the complex array of vitamins supplied by the media components and to the cumbersomeness of making single omission tests with the added vitamin mix. The concentrations of vitamins (and other substances) added to the TP-S-1 medium (Diamond, 1968) are listed in Table 3. There is evidence substantiating the value of the first six listed substances. As regards the seventh, it was possible greatly to inhibit the growth of amoebae in TP-S-1 medium by the addition of 1 or 2 units per ml of avidin (R. E. Reeves and B. West, unpublished work, 1982). There is no evidence that any of the remaining added substances are required, but there is evidence that two are not. Weinbach et al. (1980) suggested a role for ubiquinone in amoebal metabolism.

Ions of iron, magnesium, manganese, cobalt and zinc have been implicated in amoebic enzyme activities; the activities of aldolase and the NADP-linked alcohol dehydrogenase are protected by incorporating zinc in enzyme buffer solutions. The importance of iron was shown in growth studies by Latour and Reeves (1965), Diamond et al. (1978) and Smith and Meerovitch (1982). Diamond's TYI-S-33 medium includes added ferric ammonium citrate. A need for iron is apparent from the large amount of iron sulphur complexes amoebae contain. A requirement for calcium ions has not been demonstrated; amoebae grew well in the presence of an excess of sodium oxalate (R. E. Reeves, unpublished work).

TABLE 3

*The vitamins of TP-S-1 axenic growth medium[a]*

| Vitamin | 2% Panmede[c] | Addition after autoclaving | Published evidence of requirement for axenic growth |
|---|---|---|---|
| Thiamin | 0·20 | 0·17 | Susskind et al. (1982) |
| Riboflavin | 2·8 | 0·17 | Lo and Reeves (1979a,b) |
| Niacin | 9·4 | 0·42 ⎫ | Weik and Reeves (1980) |
| Nicotinamide | N.A. | 0·42 ⎭ | |
| Pantothenic acid | 30 | 0·17 ⎫ | With bacterial associate, |
| Pyridoxine | 0·36 | 0·42 ⎭ | Reeves et al. (1959a,b) |
| Biotin | 0·06 | 0·17 | None, but see text |
| Inositol | 50 | 0·88 ⎫ | Not required (see Weik |
| Calciferol | N.A. | 1·7 ⎭ | and Reeves, 1980) |
| Folic acid | 0·4 | 0·17 | None |
| Folinic acid | 0·1 | — | None |
| Vitamin $B_{12}$ | 0·12 | — | None |
| p-Aminobenzoic acid | N.A. | 0·84 | None |
| Choline chloride | N.A. | 7·65 | None |
| Vitamin A | N.A. | 1.7 | None |
| Menadione | N.A. | 0·34 | None |
| Tocopherol acetate | N.A. | 0·17 | None |

Contribution[b] to final medium by

[a] In addition to 200 μg ml⁻¹ of ascorbic acid added before autoclaving.
[b] μg ml⁻¹; N.A. = not assayed; — = not added.
[c] Calculated from supplier's assay.

The metal ions and the organic cofactors that are involved in amoebal metabolism are loosely held by those proteins that require their participation. The organic cofactors may be taken up by the relatively slow pinocytotic process; if so their concentration in the test medium may need to be greater than for organisms with specific uptake systems.

### B. SUBSTRATES

In attempts to refine axenic culture media the following substances required as cell components may not be synthesized by amoebic metabolism, or not in sufficient amounts. Cholesterol was the first of these to be recognized (Snyder and Meleney, 1943) and its requirement has been confirmed (Latour et al., 1965). All cells, presumably, require hexosamine in some form. There is no report of amoebae synthesizing glucosamine, and it was found to be a requirement for their growth (Greenberg et al., 1956). The capability of axenic amoebae to synthesize glycerol phosphate was questioned by Reeves and Lobelle-Rich (1983). Since it is abundant in amoebic lipids (Sawyer et al., 1967), it may prove to be a requirement.

Amoebae are limited in their ability to synthesize amino acids from carbohydrates. It seems likely that those aspartyl and glutamyl residues, which are incorporated into amoebic protein in the form of asparagine and glutamine, do not arise from free aspartic and glutamic acids in the culture fluids. If this is so, asparagine and glutamine, at substrate levels, may be required in a defined growth medium. These two amino acids seem not to be free in axenic growth medium, but the organism may obtain them by hydrolytic cleavage of peptides or serum proteins present in the medium.

## VI. Regulation

With a succession of colleagues I have tried to show allosteric modulation or some system for the regulation of amoebic glycolytic enzymes. The results of probing for regulatory features have been disappointing. They have yielded only thermodynamic explanations of enzyme regulation: i.e. that the rate of a reversible reaction slows as equilibrium is approached.

All the steps of the glycolytic pathway, save those catalysed by glucokinase and pyruvate synthase, are physiologically reversible reactions; even the pyruvate synthase reaction is reversible in the presence of hydrogenase and hydrogen gas (Reeves et al., 1977). It is not surprising to me that thermodynamic regulation suffices to control fermentation. The pivotal step in the regulation of the glycolytic pathway of most organisms is held to be at the phosphofructokinase step. For organisms using an ATP-dependent enzyme this is an irreversible step, one profitably subject to allosteric modulation. In E. histolytica the reversible $PP_i$-dependent phosphofructokinase is not affected by any modulator tested by us, or by fructose 2,6-bisphosphate (Van Schaftingen et al., 1982). However, in some other organisms this enzyme activity is greatly affected by modulators (Anderson and Sabularse, 1982).

Metabolic regulation, of the thermodynamic variety, is demonstrable in amoebae. Amoebae grown anaerobically and axenically cannot metabolize glucose in the absence of acetate or one of the oxidized substrates of amoebic NADP-dependent alcohol dehydrogenase (Reeves and Warren, 1977). Lacking one of these substances, when anaerobic axenic amoebae make acetate, glycolysis comes to a halt because the endogenous nicotinamide nucleotides become fully reduced. This metabolic block is overcome by incorporating acetate in the suspending fluid so that the reversible enzyme at step 15 (Table 1) makes no net acetate, or by the addition of acetone or any one of several aldehydes or ketones which are reduced by amoebic alcohol dehydrogenase. A second instance of thermodynamic regulation occurs in carbonate-containing buffers which cause the build-up of high concentrations of intracellular L-malate (Reeves *et al.*, 1974b).

## VII. Evolutionary Considerations

It is not my purpose to review the morphology of *E. histolytica*. In this paragraph, only observations that tend to underscore peculiarities in its fine structure and which have a bearing on its position in evolution will be mentioned. Lowe and Maegraith (1970) found neither mitochondria nor Golgi apparatus. Smooth endoplasmic reticulum, but not the rough variety, was seen by Griffin and Juniper (1971). A rather unconvincing 'Golgi-like complex' was seen in the HK9 strain by Proctor and Gregory (1972), but was not mentioned in their later work on the NIH:200 strain (Proctor and Gregory, 1973). Gicquaud (1979) reports seeing 30 chromosomes and a mitotic spindle in dividing cells of the Laredo strain of *E. histolytica*-like amoeba, but such structures have not been reported in dividing cells of typical *E. histolytica* strains. Workers are in agreement that the nuclear membrane remains intact during nuclear division. I have seen and photographed nuclear division many times without seeing mitotic figures. The failure of colchicine at low concentrations to modify amoebal growth (Gillen and Diamond, 1978b) is evidence which, coupled with the inadequacy of the findings offered in opposition (Pan and Geiman, 1955), leads me to the conclusion that nuclear division may be amitotic.

In his interesting discussion of the modulation of enzyme activity, Sols (1981) traces the development of basic regulatory mechanisms. The oligomeric aldolase of plants and animals had its origin no less than $1.5 \times 10^9$ years ago (Heil and Lebherz, 1978), and the 40 000 dalton monomeric unit has since been highly conserved. The lack of oligomers in amoebal aldolase places it in a very primitive evolutionary state. However, Sols (1981, p. 96) points out, 'To lose a modulatory mechanism is much easier than to gain it.' Is it possible that due to their parasitic existence the amoebae have lost regulatory mechanisms?

The natural habitat of *E. histolytica* is partially aerobic, certainly so in its tissue-invading stages. The organism not only lacks a functioning tricarboxylic acid cycle, but it transfers electrons to oxygen to form water, a step which it accomplishes without cytochromes. Did its parasitic existence cause it to shuck off without a trace such elaborate and useful features as the tricarboxylic cycle and haem proteins? Did it discard ATP-dependent kinases for less regulable $PP_i$-dependent kinases? Did it give up mitochondria? Did it give up the capability of synthesizing its own membrane lipids and a major part of its own membrane proteins? Did it discard rough endoplasmic reticulum and Golgi bodies which would assist it in inserting newly formed protein into its membranes? Unlike other eukaryotes, did it give up DNA histones? Did parasitic existence cause it to lose the ability to make recognizable chromosomes or a mitotic spindle? This line of thought presumes that too many things have been given up in order for the organism to enjoy its parasitic life. Amoebic metabolism may, instead, reflect approximately the state of the art at the time the organism adopted the parasitic mode, soon after some other species developed a functional gut. Its presently displayed characteristics seem to be those of a very primitive organism long sheltered from environmental pressures.

## VIII. Speculation

The organization of this section will informally follow that of the main body of the review.

(i) *Glycolytic enzymes.* Apart from the previously mentioned lack of regulatory features amoebic glycolytic enzymes are notable for the absence of oligomeric structure. This reinforces my view that they represent a very early stage of enzyme evolution.

Our sustained interest in amoebic glycogen metabolism was dampened by the resolute rejection of each manuscript submitted on the subject. Indeed, our observations are as troublesome to me as they were to the editors. The glycogen synthesis apparatus contains the 1,4-α-glucan branching enzyme, EC 2.4.1.18. But, I was unable to identify a debranching enzyme, or to see evidence of its functioning in glucose-starved whole cells. Freshly harvested cells contain up to 25% of their dry weight as glycogen. One-half of this glycogen serves to sustain amoebic metabolism for a few hours at room temperature. But, metabolism comes to a halt when half of the glycogen store has been consumed. Thereafter, the cells lose their integrity. This observation supports the idea of no functional debranching enzyme. Mammalian liver cells possessing the debranching enzyme are able to reduce their glycogen stores much further. In hindsight, we were remiss in not studying glycogen metabolism further. Regulation of glycogen synthesis probably occurs, but we did not discover it.

(ii) *Enzyme induction.* We have never seen enzyme repression, derepression or induction in growing cells at any stage of culture development. Admittedly, evidence offered on the last point in Section II B 2 is trivial. But, it does not constitute the full extent of our search for one or another of these phenomena.

(iii) *Electron transport*. Anaerobic metabolism proceeds from hexose to carbon dioxide and ethanol by steps in which all the carbon-containing intermediates are known. The mechanism is uncertain for the transfer to NAD of electrons from pyruvate oxidation. I believe that these electrons reduce iron–sulphur centres in the enzyme; from there, that they are transferred to a one-electron carrier having an oxidation–reduction potential in the vicinity of $-0.4$ volts. This carrier may be the ferredoxin isolated by Reeves *et al.* (1980) or it may be some other endogenous molecule. That this carrier has a low oxidation–reduction potential is indicated by the fact that amoebal metabolism will produce hydrogen gas in the presence of hydrogenase (Montalvo *et al.*, 1971).

Aerobic metabolism (respiration) has an additional unknown. Lo and Reeves (1980) showed that intact cells utilize oxygen without production of peroxide, but that disrupted cells produce peroxide. They purified to apparent homogeneity an enzyme called NADPH:flavin oxidoreductase. This enzyme has many of the properties of the NADH-linked diaphorase described by Weinbach *et al.* (1977, 1980). Lo and Reeves (1980) concluded that cell homogenates first transfer electrons from the oxidation of certain reduced substrates to NADP, then to flavin, and thence to oxygen with the production of peroxide. In intact cells the last step is different; the electrons from reduced flavin then produce water. How the organism, lacking cytochromes, is able to accomplish this feat is one of the great unsolved problems in metabolism. I believe that in amoebae this transfer does not involve the conservation of energy. The more sophisticated cytochrome-containing organisms would be able to make ATP from the energy difference between reduced flavin and water.

(iv) *Associate organisms*. In order for an enzyme activity from an associate organism to have a reasonable temporal existence free in amoebic cytoplasm it must be one that is linked to a metabolic intermediate. Possibly such substrate binding is, to a considerable extent, responsible for the protection of an organism's endogenous enzymes, and this principle would similarly apply to exogenous proteins. This might explain why bacterial 1-phosphofructokinase was never seen in amoebal homogenates, i.e. there is no fructose 1-phosphate for it to combine with. Some of the enzyme activities reported from amoebae grown with bacteria fall into the category which would find an appropriate substrate in the amoeba; however, the Entner-Doudoroff enzymes and succinate dehydrogenase do not. Studies on the duration of activity in amoebae washed free of bacteria would be important in testing this hypothesis. Such studies have been made only with hydrogenase, for which the amoebal substrate seems to be the reduced electron carrier from pyruvate oxidation.

(v) *Growth requirements.* I see no evidence that calciferol, ascorbic acid, menadione, vitamin K, vitamin $B_{12}$, vitamin A, or tocopherol enter into amoebal metabolism. Ubiquinone was said to be a component of the amoebic respiratory chain (World Health Organization, 1981, p. 30; Weinbach et al., 1980).

(vi) *Evolutionary considerations.* The principal pathway for carbohydrate metabolism by axenic amoebae (shown in Fig. 1a) could become, with some important changes, one that flows from hexose to acetate and carbon dioxide without the intervention of any nucleotide (Fig. 3). This pathway would be independent of molecular oxygen, although it does make acetate. The necessary changes are: (1) a $PP_i$-dependent phosphorylation at the first step; (2) inorganic electron acceptors (probably ferric iron or sulphur); (3) the bypassing of steps 7, 10 and 15 which, in the existing evolutionary state, require nucleotides; (4) the substitution of a simple thiol for CoA at the pyruvate synthase step; and (5) phosphorolysis of the resulting thioacetate to give acetyl phosphate. With these changes the pathway is one which may have functioned during geological ages before the origin of life. If catalysts capable of initiating the pathway of Fig. 3 were encapsulated within a membrane, the process would be self sustaining: hexose in, acetic acid and carbon dioxide out, with the $PP_i$ used at the first and third steps being reformed later. No ionically charged species pass through the membrane; two thiol esters formed in the pathway are ripe for exploitation for additional energy. Amoebae maintain in their armoury of enzymes the one capable of catalysing the final step of Fig. 3 (Reeves and Guthrie, 1975), as do certain microorganisms (Liu et al., 1982). If life had a heterotrophic origin the pathway of Fig. 3 requires consideration.

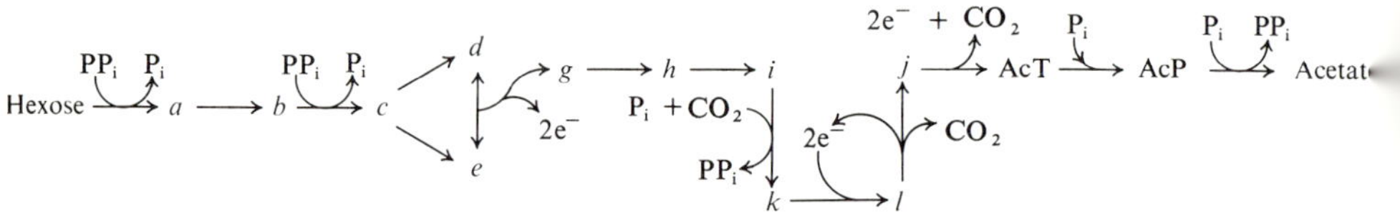

FIG. 3. A speculative pathway for the prebiotic catabolism of hexose to carbon dioxide and acetate. Inorganic electron acceptors and metal catalysts at the various steps foreshadow the later development of protein catalysts and nucleotide electron acceptors. The chemical energy of pyrophosphate ($PP_i$) at steps 1 and 3 is multiplied at the later steps where $PP_i$ is formed. The lettered reaction intermediates are presumed to be those of Table 1 and Fig. 1a, but steps 7, 10 and 15 are bypassed. AcT and AcP are a thiolacetate and acetylphosphate, respectively.

## IX. Conclusions

The glycolytic pathway which supplies biochemical energy for amoebic growth is strong and complete in both bacteria-grown and axenic amoebae. There is no evidence that amoebae are able to switch from carbohydrate to an alternate energy source. It is a dependence upon membrane-forming components which restricts trophozoite growth. Nearly all the membrane lipids and a large fraction of membrane protein are of exogenous origin. Lacking rough endoplasmic reticulum and Golgi bodies, amoebae do not possess the usual means to incorporate newly synthesized proteins into their membrane. Many of the studied membrane enzymes may be of exogenous origin. I raise this question about the amoebic phosphatases and the cryptic inorganic pyrophosphatase reported by McLaughlin et al. (1978). Only for actin is there an indication that one of the membrane enzymes was synthesized by the amoebae.

An understanding of the pathogenicity of the organism is likely to come from investigation of the interactions between amoebae, the host, and the associate organisms of the gut. The studies by Sargeaunt and his co-workers illustrate this approach (Sargeaunt and Williams, 1978; Sargeaunt et al., 1980, 1982).

Refinement of axenic culture media would permit use of the powerful tool of nutrition to define the limitations of amoebic metabolism more exactly than has been done by biochemistry. It would expedite studies of protein and lipid metabolism, two areas which share the questionable distinction of having virtually no solid scientific base in the literature at present.

Definition of the cell cycle is the most likely way work with axenic cells can contribute to the health-related aspects of amoebae. Austin and Warren (1982) are making progress in the direction of synchronized cultures. From studies of the cell cycle may evolve an understanding of the encystation process.

Apart from disease-related consideration, *E. histolytica* grown axenically may be one of the most primitive organisms, a 'metabolic fossil' (Lipmann, 1971). Its apparent versatility in sequestering useful enzyme activities from ingested associates is remarkable. Is it also able to translate into protein the messenger RNA of ingested organisms? Does it transcribe and translate useful segments of the DNA of ingested organisms? Does it incorporate, temporarily at least, part of the ingested DNA into its own genome and thus become more pathogenic? Answers to these and similar questions require the techniques of modern molecular biology. I earnestly hope to induce practitioners of that discipline to take an interest in this parasite.

## Acknowledgment

I am indebted to Professors R. A. Coulson and Lionel G. Warren for constructive criticism of this manuscript. The patient work of Wanda Santa Marina and Mildred Williams through a score of major revisions made the writing possible.

## Appendix A.

### Quantitation

A quantitative approach is essential to an understanding of intermediary metabolism. In our work packed cell volume is measured by centrifugation at 800 $g$ for 5 min at room temperature in the horizontal mode in spent culture fluid. Wet fresh cell weight in grams is 1·05 × packed cell volume in ml. and includes about 15% interstitial fluid. Total cell protein (the biochemist's fetish) is a less desirable measure of growth since it includes a highly variable amount of protein of exogenous origin. Visual inspection of the primary cell pellet is important. A layer of colourless cells at the top of the pellet signifies cells that are past their prime. Opalescence in the cell washings indicates cell rupture with a leakage of glycogen. Colourless cells may be intact by microscopic observation, but they are permeable to low molecular-weight substances and may cause misleading results in physiological experiments. Inoculum size and the time of culture incubation must be controlled to avoid opalescent washes and a colourless layer on the cell pellet. To achieve reproducible cell harvests, inocula must be taken from cultures in the logarithmic growth stage, i.e. from cultures that have not yet reached one-half maximum growth. During harvesting and washing, cells should not be subjected to osmotic stress. Cell counts are not a reliable measure of tissue mass. We have seen cell counts vary from $0·7 \times 10^8$ to $2 \times 10^8$ cells per ml packed cell volume. Meleney and Zuckerman (1948) reported sudden changes in the average size of amoebae within a single culture. Goldman and Davis (1965) found that three of the widely investigated strains contain size-variant substrains. This variability of amoebal cultures might be minimized by cloning. Suitable techniques for this have been established (Gillen and Diamond, 1978a).

Amoebae are easily homogenized by sonication, or by tissue grinding in a hypo-osmotic buffer. The latter procedure, blanketed with argon, is used to obtain those enzymes that require anaerobic conditions for activity.

## Appendix B

FREE AMINO ACID CONCENTRATIONS IN AXENIC GROWTH MEDIA (UNPUBLISHED WORK BY R. A. COULSON, W. T. MORGAN AND R. E. REEVES, 1982)

| Amino acid | TP-S-1[b] (mM) | TYI-S-33[c] (mM) |
|---|---|---|
| Asp | 1·6 | 1·8 |
| Thr | 2·4 | 2·0 |
| Ser | 3·4 | 4·0 |
| Glu | 3·2 | 4·7 |
| Pro | 3·1 | 1·9 |
| Gly | 2·4 | 1·8 |
| Ala | 4·2 | 5·6 |
| Cys[a] | — | — |
| Val | 4·4 | 4·4 |
| Met | 2·2 | 1·8 |
| Ile | 3·0 | 3·0 |
| Leu | 11·1 | 11·7 |
| Tyr | 0·85 | 0·7 |
| Phe | 4·2 | 3·7 |
| Lys | 4·0 | 4·0 |
| His | 0·75 | 0·86 |
| Arg | 1·16 | 2·3 |
| Trp | — | — |
| Total | 51·96 | 54·26 |

[a] Cystine and tryptophan were not assayed. The cysteine hydrochloride added to Diamond's media before autoclaving can be replaced by an equimolar amount of sodium thioglycolate, freshly prepared from thioglycolic acid (R. E. Reeves and B. West, unpublished results, 1982).
[b] Diamond (1968).
[c] Diamond et al. (1978).

## References

Albach, R. A. and Booden, T. (1978). Amoebae. In "Parasitic Protozoa" (J. P. Kreier, ed.), Vol. 2, pp. 455–505. Academic Press, New York.

Anderson, R. L. and Sabularse, D. C. (1982). Inorganic pyrophosphate: D-fructose-6-phosphate 1-phosphotransferase from Mung Bean. *Methods in Enzymology* **90**, 91–97.

Austin, C. J. and Warren, L. G. (1982). Induced division synchrony in *Entamoeba histolytica*. Effects of hydroxyurea and serum deprivation. *American Journal of Tropical Medicine and Hygiene* **32**, 507–511.

Baernstein, H. D., Rees, C. W. and Reardon, L. V. (1954). Symbiosis in culture of *Endamoeba histolytica* and single species of bacteria. *American Journal of Tropical Medicine and Hygiene* **3**, 839–848.

Band, R. N. and Cerrito, H. (1979). Growth response of axenic *Entamoeba histolytica* to hydrogen, carbon dioxide, and oxygen. *Journal of Protozoology* **26**, 282–286.

Becker, C. E. and Geiman, Q. M. (1954). Amino acids found in protein from *Entamoeba histolytica*. *Journal of the American Chemical Society* **76**, 3029.

Becker, C. E. and Geiman, Q. M. (1955). Utilization of glucose by two strains of *Entamoeba histolytica*. *Experimental Parasitology* **4**, 493–501.

Biagi, F., Robledo, E., Servin, H. and Martuscelli, A. (1962). The effect of cholesterol on the pathogenicity of *Entamoeba histolytica*. *American Journal of Tropical Medicine and Hygiene* **11**, 333–340.

Bird, R. G. and McCaul, T. F. (1976). The rhabdoviruses of *Entamoeba histolytica* and *Entamoeba invadens*. *Annals of Tropical Medicine and Parasitology* **70**, 81–93.

Blytt, H. J. and Reeves, R. E. (1976). Glycogen metabolism in *Entamoeba histolytica*. *Federation Proceedings* **35**, 1400.

Boonlayangoor, P., Albach, R. A., Stern, M. L. and Booden, T. (1978a). *Entamoeba histolytica*: uptake of purine bases and nucleosides during axenic growth. *Experimental Parasitology* **45**, 225–233.

Boonlayangoor, P., Otten, M., Booden, T. and Albach, R. A. (1978b). Isolation and sedimentation analysis of RNA from *Entamoeba histolytica*. *Archivos de Investigacion Medica (Mexico)* **9**, Supplement **1**, 121–128.

Boonlayangoor, P., Albach, R. A. and Booden, T. (1980). Purine nucleotide synthesis in *Entamoeba histolytica*: a preliminary study. *Archivos de Investigacion Medica (Mexico)* **11**, Supplement **1**, 83–88.

Bos, H. J. (1976). An hypothesis about the role of intestinal bacteria in the virulence of *Entamoeba histolytica*. In "Proceedings of the International Conference on Amebiasis" (B. Sepulveda and L. S. Diamond, eds), pp. 551–557. Instituto Mexicano del Serguro Social, Mexico, D. F.

Bos, H. J. and Hage, A. J. (1975). Virulence of bacteria-associated, *Crithidia*-associated, and axenic *Entamoeba histolytica*: experimental hamster liver infections with strains from patients and carriers. *Zeitschrift für Parasitenkunde* **47**, 79–89.

Bos, H. J. and Van de Griend, R. J. (1977). Virulence and toxicity of axenic *Entamoeba histolytica*. *Nature* **265**, 341–343.

Bragg, P. D. and Reeves, R. E. (1962). Pathways of glucose dissimulation in the Laredo strain of *Entamoeba histolytica*. *Experimental Parasitology* **13**, 393–400.

Carter, W. A., Levy, H. B. and Diamond, L. S. (1967). Protein synthesis by amoebal ribosomes. *Nature* **213**, 722–724.

Chang, S. L. (1945). Studies on *Entamoeba histolytica*. V. On the decrease in infectivity and pathogenicity for kittens of *E. histolytica* during prolonged *in vitro* cultivation and restoration of these characters following encystment and direct animal passage. *Journal of Infectious Diseases* **76**, 126–134.

Coulson, R. A. and Hernandez, T. (1968). Amino acid metabolism in chameleons. *Comparative Biochemistry and Physiology* **25**, 861–872.

Diamond, L. S. (1968). Techniques of axenic cultivation of *Entamoeba histolytica* Schaudinn, 1903 and *E. histolytica*-like amebae. *Journal of Parasitology* **54**, 1047–1056.

Diamond, L. S., Mattern, C. F. T. and Bartgis, I. L. (1972). Viruses of *Entamoeba histolytica*. I. Identification of transmissible virus-like agents. *Journal of Virology* **9**, 326–341.

Diamond, L. S., Mattern, C. F. T., Bartgis, I. L., Daniel, W. A. and Keister, D. B. (1976). Viruses of *Entamoeba histolytica*. VI. A study of host range. In "Proceedings of the International Conference on Amebiasis (B. Sepulveda and L. S. Diamond, eds), 334–345. Instituto Mexicano del Seguro Social, Mexico, D. F.

Diamond, L. S., Harlow, D. R. and Cunnick, C. C. (1978). A new medium for axenic cultivation of *Entamoeba histolytica* and other *Entamoeba*. *Transactions of the Royal Society of Tropical Medicine and Hygiene* **72**, 431–432.

Elsdon-Dew, R. (1976). The challenge of amebiasis. In "Proceedings of the International Conference on Amebiasis" (B. Sepulveda and L. S. Diamond, eds), pp. 773–780. Instituto Mexicano del Seguro Social, Mexico, D. F.

Entner, N. (1958). On the pathway of carbohydrate metabolism in *Entamoeba histolytica*. *Journal of Parasitology* **44**, 638.

Entner, N. and Anderson, H. H. (1954). Lactic and succinic acid formation by *Entamoeba histolytica in vitro*. *Experimental Parasitology* **3**, 234–313.

Entner, N. and Doudoroff, M. (1952). Glucose and gluconic acid oxidation of *Pseudomonas saccharophila*. *Journal of Biological Chemistry* **106**, 853–862.

Entner, N. and Grollman, A. P. (1973). Inhibition of protein synthesis: a mechanism of amebicide action of emetine and other related compounds. *Journal of Protozoology* **20**, 160–163.

Enzyme Nomenclature Committee. (1979). "Enzyme Nomenclature: A Report of the Nomenclature Committee of the International Union of Biochemistry." Academic Press, New York and London.

Eubank, W. B. and Reeves, R. E. (1982). Analog inhibitors for the pyrophosphate-dependent phosphofructokinase of *Entamoeba histolytica* and their effect on culture growth. *Journal of Parasitology* **68**, 599–602.

Feria-Velasco, A., Martinez-Zedillo, G., Manzo, N. T-G. and Guberrez-Pastrana, M. D. (1973). Investigacion de acido sialico en la cubierta exterior de trofozoites de *E. histolytica*. Estudio bioquimico y de alta resolucion. *Archivos de Investigacion Medica (Mexico)* **4**, Supplement **1**, 33–38.

Gelderman, A. H., Keister, D. B., Bartgis, I. L. and Diamond, L. S. (1971a). Characterization of the deoxyribonucleic acid of representative strains of *Entamoeba histolytica*, *Entamoeba histolytica*-like amebae, and *Entamoeba moshkovskii*. *Journal of Parasitology* **57**, 906–911.

Gelderman, A. H., Bartgis, I. L., Keister, D. B. and Diamond, L. S. (1971b). A comparison of genome sizes and thermodenaturation-derived base composition of DNA's from several members of *Entamoeba* (*histolytica* group). *Journal of Parisitology* **57**, 912–916.

Gicguaud, C. R. (1979). Etude de l'ultrastructure du noyau et de la mitose de *Entamoeba histolytica*. *Biologie Cellulaire* **35**, 305–312.

Gillen, F. D. and Diamond, L. S. (1978a). Clonal growth of *Entamoeba histolytica* and other species of *Entamoeba* in agar. *Journal of Protozoology* **25**, 539–543.

Gillen, F. D. and Diamond, L. S. (1978b). Clonal growth of *Entamoeba* in agar: some applications of this technique to the study of their cell biology. *Archivos de Investigacion Medica (Mexico)* **9**, Supplement **1**, 237–246.

Goldman, M. and Davis, V. (1965). Isolation of different-sized substrains from three stock cultures of *Entamoeba histolytica*, with observations on spontaneous size changes affecting whole populations. *Journal of Protozoology* **12**, 509–523.

Greenberg, J., Taylor, D. J. and Bond, H. W. (1956). Glucosamine in the culture of *Entamoeba histolytica* with a mixed bacterial flora. *American Journal of Tropical Medicine and Hygiene* **5**, 62–66.
Griffin, J. L. and Juniper, K. Jr. (1971). Ultrastructure of *Entamoeba histolytica* from human amebic dysentery. *Archives of Pathology* **91**, 271–280.
Guthrie, J. D. and Reeves, R. E. (1977). Iron–sulfur proteins of *Entamoeba histolytica*. *Proceedings of the Louisiana Academy of Sciences* **40**, 119.
Hallman, F. A. and DeLameter, J. N. (1953). Demonstration of amylolytic activity in *Endamoeba histolytica*. *Experimental Parasitology* **2**, 170–174.
Hallman, F. A., Michaelson, J. B., Blumenthal, H. and DeLameter, J. N. (1954). Studies on the carbohydrate metabolism of *Endamoeba histolytica*. I. The utilization of glucose. *American Journal of Hygiene* **59**, 128–131.
Harlow, D. R., Weinbach, E. C. and Diamond, L. S. (1976). Nicotinamide nucleotide transhydrogenase in *Entamoeba histolytica*, a protozoan lacking mitochondria. *Comparative Biochemistry and Physiology* **53B**, 141–144.
Hauswirth, J. W. and Warren, L. G. (1981). Preliminary studies on the incorporation of hydrogenase from *Bacteroides symbiosus* by *Entamoeba histolytica*. In "Southern Association of Parasitologists, Program and Abstracts, 14th meeting", p. 17. The University of Oklahoma Biological Station, Norman, Oklahoma.
Heebner, G. M. and Albach, R. A. (1982). High salt SDS-DEP technic for isolation of "intact" rRNA from *Entamoeba histolytica*. *Archivos de Investigacion Medica (Mexico)* **13**, *Supplement* **3**, 23–28.
Heil, J. A. and Lebherz, H. G. (1978). "Hybridization" between aldolase subunits derived from mammalian and plant origin. *Journal of Biological Chemistry* **253**, 6599–6605.
Hernandez-Velarde, R., Garuno-Rodriguez, Q. F. B. and Munoz, O. (1980). Los acidos ribonucleicos de transferencia de *Entamoeba histolytica*. I. Analisis cromologrfico de las especies isoaceptores de t-RNA de trofozoitos. *Archivos de Investigacion Medica (Mexico)* **11**, *Supplement* **1**, 89–94.
Hilker, D. M. and White, A. G. C. (1959). Some aspects of carbohydrate metabolism of *Entamoeba histolytica*. *Experimental Parasitology* **8**, 539–548.
Hilker, D. M., Sherman, H. J. and White, A. G. C. (1957). Starch hydrolysis by *Entamoeba histolytica*. *Experimental Parasitology* **6**, 459–464.
Hoshino, K. and Udagawa, K. (1960). [Organism producing isopropanol from acetone. VI. Isopropanol dehydrogenase and alcohol dehydrogenase of *Lactobacillus brevis* var. *hofuensis*.] *Nippon Nogei Kagaku Kaishi* **34**, 616–619. [In Japanese.]
Jarumilinta, R. and Maegraith, B. G. (1960). Hyaluronidase activity in stock cultures of *Entamoeba histolytica*. *Annals of Tropical Medicine and Parasitology* **54**, 118–128.
Kalra, I. S., Sabri, M. I., Dutta, G. P. and Mohan Rao, V. K. (1968). Succinate dehydrogenase activity of trophozoites of *Entamoeba histolytica*. *Indian Journal of Microbiology* **8**, 105–116.
Kalra, I. S., Dutta, G. P. and Mohan Rao, V. K. (1969). *Entamoeba histolytica*: effects of metal ions, metal binders, therapeutics, antibiotics and inhibitors on aldolase activity. *Experimental Parasitology* **24**, 26–31.
Kun, E. and Bradin, J. L. (1953). The role of sulfur in the metabolism of *Endamoeba histolytica*. *Biochimica et Biophysica Acta* **11**, 312–313.
Kun, E., Bradin, J. L. Jr. and Dechary, J. M. (1956). Correlation between $CO_2$ and $H_2S$ production by *Endamoeba histolytica*. *Biochimica et Biophysica Acta* **19**, 153–159.

Latour, N. G. and Reeves, R. E. (1965). An iron-requirement for growth of *Entamoeba histolytica* in culture, and the antiamebal activity of 7-iodo-8-hydroxyquinoline-5-sulfonic acid. *Experimental Parasitology* 17, 203–209.

Latour, N. G., Reeves, R. E. and Guidry, M. A. (1965). Steroid requirement of *Entamoeba histolytica*. *Experimental Parasitology* 16, 18–22.

Lee, E., Palacios, O. and Landa, L. (1971). Study of the enzymatic activities of *Entamoeba histolytica* axenically grown. *Archivos de Investigacion Medica (Mexico)* 2, Supplement 1, 173–179.

LeFevre, P. G. (1961). Sugar transport in the red blood cell: structure–activity relationships in substrates and antagonists. *Pharmacological Reviews* 13, 39–70.

Lindmark, D. G. (1976). Certain enzymes of the energy metabolism of *Entamoeba invadens* and their subcellular localization. In "Proceedings of the International Conference on Amebiasis" (B. Sepulveda and L. S. Diamond, eds), pp. 185–190. Instituto Mexicano del Seguro Social, Mexico, D. F.

Lindmark, D. G. and Müller, M. (1973). Hydrogenosome, a cytoplasmic organelle of the anaerobic flagellate *Tritrichomonas foetus*, and its role in pyruvate metabolism. *Journal of Biological Chemistry* 248, 7724–7728.

Lipmann, F. (1971). Attempts to map a process evolution of peptide biosynthesis. *Science* 173, 875–884.

Liu, C-L., Hart, N. and Peck, H. D. Jr. (1982). Inorganic pyrophosphate: energy source for sulfate-reducing bacteria of the genus *Desulfotomaculum*. *Science* 217, 363–364.

Lo, H-s. and Chang, C-j. (1982). Purification and properties of NADP-linked alcohol dehydrogenase from *Entamoeba histolytica*. *Journal of Parasitology* 68, 372–377.

Lo, H-s. and Reeves, R. E. (1978). Pyruvate to ethanol pathway in *Entamoeba histolytica*. *Biochemical Journal* 171, 225–230.

Lo, H-s. and Reeves, R. E. (1979a). Riboflavin requirement for the cultivation of axenic *Entamoeba histolytica*. *American Journal of Tropical Medicine and Hygiene* 28, 194–197.

Lo, H-s. and Reeves, R. E. (1979b). *Entamoeba histolytica*: flavins in axenic organisms. *Experimental Parasitology* 47, 180–184.

Lo, H-s. and Reeves, R. E. (1980). Purification and properties of NADPH: flavin oxidoreductase from *Entamoeba histolytica*. *Molecular and Biochemical Parasitology* 2, 23–30.

Lobelle-Rich, P. and Reeves, R. E. (1982). Galactose-1-phosphate uridylyltransferase in *Entamoeba histolytica*. *Methods in Enzymology* 90, 552–555.

Lobelle-Rich, P. and Reeves, R. E. (1983a). Separation and characterization of two UTP-utilizing hexose phosphate uridylyltransferases from *Entamoeba histolytica*. *Molecular and Biochemical Parasitology* 7, 173–182.

Lobelle-Rich, P. A. and Reeves, R. E. (1983b). The partial purification and characterization of adenosine kinase from *Entamoeba histolytica*. *American Journal of Tropical Medicine and Hygiene* 32, 976–979.

Lopez-Revilla, R. and Navarro, V. (1978). Composicion de aminoacidos en las proteinas totales de dos cepas de *Entamoeba histolytica* in cultivo axenico. *Archivos de Investigacion Medica (Mexico)* 9, Supplement 1, 129–132.

Lowe, C. Y. and Maegraith, B. G. (1970). Electron microscopy of *Entamoeba histolytica* in culture. *Annals of Tropical Medicine and Parasitology* 64, 283–291.

Lushbaugh, W. B. and Pittman, F. E. (1982). Correlation of cytotoxicity with proteinase activity of *Entamoeba histolytica* extracts. *Molecular and Biochemical Parasitology*, Supplement, 559–560.

Mattern, C. F. T. and Keister, D. B. (1977). Experimental amebiasis. II. Hepatic amebiasis in newborn hamsters. *American Journal of Tropical Medicine and Hygiene* **26**, 402–411.

Mattern, C. F. T., Diamond, L. S. and Daniel, W. A. (1972). Viruses of *Entamoeba histolytica*. II. Morphogenesis of the polyhedral particle (ABRM$_2$ → HK-9) → HB-301 and the filamentous agent (ABRM)$_2$ → HK-9: *Journal of Virology* **9**, 342–358.

Mattern, C. F. T., Diamond, L. S. and Keister, D. B. (1978a) Amoebal viruses and the virulence of *Entamoeba histolytica*. *Archivos de Investigacion Medica (Mexico)*, **9**, *Supplement* **1**, 165–166.

Mattern, C. F. T., Keister, D. B. and Caspar, P. A. (1978b). Experimental amebiasis. III. A rapid *in vitro* assay for virulence of *Entamoeba histolytica*. *American Journal of Tropical Medicine and Hygiene* **27**, 882–887.

Mattern, C. F. T., Keister, D. B. and Diamond, L. S. (1979). Experimental amebiasis. IV. Amebal viruses and the virulence of *Entamoeba histolytica*. *American Journal of Tropical Medicine and Hygiene* **28**, 653–657.

McLaughlin, J. and Faubert, G. (1977). Partial purification and some properties of a neutral sulfhydryl and an acid proteinase from *Entamoeba histolytica*. *Canadian Journal of Microbiology* **23**, 420–425.

McLaughlin, J., Lindmark, D. G. and Müller, M. (1978). Inorganic pyrophosphatase and nucleoside diphosphatase in the parasite protozoan, *Entamoeba histolytica*. *Biochemical and Biophysical Research Communications* **82**, 913–920.

Meerovitch, E. and Ghadrian, E. (1978). Restoration of virulence of axenically-cultivated *Entamoeba histolytica* by cholesterol. *Archivos de Investigacion Medica (Mexico)* **9**, *Supplement* **1**, 253–256.

Meleney, H. E. and Zuckerman, L. K. (1948). Note on a strain of small race *Endamoeba histolytica* which became large in culture. *American Journal of Hygiene* **47**, 187–188.

Mendoza, F., Arcos, L., Ortiz-Ortiz, L. and Diaz de Leon, L. (1982). Protein biosynthesis by *Entamoeba histolytica* in culture. *Archivos de Investigacion Medica (Mexico)* **13**, *Supplement* **3**, 71–76.

Meza, I., Sabanero, M., Cazares, F. and Bryan, J. (1982). Isolation and characterization of the actin from *Entamoeba histolytica*. *Journal of Cell Biology* **95**, 293a.

Mignani, E. and Bickel, J. (1961). Studi sulla biologia di *E. histolytica*. V. Osservazioni sull athrita' enzymatica di stipiti di *E. histolytica* in coltura. *Rivista di Parassitologia* **22**, 157–169.

Miller, J. H. and Swartzwelder, J. C. (1960). Virus-like particles in *Entamoeba histolytica* trophozoite. *Journal of Parasitology* **46**, 523–529.

Mohan Rao, V. K. and Dutta, G. P. (1966). Transaminase activity of *Entamoeba histolytica*. *Indian Journal of Microbiology* **6**, 63–76.

Montalvo, F. and Reeves, R. E. (1968). Electrophoretic characterization of amebal glucosephosphate isomerases. *Experimental Parasitology* **22**, 129–136.

Montalvo, F. E., Reeves, R. E. and Warren, L. G. (1971). Aerobic and anaerobic metabolism in *Entamoeba histolytica*. *Experimental Parasitology* **30**, 249–256.

Mukhtar, H. and Mohan Rao, V. K. (1972). α-Amylase activity in *Entamoeba histolytica*. *Indian Journal of Microbiology* **12**, 1–6.

Munoz, M. A. de L., Caldevon, J. and Rojkind, M. (1982). The collagenase of *Entamoeba histolytica*. *Journal of Experimental Medicine* **155**, 42–51.

Nakamura, M. (1953). Nutrition and physiology of *Endamoeba histolytica*. *Bacteriological Reviews* **17**, 189–212.

Nakamura, M. (1955). Growth factors for *Endamoeba histolytica*. *Proceedings of the Society for Experimental Biology and Medicine* **89**, 680–682.
Nakamura, M. and Baker, E. E. (1956). Nutritional requirements of *Endamoeba histolytica*. *American Journal of Hygiene* **64**, 12–22.
Neal, R. A. (1960). Enzymatic proteolysis by *Entamoeba histolytica*: biochemical characteristics and relationship with invasiveness. *Parasitology* **50**, 531–550.
Neal, R. A. (1983). Experimental amoebiasis and the development of antiamoebic compounds. *Parasitology* **86**, 175–191.
Nielands, J. B. (1955). Lactic dehydrogenase of heart muscle. *Methods in Enzymology* **1**, 449–454.
Noya Gonzalez, O., Warren, L. G. and Gohd, R. (1980). Serum proteins in the plasma membrane of *Entamoeba histolytica*. *Archivos de Investigacion Medica (Mexico)* **11**, Supplement **1**, 109–114.
O'Brien, W. E., Bowien, S. and Wood, H. G. (1975). Isolation and characterization of a pyrophosphate-dependent phosphofructokinase from *Propionibacterium shermanni*. *Journal of Biological Chemistry* **250**, 8690–8695.
Pan, C-T. and Geiman, Q. M. (1955). Comparative studies of intestinal amebae. I. Distribution and cyclic changes of the nucleic acids in *Endamoeba histolytica* and *Endamoebi coli*. *American Journal of Hygiene* **62**, 66–79.
Proctor, E. M. and Gregory, M. A. (1972). The ultrastructure of axenically cultivated trophozoites of *Entamoeba histolytica* with particular reference to an observed variation in structural pattern. *Annals of Tropical Medicine and Parasitology* **66**, 335–338.
Proctor, E. M. and Gregory, M. A. (1973). Ultrastructure of *E. histolytica* strain NIH 200. *International Journal for Parasitology* **3**, 457–460.
Raether, W. (1976). Pathogenicity of monoxenically and axenically grown *Entamoeba histolytica* cultures after intrahepatic infection in the golden hamster. In "Proceedings of the International Conference on Amebiasis" (B. Sepulveda and L. S. Diamond, eds), pp. 570–573. Instituto Mexicano del Seguro Social, Mexico, D. F.
Reeves, R. E. (1968). A new enzyme with the glycolytic function of pyruvate kinase. *Journal of Biological Chemistry* **243**, 3202–3204.
Reeves, R. E. (1970). Phosphopyruvate carboxylase from *Entamoeba histolytica*. *Biochimica et Biophysica Acta* **220**, 346–349.
Reeves, R. E. (1972). Carbohydrate metabolism in *Entamoeba histolytica*. In "Comparative Biochemistry of Parasites" (H. Van den Bossche, ed.), pp. 351–358. Academic Press, New York and London.
Reeves, R. E. (1974). Glycolytic enzymes in *E. histolytica*. *Archivos de Investigacion Medica (Mexico)* **5**, Supplement **2**, 411–414.
Reeves, R. E. and Bischoff, J. M. (1968). Classification of *Entamoeba* species by means of electrophoretic properties of amebal enzymes. *Journal of Parasitology* **54**, 594–600.
Reeves, R. E. and Guthrie, J. D. (1975). Acetate kinase (pyrophosphate). A fourth pyrophosphate-dependent kinase from *Entamoeba histolytica*. *Biochemical and Biophysical Research Communications* **66**, 1389–1395.
Reeves, R. E. and Lobelle-Rich, P. (1983). Absence of $\alpha$-glycerophosphate dehydrogenase in axenically-grown *Entamoeba histolytica*. *American Journal of Tropical Medicine and Hygiene* **32**, 1177–1178.
Reeves, R. E. and South, D. J. (1974). Phosphoglycerate kinase (GTP). An enzyme from *Entamoeba histolytica* selective for guanine nucleotides. *Biochemical and Biophysical Research Communications* **58**, 1053–1057.

Reeves, R. E. and Ward, A. B. (1965). Large lot cultivation of *Entamoeba histolytica*. *Journal of Parasitology* **51**, 321–324.

Reeves, R. E. and Warren L. G. (1977). Factors regulating amebal glycolysis. *Fifth International Congress of Protozoology, New York: Abstracts*, p. 223.

Reeves, R. E. and West, B. (1980). *Entamoeba histolytica*: nucleic acid precursors affecting axenic growth. *Experimental Parasitology* **49**, 78–82.

Reeves, R. E., Meleney, H. E. and Frye, W. W. (1957). A modified Shaffer-Frye technique for the cultivation of *Entamoeba histolytica* and some observations on its carbohydrate requirements. *American Journal of Hygiene* **66**, 56–62.

Reeves, R. E., Meleney, H. E. and Frye, W. W. (1959a). The cultivation of *Entamoeba histolytica* with penicillin-inhibited *Bacteroides symbiosus* cells. I. A pyridoxine requirement. *American Journal of Hygiene* **69**, 25–31.

Reeves, R. E., Meleney, H. E. and Isbelle, D. (1959b). The cultivation of *Entamoeba histolytica* with penicillin-inhibited *Bacteroides symbiosus* cells. II. A pantothenate requirement. *American Journal of Hygiene* **69**, 32–37.

Reeves, R. E., Warren, L. G. and Hsu, D-s. (1966). 1-Phosphofructokinase from an anaerobe. *Journal of Biological Chemistry* **241**, 1257–1261.

Reeves, R. E., Montalvo, F. and Sillero, A. (1967). Glucokinase from *Entamoeba histolytica* and related organisms. *Biochemistry* **6**, 1752–1760.

Reeves, R. E., Menzies, R. A. and Hsu, D-s. (1968). The pyruvate-phosphate dikinase reaction. The fate of phosphate and the equilibrium. *Journal of Biological Chemistry* **243**, 5486–5491.

Reeves, R. E., Montalvo, F. E. and Lushbaugh, T. S. (1971a). Nicotinamide adenine dinucleotide phosphate-dependent alcohol dehydrogenase: The enzyme from *Entamoeba histolytica* and some enzyme inhibitors. *International Journal of Biochemistry* **2**, 55–64.

Reeves, R. E., Lushbaugh, T. S. and Montalvo, F. E. (1971b). Characterization of deoxyribonucleic acid of *Entamoeba histolytica* by cesium chloride density centrifugation. *Journal of Parasitology* **57**, 939–944.

Reeves, R. E., South, D. J., Blytt, H. J. and Warren, L. G. (1974a). Pyrophosphate: D-fructose 6-phosphate 1-phosphotransferase. A new enzyme with the glycolytic function of 6-phosphofructokinase. *Journal of Biological Chemistry* **249**, 7737–7741.

Reeves, R. E., Warren, L. G. and Guthrie, J. D. (1974b). Studies on the intracellular concentrations of glycolytic intermediates in *E. histolytica*. *Archivos de Investigacion Medica (Mexico)* **5**, Supplement **2**, 331–336.

Reeves, R. E., Serrano, R. and South, D. J. (1976). 6-Phosphofructokinase (pyrophosphate). Properties of the enzyme from *Entamoeba histolytica* and its reaction mechanism. *Journal of Biological Chemistry* **251**, 2958–2962.

Reeves, R. E., Warren, L. G., Susskind, B. and Lo, H-s. (1977). An energy-conserving pyruvate-to-acetate pathway in *Entamoeba histolytica*. Pyruvate synthase and a new acetate thiokinase. *Journal of Biological Chemistry* **252**, 726–731.

Reeves, R. E., Guthrie, J. D. and Lobelle-Rich, P. (1980). *Entamoeba histolytica*: isolation of ferredoxin. *Experimental Parasitology* **49**, 83–88.

Reeves, R. E., Lobelle-Rich, P. and Eubank, W. B. (1982). 6-Phosphofructokinase (PPi) from *Entamoeba histolytica*. *Methods in Enzymology* **90**, 97–102.

Sabanero, M. and Meza, I. (1982). Localizacion de actina en trofozoitos de *Entamoeba histolytica*. *Archivos de Investigacion Medica (Mexico)* **13**, Supplement **3**, 37–42.

Sargeaunt, P. G. and Williams, J. E. (1978). Electrophoretic isoenzyme patterns of *Entamoeba histolytica* and *Entamoeba coli*. *Transactions of the Royal Society of Tropical Medicine and Hygiene* **72,** 164–166.
Sargeaunt, P. G., Williams, J. E. and Neal, R. A. (1980). A comparative study of *Entamoeba histolytica* (NIH: 200, HK9, etc.), *E. histolytica*-like and other morphologically identical amoebae using isoenzyme electrophoresis. *Transactions of the Royal Society of Tropical Medicine and Hygiene* **74,** 469–474.
Sargeaunt, P. G., Williams, J. R., Jackson, T. F. H. G. and Simjee, E. E. (1982). A zymodeme study of *Entamoeba histolytica* in a group of South African schoolchildren. *Transactions of the Royal Society of Tropical Medicine and Hygiene* **76,** 401–402.
Sawyer, M. K., Bischoff, J. M., Guidry, M. A. and Reeves, R. E. (1967). Lipids from *Entamoeba histolytica*. *Experimental Parasitology* **20,** 295–302.
Serrano, R. and Reeves, R. E. (1974). Glucose transport in *Entamoeba histolytica*. *Biochemical Journal* **144,** 43–48.
Serrano, R. and Reeves, R. E. (1975). Physiological significance of glucose transport in *Entamoeba histolytica*. *Experimental Parasitology* **37,** 411–416.
Serrano, R., Deas, J. E. and Warren, L. G. (1977). *Entamoeba histolytica:* membrane fractions. *Experimental Parasitology* **41,** 370–384.
Shaffer, J. G. and Frye, W. W. (1948). Studies on the growth requirements of *Entamoeba histolytica*. I. Maintenance of a strain of *E. histolytica* through one hundred transplants in the absence of actively multiplying bacterial flora. *American Journal of Hygiene* **47,** 214–221.
Sharma, R. (1959). Effect of cholesterol on the growth and virulence of *Entamoeba histolytica*. *Transactions of the Royal Society of Tropical Medicine and Hygiene* **32,** 976–979.
Singh, B. N., Srivastava, R. V. N. and Dutta, G. P. (1971). Virulence of strains of *Entamoeba histolytica* to rats and the effect of cholesterol, rat cecal and hamster liver passage on the virulence of non-invasive strains. *Indian Journal of Experimental Biology* **9,** 21–27.
Smith, J. M. and Meerovitch, E. (1982). Specificity of iron requirements for *Entamoeba histolytica*. *Archivos de Investigacion Medica (Mexico)* **13,** Supplement **3,** 63–69.
Snyder, T. L. and Meleney, H. E. (1943). Anaerobiosis and cholesterol as growth requirements for *Endamoeba histolytica*. *Journal of Parasitology* **29,** 278–284.
Sols, A. (1981). Modulation of enzyme activity. *Current Topics in Cellular Regulation* **19,** 77–101.
Sui, P. M. L. and Wood, H. G. (1962). Phosphoenolpyruvic carboxytransphosphorylase, a $CO_2$ fixation enzyme from propionic acid bacteria. *Journal of Biological Chemistry* **237,** 3044–3051.
Susskind, B. M., Warren, L. G. and Reeves, R. E. (1980). Incorporation of glucose carbon into ribonucleotides of axenic *Entamoeba histolytica*. *Journal of Parasitology* **66,** 759–764.
Susskind, B. M., Warren, L. G. and Reeves, R. E. (1982). A pathway for the interconversion of hexose and pentose in the parasitic amoeba *Entamoeba histolytica*. *Biochemical Journal* **204,** 191–196.
Takeuchi, T., Weinbach, E. C. and Diamond, L. S. (1975). Pyruvate oxidase (CoA acetylating) in *Entamoeba histolytica*. *Biochemical and Biophysical Research Communications* **65,** 591–596.

Takeuchi, T., Weinbach, E. C. and Diamond, L. S. (1977a). *Entamoeba histolytica*: localization and characterization of phosphorylase and particulate glycogen. *Experimental Parasitology* **43**, 107–114.

Takeuchi, T., Weinbach, E. C. and Diamond, L. S. (1977b). *Entamoeba histolytica*: localization and characterization of phosphoglucomutase, uridine diphosphate glucose pyrophosphorylase, and glycogen synthase. *Experimental Parasitology* **43**, 115–121.

Takeuchi, T., Weinbach, E. C., Gottlieb, M. and Diamond, L. S. (1979). Mechanisms of L-serine oxidation in *Entamoeba histolytica*. *Comparative Biochemistry and Physiology* **62B**, 281–285.

Takeuchi, T., Kobayashi, S., Tanabe, M., Kanada, Y. and Asami, K. (1980). $NADP^+$-dependent α-glycerolphosphate dehydrogenase activity in *Entamoeba histolytica*. *Japanese Journal of Parasitology* **29**, 39–43.

Tanowitz, H. B., Wittner, M., Rosenbaum, R. M. and Kress, Y. (1975). In vitro studies on the differential toxicity of metronidazole in protozoa and mammalian cells. *Annals of Tropical Medicine and Parasitology* **69**, 19–28.

Van Schaftingen, E., Ledever, B., Bartrons, R. and Hers, H-G. (1982). A kinetic study of pyrophosphate: fructose-6-phosphate phosphotransferase from potato tubers. *European Journal of Biochemistry* **129**, 191–195.

Villegas Gonzalez, J., Fastag de Shor, A., Villegas Silva, R. and De la Torre, M. (1976). Observations on the virulence of *E. histolytica* proceeding from an axenic medium and incubated with herpes simplex viruses. *In* "Proceedings of the International Conference on Amebiasis" (B. Sepulveda and L. S. Diamond, eds), pp. 152–157. Instituto Mexicano del Seguro Social, Mexico, D. F.

Von Brandt, T. (1973). *In* "Biochemistry of Parasites," 2nd edition, pp. 268–273. Academic Press, New York.

Weik, R. R. and Reeves, R. E. (1980). Niacin requirement for growth of axenic *Entamoeba histolytica*. *American Journal of Tropical Medicine and Hygiene* **29**, 1201–1204.

Weinbach, E. C. and Diamond, L. S. (1974). *Entamoeba histolytica*: I. Aerobic metabolism. *Experimental Parasitology* **35**, 232–243.

Weinbach, E. C., Diamond, L. S., Claggett, C. E. and Kon, H. (1976a). Iron–sulfur proteins of *Entamoeba histolytica*. *Journal of Parasitology* **62**, 127–128.

Weinbach, E. C., Harlow, D. R., Takeuchi, T., Diamond, L. S., Claggett, C. E. and Kon, H. (1976b). Anaerobic metabolism of *Entamoeba histolytica*: facts and fallacies. *In* "Proceedings of the International Conference on Amebiasis" (B. Sepulveda and L. S. Diamond, eds), pp. 190–203. Instituto Mexicano del Seguro Social, Mexico, D. F.

Weinbach, E. C., Harlow, D. R., Claggett, C. E. and Diamond, L. S. (1977). *Entamoeba histolytica*: diaphorase activities. *Experimental Parasitology* **41**, 186–197.

Weinbach, E. C., Takeuchi, T., Claggett, C. E., Inohue, F., Kon, H. and Diamond, L. S. (1980). Role of iron–sulfur proteins in the electron transport system of *Entamoeba histolytica*. *Archivos de Investigacion Medica (Mexico)* **11**, Supplement **1**, 75–81.

Wessenberg, H. (1974). The pathogenicity of *Entamoeba histolytica*. Is heat stress a factor? *Perspectives in Biology and Medicine* **17**, 250–266.

Wittner, M. and Rosenbaum, R. M. (1970). Role of bacteria in modifying virulence of *Entamoeba histolytica*. Studies of amebae from axenic cultures. *American Journal of Tropical Medicine and Hygiene* **19**, 755–761.

World Health Organization (1981). Intestinal protozoan and helminthic infections. *WHO Technical Report Series* No. 666, pp. 29–45.

# Cell-Mediated Damage to Helminths

A. E. BUTTERWORTH

*Department of Pathology, University of Cambridge, Tennis Court Road, Cambridge, UK*

| | | |
|---|---|---|
| I. | Introduction | 144 |
| II. | Methods for Studying Cell-Mediated Damage to Helminths *In Vitro* | 145 |
| | A. Biological Assays | 145 |
| | B. Isotopic Assays | 146 |
| | C. Microscopical Assays | 146 |
| III. | Characteristics of Cellular Effector Mechanisms Active Against Helminths | 148 |
| | A. Introduction | 148 |
| | B. Granulocyte-Mediated Damage | 149 |
| | C. Macrophage-Mediated Damage | 172 |
| | D. Lymphocyte-Mediated Damage | 178 |
| | E. Conclusions | 181 |
| IV. | Effector Mechanisms Active Against Particular Helminth Species | 182 |
| | A. Introduction | 182 |
| | B. Platyhelminthes: Trematoda | 182 |
| | C. Platyhelminthes: Cestoda | 187 |
| | D. Nematoda | 189 |
| V. | The Role of Cellular Effector Mechanisms in Immunity *In Vivo* | 198 |
| | A. Introduction | 198 |
| | B. Histological Studies | 199 |
| | C. Manipulation of Experimental Animal Models | 200 |
| | D. Correlative Studies in Experimental Animals and Man | 203 |
| | E. Conclusions | 205 |
| VI. | Summary and Conclusions | 205 |
| | Acknowledgements | 207 |
| | References | 207 |

## I. Introduction

Different helminth parasites, and even different stages of the same parasite species, vary markedly in their surface structure. However, they have one common feature that distinguishes them from most other pathogenic organisms; namely that they present to the host's defences a large non-phagocytosable surface. This imposes a severe restriction on the cellular effector mechanisms that will be active against such organisms. It might be predicted that some effector cells, such as the various types of cytotoxic lymphocyte, will be relatively inactive in mediating damage. Equally, it might be predicted that the mammalian host will have evolved one or more specialized defences that are particularly well adapted to damaging these large tissue-invasive organisms, and an example that will be considered in detail is the antibody- or complement-dependent damage mediated by eosinophils. After a brief summary of methodology, therefore, the first main section of this review deals in general terms with the nature and mode of action of the main categories of effector cells, and their capacity to act against selected helminth parasites *in vitro*.

Many of the most recent detailed studies of the damage attributable to such effector mechanisms have involved as target organisms the larval stages of *Schistosoma mansoni* or *Trichinella spiralis*, tested in the presence of human or rodent effector cells. It is now clear, however, that there are marked differences not only in the effects of cells from different host species but also, more importantly, in the relative susceptibility of different groups, species and stages of target organism. The next main part of this review, therefore, summarizes briefly the distinctive features of the most important or best-studied host–parasite combinations.

Finally it should be emphasized that studies *in vitro* offer two main advantages. First, they allow a detailed analysis of the nature and mode of action of those cellular effector mechanisms that *may* be operative *in vivo*, and of their interactions with other systems. Secondly, apart from anecdotal studies on individuals with innate or acquired defects in specific cell functions, they provide the only approach to the identification of the mechanisms of human immunity. However, the fact that a particular effector mechanism is demonstrable *in vitro* does not necessarily imply that it is active *in vivo*. The final section of this review, therefore, summarizes briefly the evidence for a role for cellular immune effector mechanisms *in vivo*, both in experimental animals and in man.

## II. METHODS FOR STUDYING CELL-MEDIATED DAMAGE TO HELMINTHS *In Vitro*

### A. BIOLOGICAL ASSAYS

A problem that is immediately encountered in studies on cell-mediated damage to helminths, in contrast to unicellular organisms or virus-infected, neoplastic or normal mammalian cells, is to know what constitutes irreversible damage to the target. Damage to tumour cells mediated by cytotoxic T lymphocytes, for example, follows a precise and clearly-defined course during which, at a given time, damage becomes irreversible (Martz, 1977; Sanderson, 1982). Such damage is then necessarily followed by disruption of the cell, and the loss of its capacity to divide *in vitro* or *in vivo*. The situation for a multicellular helminth is usually not so straightforward. On the one hand, the surface layer may be severely damaged, yet the organism may have the capacity to repair this damage and remain viable: and, on the other hand, relatively inconspicuous damage to the surface may be associated with a complete loss of viability.

In principle, therefore, the most useful assays would be those that measure some biological property of the target that can be regarded as a necessary condition for viability. Such assays have included the following.

(a) Measurement of the capacity of schistosomula of *Schistosoma mansoni* to mature into adult worms after intravenous or intraperitoneal injection into mice (Kassis *et al.*, 1979; Dessein *et al.*, 1983b).

(b) Measurement of the capacity of ova of *Schistosoma mansoni* to hatch in water, releasing viable miracidia (James and Colley, 1976).

(c) Measurment of the capacity of newborn larvae of *Trichinella spiralis* to encyst in muscles, after intravenous injection (Bass and Szejda, 1979a,b).

(d) Assessment of the capacity of *Fasciola hepatica* to mature into adult worms after intraperitoneal inoculation (Doy and Hughes, 1982b).

(e) Assessment of the capacity of *Nematospiroides dubius* infective larvae to mature after injection into mice (Chaicumpa and Jenkin, 1978).

In practice, however, it would not be feasible to use such assays on a regular basis, because of the limitations imposed by time and by the numbers of samples that can be handled. Most assays in regular use, therefore, involve either measurement of the release of radioisotopes from labelled targets, or microscopical estimates of damage, and these are usually satisfactory in terms of speed, convenience and reproducibility. However, unless a strict correlation has been observed between the results of such assays and those of a biological test system, it is preferable to describe the observed effects as representing damage to, rather than death of, the target organism.

B. ISOTOPIC ASSAYS

Assays that involve the measurement of the release of radioisotopes, such as $^{51}$Cr, from prelabelled organisms, have the advantages of reproducibility, sensitivity, objectivity and the capacity to handle very large numbers of samples. Such assays have been developed for schistosomula of *Schistosoma mansoni* (Butterworth *et al.*, 1974, 1975, 1976a; David *et al.*, 1977), microfilariae of *Litomosoides carinii* (Subrahmanyam *et al.*, 1976), and excysted metacercariae of *Fasciola hepatica* (Duffus and Franks, 1980). However, their use is somewhat limited, since release of isotope may not reflect severe damage to the target, and since the correlation with microscopical assays may not be good (Vadas *et al.*, 1979a; Butterworth and Vadas, 1979). Their main advantage now may be less in the detailed analysis of cellular effector mechanisms, and more in the routine screening of large numbers of serum or cell samples for anti-parasite activity (Butterworth *et al.*, 1976a; Sturrock *et al.*, 1978, 1981, 1983).

C. MICROSCOPICAL ASSAYS

1. *Light microscopy*

Most assays currently in use involve some form of quantitative light microscopy. Originally developed by Clegg and Smithers (1972) for the investigation of the effects of antibody and complement on schistosomula of *Schistosoma mansoni*, there are now a number of variations, depending on the target organism and the test system.

The length of time required for damage to be detectable microscopically is rather greater for helminths than for most unicellular targets. Although incubation times as short as 4 hours have sometimes been used, most assays involve periods of 18 to 48 hours, whereas in the assay originally developed Clegg and Smithers (1972) periods of 3–4 days were needed for the full expression of damage. However, it should be emphasized that damage may be initiated during the very early stages of the reaction: the long culture periods may solely be required for such damage to become detectable microscopically.

Cell-mediated damage can be observed either *in situ*, if flat-bottomed microtitre wells are used for the assay, or after transfer of the contents of the reaction tubes to microscope slides. The criteria subsequently used for defining damage are various, and have included the following.

(a) Loss of motility.

(b) Gross disruption of the organism, for example of newborn larvae of *Trichinella spiralis* (Kazura and Grove, 1978; Kazura, 1981; Grover *et al.*,

1983). In this case, the numbers of remaining organisms are determined as a percentage of the starting population.

(c) Other gross changes to the parasite, including swelling, changes in refractive properties, ballooning or blebbing of the tegument, distortion of shape or extrusion of internal contents.

(d) Changes in permeability to various dyes. This is the preferable approach, since it offers a greater degree of objectivity than the others listed above. Apart from trypan blue (Mehta et al., 1980), conventionally used for determining the viability of mammalian cells, the dyes used have included methylene blue and toluidine blue, both of which are somewhat more sensitive for helminths (Clegg and Smithers, 1972; Mackenzie et al., 1977; Kassis et al., 1979; Anwar et al., 1979; Vadas et al., 1979a). Less widely used dyes have been ethidium bromide in combination with fluorescein dibutyrate or other fluorescein esters. The ethidium bromide stains dead organisms with an orange fluorescence, whereas the fluorescein esters are taken up by live parasites and converted into fluorescein, which fails to diffuse out of the organism (B. Cottrell, personal communication). In all such assays, problems are encountered if the number of effector cells adhering to the target is so great that the underlying organism is obscured. Such organisms may be intact and viable, even though enclosed within a 'coffin' of effector cells.

Although each assay is standardized within each laboratory, there are usually differences in procedure between different groups. Apart from relevant and testable variables such as the nature, source and amount of antibody, complement and effector cells, other variables that may not be consistent between different groups, and that may influence the results include: (a) the method of preparation of the target organism; (b) the culture medium used, and the presence or absence of foetal calf serum or other serum supplements; (c) the shape of the reaction tube or well (round-bottomed or flat), and the volumes of reagents added; and (d) the criteria chosen to represent damage.

Differences in results between different groups, discussed in more detail below, may be attributable to such factors. In such cases, the most appropriate assay is that which shows the closest correlation with the biological assays described above.

2. *Electron microscopy*

Electron microscopical techniques cannot be used to quantify the extent of damage to a target population, but they are invaluable in studies on the mechanism whereby damage is induced. They have proved particularly useful, for example, in demonstrating the extensive degranulation of eosinophils, with release of granule contents onto the surface of the parasite, that occurs during eosinophil-mediated damage to schistosomula of *Schistosoma mansoni* (Glauert and Butterworth, 1977; McLaren *et al.*, 1977, 1978a; Glauert *et al.*, 1978) and other helminths (McLaren *et al.*, 1977; McLaren, 1980a), and in identifying the nature of the damage. Examination of thin sections by transmission microscopy has usually proved to be the most useful technique, but studies by scanning electron microscopy (Baron and Tanner, 1977; Greene *et al.*, 1981) and by freeze-fracture techniques (Caulfield *et al.*, 1980a,b, 1982; Torpier *et al.*, 1979; Torpier and Capron, 1980) have also yielded useful results. Their main limitation is in the number of experimental conditions that can reasonably be tested.

### III. Characteristics of Cellular Effector Mechanisms Active Against Helminths

#### A. INTRODUCTION

Various cell types, active in other cytotoxic systems, have been studied for their capacity to damage helminths *in vitro*. These include: (a) cells of the granulocyte series, classically noted for their capacity to cause rapid intracellular killing of bacteria of acute pyogenic infections; (b) cells of the monocyte–macrophage series, usually noted for their capacity to undergo T lymphocyte-dependent activation and to cause a rather slower death of resistant obligatory intracellular parasites, including both bacteria (e.g. North, 1981) and protozoa (reviewed by Thorne and Blackwell, 1983); and (c) several types of 'executive' lymphocyte, usually considered in the context of extracellular cytotoxicity to virus-infected (Blanden, 1974; Zinkernagel and Doherty, 1979; Zinkernagel and Rosenthal, 1981), neoplastic or normal mammalian target cells.

There is no reason for assuming *a priori* that any one cell type will be more active against helminths than any other, nor that the mechanism of attachment and damage will be the same as in other systems. In this section, the general properties of each cell type will be considered, with emphasis on the nature of the ligands, if any, that mediate attachment of the effector cell to the parasite; the mechanisms of damage; the functional state of the effector cell, and the ways in which functional activity may be altered; and differences in activity of cells from different host species. Particular emphasis is laid at this stage on schistosomula of *Schistosoma mansoni* and newborn larvae of *Trichinella spiralis*, which have been the most extensively studied targets.

## B. GRANULOCYTE-MEDIATED DAMAGE

### 1. *Introduction*

Until quite recently, it was assumed that the eosinophil was a close relation of the neutrophil, with similar but less effective functional properties: it could, for example, phagocytose antigen–antibody complexes (Sabesin, 1963; Archer and Hirsch, 1963; Litt, 1964; Ishikawa *et al.*, 1974) and phagocytose and kill antibody-coated spores of *Candida albicans* (Ishikawa *et al.*, 1972), but it was less active in killing other intracellular microorganisms (Cline *et al.*, 1968; Cohen and Sapp, 1969; Baehner and Johnston, 1971; Mickenberg *et al.*, 1972). More recently, however, it has been recognized that eosinophil precursors are distinguishable from neutrophil and macrophage precursors very early during differentiation in the bone marrow, and that the development of eosinophil colonies in bone marrow cultures *in vitro* may be stimulated by mediators that are clearly separable from those that stimulate the formation of granulocyte and macrophage colonies (Rabellino and Metcalf, 1975; Nicola *et al.*, 1978, 1979). In addition, mature eosinophils and neutrophils differ markedly in their content of enzymes and other proteins. For example, both eosinophils and neutrophils contain a peroxidase, but the enzyme in the two cells differs antigenically (Salmon *et al.*, 1970), in substrate specificity and in susceptibility to inhibitors (Klebanoff *et al.*, 1980): and genetic defects in one enzyme are not associated with similar defects in the other (Presentey and Szapiro, 1969; Salmon *et al.*, 1970). Eosinophils do not bear, within their granules, several of the enzymes that are characteristic of the neutrophil: and instead they contain their own characteristic moieties. Unusual eosinophil enzymes include arylsulphatase B (Tanaka *et al.*, 1962; Wasserman *et al.*, 1975), phospholipase D (Kater *et al.*, 1976) and lysophospholipase (Ottolenghi, 1969; Weller *et al.*, 1980). Histaminase is present in both eosinophils and neutrophils (Zeiger *et al.*, 1976). Other characteristic proteins, with no demonstrable enzymatic activity, are also present in large amounts in the eosinophil granule. The eosinophil major basic protein (MBP) (Gleich *et al.*, 1973, 1974), which is localized in the crystalloid core of the eosinophil granule (Lewis *et al.*, 1978), may account for over 50% of the total granule protein. It is a small polypeptide, with a molecular weight of 9300 in man and 11 000 in the guinea pig (Gleich *et al.*, 1974, 1976) and it has a high arginine content and a pI of greater than 11. A second arginine-rich protein, the eosinophil cationic protein (ECP), has also been described as a major component of human eosinophil granule (Olsson and Venge, 1974; Olsson *et al.*, 1977): it has a molecular weight of 21 000, and the amino acid contents of MBP and ECP respectively are such that neither can be derived directly from the other. The relationship between these two proteins, and the relative amounts of each in the cell, are not yet clearly understood. Apart from these two major proteins, various minor proteins are also found in the eosinophil, including particularly an eosinophil-derived neurotoxin, thought to be responsible for the Gordon phenomenon seen after intrathecal injection of eosinophils into rabbits (Ackerman et al., 1983).

It has become apparent, therefore, that the eosinophil differs fundamentally from the neutrophil, and might be expected to have some distinctive functional properties: and, during the last 10 years, two such properties have been ascribed to these cells. First, they may show a particular capacity to modulate, or dampen down, immediate hypersensitivity reactions attributable to immunoglobulin E (IgE)-dependent mast cell degranulation. The suggested sequence of events starts with the release, from the mast cell, of a variety of mediators that are more or less selectively chemotactic for eosinophils. The first of these mediators to be described was the eosinophil chemotactic factor of anaphylaxis (ECF-A) (Kay and Austen, 1971; Kay et al., 1971), subsequently identified as two acidic tetrapeptides, Ala-Gly-Ser-Glu and Val-Gly-Ser-Glu (Goetzl and Austen, 1975, 1976). More recently, however, it has been recognized that a variety of other mast cell mediators are also chemotactic for eosinophils, including histamine (Clark et al., 1975) and its oxidative metabolite imidazole acetic acid (Turnbull and Kay, 1976), some lipoxygenase and cyclooxygenase derivatives of arachidonic acid, especially the hydroxyeicosatetraenoic acids (Goetzl et al., 1977; Goetzl and Gorman, 1978), and some uncharacterized polypeptides of intermediate molecular weight (Boswell et al., 1978). It is suggested that these mediators are released at the site of immediate hypersensitivity reactions, and that they lead to a more or less selective localization of eosinophils in the reaction, especially when such reactions are persistent or repetitive.

Having localized at the site of the immediate hypersensitivity reaction, the eosinophil may then participate in the degradation or neutralization of other mast cell mediators, by virtue of some of the unusual enzymes and other proteins that it contains. For example, the histaminase may assist in histamine breakdown, the phospholipase D may catabolize platelet activating factor (Benveniste, 1974), and especially its platelet lytic component (Kater et al., 1976; Weller et al., 1980), and the arylsulphatase may catabolize slow-reacting substance of anaphylaxis (SRS-A) (Orange et al., 1974; Wasserman et al., 1975). This last point is now somewhat doubtful, since human SRS-A has now been identified as the leukotrienes C4 and D4, and purified arylsulphatase may have no effect on these compounds. In addition, the various eosinophil basic proteins may interact with and neutralize the acidic heparin that is released in large amounts from the mast cell granule: and the generation by the eosinophils of the prostaglandins $E_1$ and $E_2$ may prevent further mast cell degranulation. The net effect of these processes would be to accelerate the degradation or inactivation of mast cell mediators, and to prevent further mediator release; and hence to dampen down a reaction that is deleterious to the host (Goetzl *et al.*, 1975).

This hypothesis, as it stands, is attractive, but suffers from a conceptual problem. What it seems to suggest is that the organism has evolved a reaction, mediated by IgE and mast cells, that is solely deleterious to itself; and that it has also evolved a particular cell, the eosinophil, whose main specialized attribute is its ability to switch off this otherwise deleterious reaction. This would present no selective advantage to the organism. Instead, an alternative hypothesis has been developed by several workers, which suggests that the eosinophil has evolved as a specialized form of defense against large, non-phagocytosable tissue-stage parasites such as helminths (Butterworth, 1977; McLaren, 1980b); and that the primary role of the IgE-dependent mast cell reaction is to localize the eosinophils to the site of the invading helminth, as well as to enhance their functional properties. The proposed sequence of events, which will be considered in detail in later sections, is: (a) that helminth infections lead to a thymus-dependent increase in the numbers of circulating eosinophils, and to an enhancement of the functional properties of the individual cells; (b) that antigens released from the invading helminths elicit a local IgE-dependent degranulation of mast cells; (c) that the release of mast cell mediators leads to a selective localization of eosinophils at the site of the invading helminth, together with an accumulation in the tissue fluids of immunoglobulins and complement; (d) that the mast cell mediators cause a further enhancement of eosinophil functional activity; (e) that the localized eosinophils, in the presence of various combinations of antibody and complement, then show a more or less selective capacity to damage the invading helminth; and (f) finally, that the eosinophils show a secondary capacity to switch off the mast cell reaction that elicited their arrival.

This scheme is clearly speculative, and the evidence for it, which will be described in the next section, is incomplete. Furthermore, the suggestion that the eosinophil may be particularly adapted for protection of the host against helminth infections does not preclude the possibility that more conventional neutrophil-mediated mechanisms are also involved, and these will also be described.

## 2. *Early demonstrations of granulocyte-mediated damage to helminths*

Most of the early work on antibody-dependent cell-mediated damage to helminths involved the use of the larval stages (schistosomula) of *Schistosoma mansoni*. These may be prepared by allowing the infective cercariae to penetrate an isolated preparation of rat or mouse skin *in vitro* (Stirewalt and Uy, 1969), and maintained in complex media for prolonged periods (Clegg and Smithers, 1972). Such organisms were found to be susceptible to damage by IgG antibodies from hyperimmunized rhesus monkeys, acting in the presence of a heat-labile component, presumably complement, present in normal monkey serum. The effect, however, was relatively weak, in that high concentrations both of immune serum and of the heat-labile component were required for the induction of damage, which was not fully manifest until the fourth day of culture. Thereafter, an accessory action of effector cells was sought. Dean *et al.* (1974, 1975) demonstrated that IgG antibodies from immunized rats or guinea pigs, in the presence of homologous neutrophils, caused a rapid death of schistosomula, associated with adherence of cells and the reduction of nitroblue tetrazolium at the surface. In these reactions, however, there was a requirement for high levels of complement, and it is possible that the effect of neutrophils was simply to accelerate the rate at which damage already initiated by complement activation could be demonstrated. Subsequently, it was shown that normal human and baboon peripheral blood leukocytes (Butterworth *et al.*, 1974, 1976a), and especially their granulocyte-rich fractions (Butterworth *et al.*, 1975) would induce an antibody-dependent release of $^{51}$Cr from prelabelled schistosomula, this release being independent of added complement. Similar studies, in which unpurified granulocytes have been used, have subsequently been carried out with a variety of different helminths, including *Litomosoides carinii* (Subrahmanyam *et al.*, 1976), *Trichinella spiralis* (Kazura and Grove, 1978), *Dipetalonema viteae* (Rudin *et al.*, 1980), *Brugia malayi* (Sim *et al.*, 1982), *Brugia pahangi* (Piessens and Dias da Silva, 1982) and *Dirofilaria immitis* (El-Sadr *et al.*, 1983).

More recently, the development of techniques for the preparation of highly purified granulocyte populations has allowed a comparison of the effects of eosinophils and neutrophils, and a detailed analysis of their mode of action. In subsequent sections, these will be considered separately.

3. *Eosinophil-mediated damage to helminths*

(*a*) *Effects of purified eosinophils.* Eosinophils may be purified from the peripheral blood or peritoneal cavity of several species by density fractionation. Early experiments involved the use of metrizoate (Day, 1970), and it was found that human blood eosinophils at up to 98% purity would induce release of $^{51}$Cr from labelled schistosomula in the presence of anti-schistosomular antibodies (Butterworth *et al.*, 1977a). Metrizoate, however, is somewhat toxic to cells, especially at high concentrations, and centrifugation on gradients of hypaque, metrizamide or percoll is now preferred (Grover *et al.*, 1978; Vadas *et al.*, 1979a; Gärtner, 1980). Under appropriate conditions, preparations of more than 95% purity may routinely be obtained, even from normal peripheral blood containing less than 5% eosinophils.

Such eosinophil-enriched cell preparations have been found to damage a wide variety of target helminths, in the presence of antibody, complement, or both together. Susceptible targets include: schistosomula of *Schistosoma mansoni* (Butterworth *et al.*, 1975, 1977a; McLaren *et al.*, 1977, 1978a; Glauert and Butterworth, 1977; Glauert *et al.*, 1978; Capron *et al.*, 1978a,b, 1979; Vadas *et al.*, 1979a; Anwar *et al.*, 1979, 1980; Kassis *et al.*, 1979; Kazura *et al.*, 1981), eggs of *Schistosoma mansoni* (James and Colley, 1976, 1978a,b,c), microfilariae of *Litomosoides carinii* (Mehta *et al.*, 1981b), microfilariae of *Onchocerca volvulus* (Greene *et al.*, 1981), newborn larvae of *Trichinella spiralis* (Kazura and Grove, 1978; Kazura, 1981; Bass and Szejda, 1979a,b; Grover *et al.*, 1983), infective larvae of *Dictyocaulus viviparus* (Knapp and Oakley, 1981), and both microfilariae and infective larvae of *Dipetalonema viteae* (Haque *et al.*, 1981, 1982),

In some cases, damage induced by eosinophils *in vitro* has been found to be associated with a reduced capacity of the target organisms subsequently to survive and mature after introduction into an appropriate experimental host (Kassis *et al.*, 1979; Bass and Szejda, 1979a; Dessein *et al.*, 1983b).

(*b*) *Ligands mediating eosinophil-dependent damage.* In most cases so far studied, damage to helminths induced by eosinophils depends on a direct interaction between the eosinophil and the surface of the target organism, this interaction being mediated by a ligand. Eosinophils bear on their surface receptors for the Fc piece of IgG (Rabellino and Metcalf, 1975; Gupta *et al.*, 1976; Tai and Spry, 1976; Butterworth *et al.*, 1976b; Anwar and Kay, 1977), and for C3 (Tai and Spry, 1976; Sher and Glover, 1976; Anwar and Kay, 1977). In the case of C3, receptors for both C3b and C3d have been identified (Anwar and Kay, 1977). Adherence of eosinophils may therefore be mediated by IgG antibodies alone, complement alone, or combinations of the two. In addition, an early report that eosinophils bear receptors for IgE (Hubscher, 1975) has recently been confirmed (Capron *et al.*, 1981a) and extended to show that IgE can serve as a ligand for eosinophil-mediated damage to schistosomula of *Schistosoma mansoni* (Capron *et al.*, 1981b) and microfilariae of *Dipetalonema viteae* (Haque *et al.*, 1981).

The subclasses of IgG mediating the attachment to helminths of human eosinophils have not yet been adequately characterized, although $IgG_1$ is known to be effective (M. Vadas and A. E. Butterworth, unpublished observations). In the rat, $IgG_{2a}$ antibodies are effective in mediating eosinophil adherence to schistosomula (Capron *et al.*, 1978a,b), and a rat monoclonal $IgG_{2a}$ antibody with specificity for a schistosomulum surface antigen both mediates eosinophil killing of schistosomula and confers protection against cercarial challenge *in vivo* (Verwaerde *et al.*, 1979; Grzych *et al.*, 1982). Rat $IgG_{2a}$, like IgE, is homocytotropic, and will sensitize mast cells for subsequent degranulation in the presence of antigen, an aspect discussed in more detail below. The analogous subclass of heat-stable homocytotropic antibodies in the mouse, IgG$_1$, will also mediate eosinophil killing by rat eosinophils (Ramalho-Pinto et al., 1979), and is capable of conferring passive protection against cercarial challenge *in vivo*.

Several helminths can activate complement by the alternative pathway, in the absence of antibody. In most cases, alternative pathway activation does not lead to direct damage to the organism, except in the case of schistosomula of *Schistosoma mansoni* that have been prepared by the unphysiological process of cercarial disruption and incubation (Santoro et al., 1979). However, alternative pathway activation leads to the binding of C3 to the surface of the organism, and such organisms are now susceptible to eosinophil attachment and damage, although the extent of the effect is usually less marked than in the situation in which complement is activated by the classical pathway, in the presence of antibody (Ramalho-Pinto et al., 1978; Anwar et al., 1979; McKean et al., 1981; Dessein et al., 1981b).

In most studies, the presence of complement enhances the extent of eosinophil-mediated damage that is observed in the presence of antibody alone (McLaren and Ramalho-Pinto, 1979; Anwar et al., 1979; McKean et al., 1981), especially when antisera with a low activity are used. However, this enhancement is not invariably seen (Butterworth et al., 1982), and care must be exercised in the interpretation of the phenomenon. The usual interpretation is that IgG and complement are synergistic and that, as with phagocytosis by neutrophils (Mantovani, 1975; Ehlenberger and Nussenzweig, 1977), the attachment of eosinophils by both Fc and C3 receptors allows a more effective expression of the cell's killing mechanisms. Two alternative explanations, however, must be considered. First, the fixation of small amounts of complement may cause a degree of damage to the target that is undetectable in the particular assay system being used, but which will cause an apparent enhancement of a separate cell-mediated effect. Secondly, especially when the killing assays are carried out in flat-bottomed chambers as opposed to round-bottomed tubes, the generation of chemotactic products of complement activation, especially C5a and $\overline{C567}$, may cause a more rapid or more effective localization of the eosinophils around the parasite. Therefore, it cannot always be assumed that an enhancing effect of complement is attributable to C3 binding and to attachment of eosinophils through C3 receptors. However, experiments involving the use of purified complement components activated in series have shown that such attachment can certainly occur in some circumstances (Anwar et al., 1979).

Certain lectins, acting as non-specific ligands, can also mediate attachment of eosinophils to target helminths (Butterworth et al., 1979a). Such attachment, however, is not followed by damage to the parasite. This observation has proved useful in understanding the mechanisms of attachment and of damage, and is discussed in more detail below.

(c) *Mechanisms of eosinophil-mediated damage.*

(i) Electron microscopical studies. Studies by electron microscopy have shown that, after an initial tight interaction of the eosinophil with the surface of the antibody- or complement-coated helminth, the cell then degranulates, and releases its granule contents as electron-dense deposits on the surface of the target. This event was initially observed with schistosomula of *Schistosoma mansoni* (McLaren *et al.*, 1977, 1978a; Glauert and Butterworth, 1977; Glauert *et al.*, 1978; Caulfield *et al.*, 1980a), but has been extended to cover a wide range of different helminths (McLaren *et al.*, 1977; McLaren, 1980b), and appears to be a general feature of the interaction of eosinophils with large non-phagocytosable surfaces (Glauert *et al.*, 1980). Such degranulation can also be observed by phase-contrast cinemicrography (Densen *et al.*, 1978). Neutrophils, in contrast, fail to degranulate, and their interactions are discussed in more detail below.

The sequence of events that occurs during degranulation has varied to a certain extent in different studies, depending on the nature of the target organism, the nature of the ligand, the assay condition tested, and the host species from which the eosinophils were recovered. In the case of human eosinophils, adhering either to antibody-coated schistosomula (Glauert *et al.*, 1978) or to a model non-phagocytosable target consisting of antigen-antibody complexes incorporated into an agar layer (Glauert *et al.*, 1980), the granules usually fuse directly with the plasma membrane in areas of attachment to the parasite, and granule contents are released directly onto the surface of the organism. In the case of rat eosinophils, granules fuse initially with each other, forming large intracytoplasmic vacuoles which subsequently discharge onto the surface (McLaren *et al.*, 1977; McLaren, 1980a,b). In both cases, the consequences of degranulation are that the eosinophil membrane is separated from the schistosomulum membrane by a dense deposit of granule contents which, in the case of rat eosinophils (McLaren *et al.*, 1977; McLaren, 1980a,b), can be demonstrated to contain peroxidase.

Subsequent damage depends on the nature of the target parasite. In the case of schistosomula, for example, the first sign of damage is the appearance of small vacuoles or blebs within the syncytial tegument (Glauert *et al.*, 1978; McLaren *et al.*, 1978a). These coalesce, and the surface membrane is lifted up. Breaks in the surface membrane appear, and eosinophils may be observed infiltrating through the gaps in the tegumental membrane, which is lifted off the organism and phagocytosed. Subsequently, eosinophils are observed in close contact with schistosomulum bodies that have been denuded of their tegumental membrane.

In preparations containing eosinophils and schistosomula without antibody or complement, contact between the cells and the parasite may sometimes be observed (Glauert *et al.*, 1978). In such cases, however, the cell fails to flatten onto the surface or to release its granule contents, and no damage to the underlying parasite is seen. These findings, together with the localized nature of the initial lesions, suggest a primary role for eosinophil degranulation in the induction of damage.

(ii) The role of released granule contents in mediating damage. The extent of degranulation observed by electron microscopy during eosinophil-mediated damage to schistosomula and other helminth targets suggested that damage might be directly attributable to the release of toxic granule contents. Isolated granule components were therefore tested for their ability to cause direct damage to schistosomula, and it was found that both human and guinea-pig eosinophil major basic protein (MBP) were toxic to schistosomula in low concentrations (Butterworth et al., 1979b). Toxicity was associated with an uptake of MBP onto the surface of the target organism, detectable by immunofluorescence, and both uptake and toxicity could be blocked by heparin. In addition, it could be demonstrated directly that MBP was released onto the surface of schistosomula during an eosinophil-mediated killing reaction: and more recent observations (A. E. Butterworth, unpublished results) indicate that heparin will also inhibit the effects of intact eosinophils. Although these observations must be interpreted with caution, they suggest that degranulation and the release of MBP may be one of the mechanisms whereby eosinophils damage schistosomula.

A second basic protein within the eosinophil, the eosinophil cationic protein (ECP) (Olsson and Venge, 1974; Olsson et al., 1977) is also directly toxic for schistosomula at low concentrations (McLaren et al., 1981). ECP has been reported to be highly toxic at concentrations of $1\mu$M, 10-fold less in molar terms than the concentrations required for MBP toxicity (Butterworth et al., 1979b). Other cationic polypeptides, such as protamine and polyarginine, are also toxic for schistosomula, but the effect differs microscopically from that induced by MBP or ECP, suggesting that the effects of the eosinophil proteins may be attributable to some property other than simply their basic nature. With both ECP and MBP, damage is associated with a marked ballooning and thinning of the schistosomular tegumental membrane (McLaren et al., 1981; A. E. Butterworth, unpublished observations). The relative role of MBP and ECP in inducing damage is uncertain; recent data indicate that, although ECP is 10-fold more active than MBP in inducing damage, there is 40-fold less within the eosinophil granule; and that, when granule contents are fractionated by column chromatography, most of the toxicity for schistosomula resides within the MBP fraction (S. J. Ackerman, D. A. Loegering, A. E. Butterworth and G. J. Gleich, unpublished observations.) However, this finding, and the exact relationship of MBP to ECP, remains to be confirmed.

Major basic protein is also toxic for newborn larvae of *Trichinella spiralis*, in contrast to other basic polypeptides such as protamine and polyarginine (Wassom and Gleich, 1979). In addition, a basic protein has also been partially purified from bovine eosinophils (Duffus et al., 1980). This material has a molecular weight of 16 000, intermediate between that of human MBP (9200) and human ECP (21 000), and has been found to be directly toxic for the larval stages of *Fasciola hepatica*.

(iii) Adherence and degranulation: a two-stage reaction. Early observations in one model of antibody-dependent eosinophil-mediated killing of schistosomula had indicated that the high levels of eosinophil-mediated killing that were observed were preceded by striking levels of eosinophil adherence to the antibody-coated larvae, when compared with neutrophils (Vadas et al., 1980a). This finding was unexpected, since it was known that neutrophils have more or stronger Fc receptors for bound IgG than neutrophils (Tai and Spry, 1976; Butterworth et al., 1976b; Ottesen et al., 1977). The observation that the initial attachment of eosinophils was rapidly followed by degranulation and the release of granule contents suggested that the released granule contents, as well as causing direct damage to the target, might also serve to maintain the adherence of cells already attached, and to promote the adherence of fresh cells. This hypothesis was tested in a series of experiments (Butterworth et al., 1979a; Vadas et al., 1980a). Both eosinophils and neutrophils initially adhere to schistosomula by a temperature-independent reaction that is blocked by preincubation of the cells with aggregated gamma-globulin or *Staphylococcus aureus* protein A (SPA), implying the involvement of the cells' Fc receptors (Vadas et al., 1980a). At 4°C, the binding of neutrophils, both to opsonized erythrocytes and to antibody-coated schistosomula, is more marked than that of eosinophils. At 37°C, the adherence of neutrophils is unchanged, whereas that of eosinophils is markedly enhanced. With prolonged incubation at 37°C, neutrophils show a peak of maximum adherence after 4 hours of incubation, and subsequently become detached from the schistosomula. Eosinophils, in contrast, progressively accumulate on the surface of live organisms. The established adherence of neutrophils after 1 hour of incubation can be reversed by the addition of SPA or aggregated gamma-globulin, whereas established eosinophil adherence cannot be reversed in this way (Vadas et al., 1980a). These findings suggest that both eosinophils and neutrophils show an initial temperature-independent binding through their Fc receptors, but that the eosinophils then undergo an additional temperature-dependent step that renders their binding progressive and irreversible; the likely nature of this second step being the degranulation of the eosinophils that occurs soon after adherence. In order to test this possiblity, eosinophils and neutrophils were bound to schistosomula by concanavalin A (Con A) instead of antibody (Butterworth et al., 1979a). Under these conditions, the binding of eosinophils and neutrophils was similar, and no differences were observed between preparations incubated at 4°C and at 37°C. No degranulation of the eosinophils occurred, and there was no damage to the parasite: and the established adherence of both eosinophils and neutrophils could be reversed by the addition of α-methylmannoside. However, the addition of calcium ionophore A23187 to Con A-bound cells induced eosinophil degranulation: under these conditions, the established adherence of eosinophils, but not of neutrophils, became irreversible by α-methylmannoside, and eosinophil-mediated damage to the target could subsequently be observed. These findings support the hypothesis that, in the antibody-dependent reaction, both the progressive adherence of eosinophils and their capacity to induce damage is attributable to the degranulation that follows the initial adherence step.

(iv) *The role of oxidative mechanisms in eosinophil-mediated damage.* The findings described in previous sections strongly support the idea that at least one of the mechanisms whereby eosinophils damage helminths involves degranulation and the release of toxic granule contents onto the surface of the parasite. Further evidence for non-oxidative mechanisms comes from the observations of Pincus *et al.* (1981), who have shown that eosinophils cultured under strictly anaerobic conditions, such that they fail to mount a respiratory burst even in response to a strong membrane stimulus, phorbol myristate acetate, show a normal capacity to kill schistosomula in the presence of antibody or antibody and complement. In addition, cells from patients with chronic granulomatous disease, who lack the capacity to generate superoxide and subsequently other active oxygen species as a result of a respiratory burst (reviewed by Babior, 1978), are still able to kill schistosomula (Butterworth *et al.*, 1980), although to a lesser degree than cells from normal individuals (Kazura *et al.*, 1981). However, under normal circumstances, eosinophils mount a strong respiratory burst after appropriate stimulation (Baehner and Johnston, 1971; Mickenberg *et al.*, 1972; De Chatelet *et al.*, 1977), and there is evidence that oxidative mechanisms may also be involved in eosinophil-mediated damage. In cell-free systems, eosinophil peroxidase, together with hydrogen peroxide and iodide, is toxic for schistosomula (Klebanoff *et al.*, 1980; Jong *et al.*, 1981), in a fashion comparable to the toxicity to many species of bacteria that is mediated by neutrophil myeloperoxidase (Klebanoff, 1975). Kazura *et al.* (1981) have demonstrated a rapid and long-lasting increase in hydrogen peroxide generation by both eosinophils and neutrophils, incubated with schistosomula in the presence of antibody, complement, or both together. Killing of parasites was correlated with hydrogen peroxide generation and could be partially inhibited with catalase at high concentrations (5000 units ml$^{-1}$). Eosinophil killing could also be partially inhibited by azide and aminotriazole, but not by cyanide (which fails to affect eosinophil peroxidase): neutrophil killing could be inhibited by all three reagents.

There is no clear explanation for the differences between Pincus et al. (1981), who attribute a major role to non-oxidative mechanisms, and Kazura et al. (1981), who consider that oxidative mechanisms are more important. A possible explanation is simply that the assay conditions were different, involving respectively round-bottomed tubes and flat-bottomed wells. Until further evidence emerges, a reasonable conclusion may be that both oxidative and non-oxidative mechanisms are involved in killing of schistosomula.

One interesting alternative possibility, however, is that oxidative effects are particularly marked in 'activated' eosinophils from patients with eosinophilia (p. 163). After stimulation with opsonized particles, eosinophils from eosinophilic individuals show increased oxygen consumption, hexose monophosphate shunt activity, superanion production and hydrogen peroxide generation in comparison with neutrophils (Baehner and Johnston, 1971: Mickenberg et al., 1972; De Chatelet et al., 1977; Klebanoff et al., 1977; Tauber et al., 1979). In addition, a direct comparison of eosinophils from eosinophilic and normal individuals has shown marked differences between eosinophils from eosinophilic and normal individuals, including a decreased surface charge, increased hexose transport and activation of the latent enzyme acid phosphatase (Bass et al., 1980). It is possible, although there is no direct evidence, that eosinophils have two distinct killing mechanisms—a 'fast' reaction, dependent on the early generation of active oxygen metabolites, and a 'slow' reaction, dependent on progressive degranulation and the release of granule contents—and that the 'fast' reaction is enhanced in conditions of eosinophil stimulation or activation. In this context, it may be noted that Rand and Colley (1982) have recently shown that normal eosinophils stimulated with lymphokine-containing lymphocyte culture supernatants show an enhanced respiratory burst activity with no increased propensity to degranulate. This aspect is considered in more detail below.

(d) *Enhancement of eosinophil functional activity*

(i) Eosinophilia and IgE responses in helminth infections. Many of the experiments described in the previous section have involved the use of eosinophils recovered from the blood of normal human subjects, or from the blood or peritoneal cavity of normal experimental animals. Helminth infections, however, are commonly associated with an increase in circulating eosinophil numbers: and the effect of eosinophils *in vivo* will not be manifest in the blood or peritoneal cavity, but rather at the site of IgE-dependent immediate hypersensitivity reactions to invading tissue-stage larvae. It is therefore important to determine how these processes—eosinophilia and immediate hypersensitivity reactions—may affect eosinophil function.

In several experimental animal models, the eosinophilia of helminth infections has been found to be a T lymphocyte-dependent event. This was originally demonstrated in the classical experiments of Basten, Beeson and their colleagues on the responses of rats to *Trichinella spiralis* (Basten and Beeson, 1970; Basten *et al.*, 1970; Walls and Beeson, 1972; Walls *et al.*, 1971, 1974). After intravenous injection of muscle-stage larvae, a sharp burst of eosinophilia was observed, reaching a peak 10 days after injection. Depletion of T lymphocytes by neonatal thymectomy, thoracic duct drainage or treatment with anti-lymphocyte serum ablated this response and, after transfer of lymphocytes from infected animals to normal syngeneic recipients an enhanced and accelerated response was observed. Such lymphocytes could also stimulate eosinophilopoiesis after implantation in millipore diffusion chambers, suggesting the action of soluble mediators. The induction of eosinophilia depended on the introduction of an intact parasite: fragmented larvae caused no eosinophilia. This suggested that the induction of eosinophilia depended on the physical state of the introduced material, rather than any particular constituents, a suggestion that is supported by the observation that the intravenous injection of dextran beads or antigen-coated inert particles cause a similar eosinophilia (Walls and Beeson, 1972; Schriber and Zucker-Franklin, 1975).

Since these early observations, a similar T lymphocyte-dependent eosinophilia has been noted in several other helminth infections, including particularly schistosomiasis (Colley, 1972, 1974; Fine *et al.*, 1973), and ascariasis (Walls, 1976), a condition in mice associated with very high peripheral blood eosinophil counts and extensive bone marrow changes (Sakai *et al.*, 1981). The expression of eosinophilia in mice has been shown in radiation chimaera studies to be under genetic control at the level of bone marrow-derived cells (Vadas, 1982). However, not all helminth-associated eosinophilias are T lymphocyte-dependent: athymic (nude) rats, for example, are capable of mounting an eosinophilic response to *Ascaris suum* (Pritchard and Eady, 1981) and *Fasciola hepatica* (Doy and Hughes, 1982a) infections.

Early evidence had indicated that eosinophilia, whether T lymphocyte-dependent or not, might be attributable to the action of humoral mediators (Basten and Beeson, 1970; McGarry and Miller, 1974; Miller and McGarry, 1976; Miller *et al.*, 1976). Since then, a variety of mediators have been isolated and partially characterized, that either elicit an eosinophilia *in vivo* or induce the formation of eosinophil colonies in bone marrow cultures *in vitro* (Metcalf *et al.*, 1974; Rabellino and Metcalf, 1975; Ruscetti *et al.*, 1976). Partially characterized mediators include: (a) eosinophilopoietin, a low molecular-weight oligopeptide-like material (Mahmoud *et al.*, 1977, 1979a); and (b) a colony-stimulating factor (CSF) derived from human placental conditioned medium (HPCM), and recently separated into two components (Johnson et al., 1977; Johnson and Metcalf, 1978; Nicola et al., 1978, 1979). Both have a molecular weight of about 30 000 daltons, but one (CSF-α) is less hydrophobic and stimulates eosinophil as well as granulocyte/macrophage colonies, whereas the second (CSF-β) is more hydrophobic and stimulates only granulocyte/macrophage colonies. Their effects on the capacity of mature eosinophils to kill helminths are discussed below (p. 163).

Other T cell mediators have been shown to affect the function of mature eosinophils. In particular, eosinophil stimulation promoter (ESP) (Colley, 1973, 1980) enhances the capacity of eosinophils to migrate out of agarose droplets, an effect dependent on altered arachidonic acid metabolism (Rand et al., 1982). It is a heat-stable, trypsin-sensitive moiety of 35 000 to 55 000 daltons (Greene and Colley, 1974) that is produced both by T lymphocytes responding to mitogen or antigen (Colley, 1976; Greene and Colley, 1976) and by isolated egg granulomas (James and Colley, 1975). ESP-containing lymphocyte culture supernatants enhance the capacity of eosinophils to destroy schistosome eggs (James and Colley, 1978b); and ESP-stimulated eosinophils show an increased respiratory burst, as reflected by hexose monophosphate shunt activity and nitroblue tetrazolium reduction, in response to opsonized zymosan, with no increased propensity to degranulate (Rand and Colley, 1982). ESP-like activities have also been described in man (Kazura et al., 1975).

In addition to an increase in circulating eosinophil levels, helminth infections commonly elicit high IgE responses (Johansson et al., 1968; Jarrett, 1973; Jarrett and Miller, 1982); and, during repeated or prolonged infections, there may be evidence of extensive IgE-mediated reactions to helminths or helminth products. Striking examples include: (a) the anaphylactic shock that accompanies rupture of a hydatid cyst (Jakubowski and Barnard, 1971) or treatment of onchocerciasis with diethylcarbamazine (Bryceson et al., 1977); (b) asthma-like reactions in tropical eosinophilia, associated with filarial migration through the lungs (Gentilini et al., 1975; Neva et al., 1975; Ottesen et al., 1979), and in *Toxocara canis* infections (Woodruff, 1970); and (c) eosinophilic meningitis associated with *Angiostrongylus cantonensis* infection of the central nervous system (Rosen et al., 1962; John and Martinez, 1975; Ottolenghi et al., 1977; Pascual et al., 1981; Enzenauer and Yamaoka, 1982).

The regulation of IgE responses to helminth and other antigens has been studied extensively, and readers are referred to reviews by Ishizaka and Ishizaka (1978), Jarrett and Miller (1982), Lehrer and Bozelka (1982), Bazin and Pauwels (1982), Katz (1982) and Kishimoto (1982). Briefly, it is found that both the induction and the suppression of such responses are controlled by a series of isotype-specific regulatory mechanisms, separate from those that control other isotype responses. It is still not clear why helminth infections, in particular, should elicit such strong IgE responses: but they are likely to be attributable to the physical nature and mode of presentation of the helminth antigens, rather than to the chemical structure of these antigens. A particularly useful model for studying IgE responses in helminth infections has been infection of the rat with *Nippostrongylus braziliensis* (Ishizaka and Ishizaka, 1978). In this model, a subpopulation of T lymphocytes has been demonstrated which, through the release of soluble mediators detectable in serum, lymph or cell-free supernatants of lymph-node cell cultures, selectively induce the expression and production by B lymphocytes of IgE antibodies with specificity for irrelevant antigens. This IgE-selective but antigen-non-specific helper effect is of considerable general interest, and may help to explain the 'potentiation' of irrelevant IgE responses that is observed in helminth infections (Jarrett, 1973; Jarrett and Miller, 1982). The IgE-potentiating factor is produced by a population of T cells with Fc receptors for IgE (Suemara *et al.*, 1980), and is itself capable of binding IgE (Yodoi *et al.*, 1980). In other situations, IgE suppressive factors, also capable of binding IgE, have been demonstrated (Hirashima *et al.*, 1980); this factor may be an unglycosylated form of the IgE potentiating factor (Yodoi *et al.*, 1981). From these and other very interesting experiments, a picture is emerging of a class of antibody, selectively elicited by helminth infections, and under a complicated and highly selective system of control.

Finally, it should be noted that although helminth infections elicit both high IgE responses and a peripheral blood eosinophilia, there is no evidence that the two processes are causally linked, whereas the subsequent influx of eosinophils into the site of immediate hypersensitivity reactions is directly dependent on IgE-mediated reactions. For example, Dessein *et al.* (1981a) have shown that suppression of the IgE response of rats to *Trichinella spiralis* infections, by treatment of neonatal animals with anti-epsilon serum before infection, causes a marked reduction in local eosinophil accumulation around the encysted larvae, but has relatively little effect on the development of blood eosinophilia. In addition, there is no direct relationship between the development of eosinophilia in different inbred strains of mice and their capacity to mount an IgE response. It would appear, therefore, that the eosinophilia and the IgE responses seen in helminth infections are separate phenomena, and their possible effects on eosinophil functional properties should be considered in isolation.

(ii) *Enhancement of eosinophil functional activity in conditions of eosinophilia.* Early studies showed that eosinophils recovered from patients with very high circulating eosinophils were relatively inactive in killing schistosomula (Butterworth *et al.*, 1977a), and it was suggested that this might be attributable to a blocking of eosinophil Fc receptors by circulating immune complexes (Butterworth, 1977). It has also been noted, however, that in such patients the eosinophils show gross morphological abnormalities, including vacuolation and degranulation (Connell, 1968; Saran, 1973; Spry and Tai, 1976), and it was later found that sera from hypereosinophilic individuals contained high levels of MBP (Ackerman et al., 1981), suggesting that the cells were spontaneously degranulating in the circulation.

Later studies, therefore, concentrated on eosinophils recovered from patients with a mild or 'physiological' eosinophilia, attributable either to helminth infections or to other allergic disorders. Cells from such individuals show several functional differences from those from normal individuals, including an increased expression of Fc receptors (Tai and Spry, 1976; Spry and Tai, 1976; Ottesen et al., 1977; Parillo and Fauci, 1978), a decrease in surface charge, an increase in hexose transport (Bass et al., 1980) and probably alterations in oxidative metabolism (Baehner and Johnston, 1971; Mickenberg et al., 1972; Bass et al., 1980). Such cells also show a strikingly enhanced capacity to kill schistosomula, in the presence of suboptimal concentrations of anti-schistosomular antibody (Vadas et al., 1979b, 1980b; David et al., 1980; Butterworth et al., 1982; Veith and Butterworth, 1983).

Three possible explanations for this alteration in functional activity may be considered. First, mature eosinophils in eosinophilic individuals may be 'activated', in a manner analogous to the activated macrophage (Bass et al., 1980). Secondly, there may be different subpopulations of eosinophils, with different functional properties, whose proportions alter during eosinophilia. Thirdly, eosinophils in eosinophilic individuals may be at a different state of maturation from those from normal individuals.

In order to test the first possibility, namely that normal eosinophils can undergo activation, several groups have investigated the effects of various mediators that may have eosinophilopoietic activity on the functional properties of normal mature cells. Vadas, Dessein and their colleagues (Vadas et al., 1981; Dessein et al., 1982, 1983a; Vadas, 1983) have shown that human placenta conditioned medium (HPCM), containing both eosinophil and granulocyte/macrophage colony-stimulating activities (Johnson and Metcalf, 1978), enhance the capacity of normal eosinophils to kill schistosomula in the presence of antibody or complement. This effect, which can also be achieved with partially-purified eosinophil colony-stimulating factor (CSF-$\alpha$) (Nicola et al., 1978, 1979), is associated with increased eosinophil autofluorescence, superoxide production, chemotactic responses to casein and killing of antibody-coated mammalian target cells (Vadas, 1983). There are no increases in Fc receptors or in the initial temperature-independent adherence of eosinophils to schistosomula, but the subsequent temperature-dependent step is markedly enhanced (Dessein et al., 1983a).

Eosinophilia is frequently a T lymphocyte-dependent event and a T cell product, eosinophil-stimulation promoter, has been shown to affect several functional properties of mature eosinophils, including their capacity to destroy schistosome eggs (James and Colley, 1978b) (p. 161). In an attempt to test for a role for such mediators in promoting eosinophil function during conditions of eosinophilia, two groups have investigated the production of eosinophil-stimulating activities by mononuclear cells from patients with eosinophilia (Veith and Butterworth, 1983; Veith et al., 1983; Dessein et al., 1983a), with similar results. Peripheral blood mononuclear cells, cultured for 2 to 24 hours in the absence of antigen or other stimulating agents, produce spontaneously a heat-stable, trypsin-sensitive moiety that markedly enhances the capacity of normal eosinophils to kill schistosomula in the presence of low concentrations of anti-schistosomular serum. The activity differs from ESP, in that it is active at very high dilutions, is produced by unstimulated mononuclear cells, and is made not by a T lymphocyte but by an adherent esterase-positive monocyte that is resistant to damage by a lytic monoclonal anti-lymphocyte antibody (Veith and Butterworth, 1983). Although mononuclear cells from most individuals produce activity detectable at high concentrations (Dessein et al., 1983a; Veith and Butterworth, 1983), there is a tendency for more activity to be produced by cells from eosinophilic individuals whose own eosinophils show high killing activity (Veith and Butterworth, 1983), suggesting a causal association between eosinophilia and eosinophil stimulation. However, Dessein et al. (1983a) have also reported that cells from patients with *Schistosoma mansoni* infection are unable to produce an eosinophil-stimulating activity.

The extent of stimulation that is observed with colony-stimulating factors and with mononuclear cell supernatants is very striking when compared, for example, with that induced by mast cell mediators (described below). This topic needs further study, since it is possible that immunity *in vivo* depends not only on the presence of specific anti-helminth antibodies and of immediate hypersensitivity reactions, but also on the numbers and functional properties of the circulating eosinophils.

(iii) Enhancement of eosinophil functional activities by mast cell mediators, and the involvement of anaphylactic antibodies in eosinophil-mediated killing. Since it is not possible to recover eosinophils directly from the site of IgE-dependent immediate hypersensitivity reactions, several groups have examined the effects of mast cell mediators on normal eosinophil function. Anwar and Kay (1978) and Kay (1982), using human eosinophils, have demonstrated that the expression of C3 receptors can be enhanced by a variety of mast cell mediators, including histamine and the chemotactic tetrapeptides of ECF-A, whereas Fc receptors were unaffected. The enhancement of C3 receptor expression was associated with an enhancement of complement-dependent, but not of antibody-dependent, killing of schistosomula (Anwar et al., 1980).

In contrast, Capron and his colleagues, using rat eosinophils, have reported that antibody-dependent killing can also be enhanced, and have demonstrated an interesting series of interactions between eosinophils and anaphylactic antibodies (Capron et al., 1978a,b, 1979, 1981a,b,c). Initially, they showed that normal rat eosinophils could damage schistosomula in the presence of mast cells and $IgG_{2a}$ antibodies. Damage was markedly reduced after mast cell depletion, but the requirement for mast cells could be replaced either by supernatants from mast cells that had been induced to degranulate by drugs or by anaphylactic reactions, or by purified mast cell mediators. The enhancement of eosinophil-mediated damage was particularly marked with the chemotactic tetrapeptides of ECF-A, and the effect of synthetic analogues on eosinophil-mediated killing was related to their chemotactic activity. Enhanced killing, in this case, was associated with an enhanced expression of Fc receptors, and a similar enhancement was seen with human eosinophils (Capron et al., 1981c). The reasons for this difference in results from those of Anwar and Kay (1978) are not clear, but could be related to the extent of Fc receptor expression on the starting eosinophil population. The differences, however, are less important than the main finding common to both groups, namely that mast cell mediators can enhance eosinophil functional properties: and other studies have shown that mast cells bind to schistosomula under appropriate conditions (Sher, 1976; Caulfield et al., 1981).

Capron and his colleagues (1979) have also shown that eosinophils from infected rats bear cytophilic antibodies on their surface, and can directly mediate killing of schistosomula in the absence of added antibody. Subsequently, they showed that both rat and human eosinophils have Fc receptors for IgE, and that IgE can mediate eosinophil killing of schistosomula (Capron et al., 1981a,b,c). This effect was particularly marked with sera from chronically infected rats: absorption of IgE from the serum, and preincubation of eosinophils with aggregated IgE, abrogated the effect.

These important observations by both Capron and colleagues and Kay and colleagues indicate a complex involvement between eosinophils and anaphylactic antibodies. Both types of rat anaphylactic antibody, IgE and $IgG_{2a}$, can mediate eosinophil killing. In addition, both types can sensitize mast cells, and the subsequent release of mast cell mediators leads to an enhancement of eosinophil function, associated with an increased expression of eosinophil cell surface receptors. Finally, as described below (p. 173), IgE is also involved in macrophage-mediated killing of several helminth targets.

(iv) *Enhancement of eosinophil functional activities by parasite products.* Eosinophil functional activities may be modified not only by host responses but also by parasite products. Eosinophil chemotactic activities have been detected in extracts of various stages of *Ascaris suum* and *Echinococcus granulosus* (Archer *et al.*, 1977), *Anisakis* (Tanaka and Torisu, 1978; Tanaka *et al.*, 1979; Torisu *et al.*, 1983), *Nippostrongylus brasiliensis* (Czarnetski, 1978) and *Schistosoma japonicum* eggs (Owhashi and Ishii, 1982), and these may help to localize eosinophils at the site of parasite invasion. In addition, Auriault *et al.* (1982) have demonstrated a low molecular-weight product, released from schistosomula of *Schistosoma mansoni*, that enhances eosinophil killing activity, an effect associated with an increased expression of eosinophil Fc receptors.

(*e*) *Species differences between eosinophils.* The findings described in the preceding sections reflect general properties of the eosinophil, common to more than one species. However, eosinophils from different species differ from each other morphologically, biochemically and functionally, and it should not be assumed that an eosinophil-mediated effect demonstrable in one species will necessarily occur in another. Human eosinophils, for example, are characterized by a great number of large granules, bearing a well-defined crystalloid core, and with a high content of basic proteins (Zucker-Franklin, 1980; Gleich *et al.*, 1976). Bovine eosinophils, in contrast, have no crystalloid cores in their granules, and relatively little MBP in relation to peroxidase (Duffus *et al.*, 1980). With bovine cells, therefore, it may be found that killing mechanisms attributable to the release of basic proteins are less important than those dependent on oxidative metabolism. Rat and human eosinophils are superficially similar, but appear to differ in their mode of degranulation: human eosinophil granules fuse directly with the plasma membrane and discharge into the exterior (Glauert *et al.*, 1978), whereas rat eosinophils discharge initially into intracellular vacuoles (McLaren *et al.*, 1977). Mouse eosinophils are noteworthy for the paucity of their granules and their relative lack of functional activity in most *in vitro* assays. This list of species differences could be extended: the point that is emphasized is that care must be taken in the extrapolation from one host–parasite combination to a second.

4. *Neutrophil-mediated damage to helminths*

(*a*) *Comparison with eosinophils.* Although much recent work has concentrated on the marked and apparently preferential damage to some helminths that is induced by eosinophils, this does not mean that neutrophils are ineffective: and in many systems, the effect of neutrophils may be equal to or greater than that of eosinophils. The relative effect of the two cell types may depend on the nature of the target helminth, the species from which the effector cells and antibodies are derived, and the conditions of the killing assay and the state of the target organism. It is sometimes difficult to determine whether recorded differences between different groups in the relative activity of the two cell types reflects a genuine biological difference or a trivial artefact of the assay.

For example, several groups have recorded an antibody- or complement-dependent killing of schistosomula of *Schistosoma mansoni* by rat or human neutrophils (Dean *et al.*, 1974, 1975; Anwar *et al.*, 1979; Moser and Sher, 1981; Kazura *et al.*, 1981). In contrast, Incani and McLaren (1981) describe a complement-dependent but not an antibody-dependent effect of rat neutrophils, while Vadas *et al.* (1979a) and Butterworth *et al.* (1982) have failed to detect antibody- or complement-dependent damage by human neutrophils, except under conditions in which marked complement-induced changes to the parasite were observed. This failure of neutrophils to kill schistosomula was not attributable to an inadequacy of the neutrophil preparation, the method of preparation of the schistosomula, or the use of round-bottomed instead of flat-bottomed culture vessels (Butterworth *et al.*, 1982; Veith *et al.*, 1983). Two further possible explanations have not yet been excluded. First, if schistosomula are undetectably damaged during preparation, they may fail to form antibody- or complement-dependent fusions with neutrophils (Caulfield *et al.*, 1980a,b), and so become susceptible to neutrophil-mediated attack. Secondly, the use of high-titre antibodies may permit an extensive smothering of the parasite by the neutrophils thus allowing them rapidly to overcome the parasite's mechanisms for evading such attack. This may be particularly true under the artificial circumstances in which haptens are attached to the parasite surface, and an anti-hapten antibody is used (Moser and Sher, 1981).

As a tentative hypothesis, it may be suggested that the interaction between schistosomula and effector cells is a kinetic situation, whose balance may be tipped in favour either of the cells or of the parasite. In the case of eosinophils, there is a progressive and cumulative interaction of cells which eventually leads to parasite killing. In the case of neutrophils, killing may occur, but only under conditions in which either the parasite is not functionally intact, or the cells are able to deliver a massive toxic signal very early during the interaction. Otherwise, the parasite escapes from neutrophil attack, by the mechanisms described below, and is subsequently refractory.

With other helminth parasites, and especially the sheathed microfilariae of several filarial nematode species, there is no disagreement about the greater effect of neutrophils, in comparison with eosinophils (Mehta *et al.*, 1982): and it is possible that eosinophil degranulation on microfilarial sheaths simply fails to affect the underlying parasite. Even so, it should be borne in mind in many studies that there is little information about the functional state of the cells tested: although normal neutrophils may be more active than normal eosinophils, the situation may not be the same when eosinophil activation has occurred, either as a result of eosinophilia or in response to mast cell or other mediators. In this context, Bass and Szejda (1979a,b) found no differences in the killing of newborn larvae of *Trichinella spiralis* between normal human neutrophils and activated eosinophils from eosinophilic individuals, when tested in the presence of hyperimmune rabbit serum. However, a different pattern emerged when human cells were tested in the presence of baboon sera, taken at different times after *Trichinella* infection (Grover *et al.*, 1983). When sera from chronically infected and rechallenged animals were tested, both eosinophils from eosinophilic individuals and normal neutrophils showed greater killing activity than normal eosinophils. In contrast, in the presence of sera from recently infected animals, only eosinophils from eosinophilic individuals showed a marked effect, normal eosinophils and normal neutrophils showing little activity. These findings emphasize the importance, when comparing eosinophils with neutrophils, of both the functional state of the cells and the source of antibody.

(*b*) *Ligands mediating neutrophil-mediated killing of helminths.* Neutrophils have surface receptors both for IgG (Swisher, 1956; Henson, 1969; Wong and Wilson, 1975) and for C3. The C3 receptors of neutrophils bind C3b only (Ehlenberger and Nussenzweig, 1977), and are identical to the C3b (CR$_1$) receptors on human erythrocytes: an antiserum against the purified 205 000 dalton receptor glycoprotein on human erythrocytes blocks the formation of neutrophil C3 rosettes by binding to an externally orientated membrane protein, also of 205 000 daltons (Fearon, 1980; Dobson *et al.*, 1981). Experiments on rosette formation also suggest that neutrophils have either larger numbers or a higher affinity of Fc and C3 receptors than eosinophils (Ottesen *et al.*, 1977; Vadas *et al.*, 1980a). The relative role of the Fc and C3 receptor in promoting adherence and phagocytosis has been extensively studied, and it has been concluded that the binding of C3 receptors in general promotes adherence of the cell to the particle, whereas the binding of Fc receptors, less effective at promoting adherence, serves to stimulate both endocytosis of the particle and the respiratory burst (Mantovani, 1975; Ehlenberger and Nussenzweig, 1977; Newman and Johnston, 1979). As would be expected, both antibody and complement can serve as ligands for neutrophil-mediated attachment to helminths (Dean *et al.*, 1974, 1975; Bass and Szejda, 1979a,b; Weiss and Tanner, 1979; Anwar *et al.*, 1979, 1980; McKean *et al.*, 1981; Incani and McLaren, 1981), although rat neutrophils fail to mediate damage in the presence of antibody alone (Dean *et al.*, 1974; Incani and McLaren, 1981). In addition, one report has indicated that IgE can act as a ligand for neutrophil-mediated damage (Mehta *et al.*, 1982), although there have been no direct demonstrations of neutrophil IgE receptors.

(*c*) *Mechanisms of neutrophil-mediated damage.* In contrast to the eosinophil, there is little evidence that neutrophils degranulate extensively upon contact with antibody- or complement-coated helminths. For example, Caulfield *et al.* (1980a,b) have studied in detail the interaction of human neutrophils with *S. mansoni* schistosomula by transmission and freeze-fracture electron microscopy. In confirmation of previous findings (McLaren *et al.*, 1977, 1978a; Glauert *et al.*, 1978), they found that eosinophils degranulate extensively on the surface of antibody-coated schistosomula. Neutrophils, instead of degranulating, show areas of tight apposition to the worm's surface. In the presence of antibody, these areas of close attachment involve small pseudopod-like processes (Glauert *et al.*, 1978), whereas when complement is present much wider areas of attachment are observed. At high resolution, fusions can be observed between the neutrophil membrane and the outer of the two lipid bilayers (Caulfield *et al.*, 1980a,b). These fusions have been studied by freeze-fracture experiments. The normal schistosomulum shows in both the E and P faces an inner membrane rich in intramembranous particles and an outer membrane poor in such particles: the fracture plane usually passes through the outer membrane, with an occasional step into the inner membrane. Areas of interaction with neutrophils are bounded by a characteristic edge, consisting of a regular linear step into the inner membrane. Within this edge, two patterns are observed. 'Particle-rich' areas appear elevated in the P face view, and fractures through such areas reveal both the outer and the inner schistosomulum membrane underneath. These areas represent fractures through unfused neutrophil membrane. In contrast, 'particle-poor' areas show a density of intramembranous particles intermediate between that of the neutrophil membrane and the schistosomulum outer membrane. Fractures through such membranes reveal only the schistosomulum inner membrane underneath. Particle-poor areas are continuous with the normal neutrophil membrane, but show the same spatial relationship to the schistosomulum inner membrane as does the normal schistosomulum outer membrane. They are therefore interpreted as representing areas of fusion between the neutrophil membrane and the schistosomulum outer membrane. When the neutrophils become detached from the schistosomula either spontaneously or after osmotic or physical stress, the edge zone disintegrates: there is now no discontinuity between the normal schistosomulum outer membrane and the previously fused area, and the intramembranous particles become evenly redistributed. These findings are interpreted as showing that, following detachment of the neutrophil, a part of its membrane remains behind, becoming permanently incorporated into the schistosomulum outer membrane. This may represent a mechanism whereby host-derived molecules are incorporated firmly into the schistosomulum (p. 184): and the rapid fusion of the neutrophil membrane may serve to prevent the neutrophil from degranulating. Similar fusions may be observed with eosinophils, although more rarely, and with erythrocytes: and they may be elicited by binding the neutrophils to the schistosomulum with concanavalin A instead of antibody (Caulfield et al., 1982).

These findings indicate that, although neutrophils contain within their granules highly toxic cationic proteins (Weiss et al., 1978) which, when isolated, are capable of damaging schistosomula of *Schistosoma mansoni* (McLaren et al., 1981), these proteins may not be released in large amounts when neutrophils adhere to target helminths, and therefore may not account for any observed damage. Instead, most studies have provided evidence for a major role of oxidative killing mechanisms. Neutrophils mount a strong and rapid respiratory burst in response to appropriate stimuli, leading to the production of superoxide anions, hydrogen peroxide, and other active oxygen species including singlet oxygen and hydroxyl radicals (reviewed by Babior, 1978). Some or all of these oxidative metabolites may be released extracellularly, and may therefore be available to damage non-phagocytosable extracellular targets such as helminths. A possible role for such reactions was first demonstrated by Dean et al. (1974), who showed that nitroblue tetrazolium was reduced at the site of neutrophils binding to schistosomula in the presence of antibody and complement. This observation was subsequently extended to several nematode parasites (Mackenzie et al., 1980, 1981), and would now be accepted as evidence for the local generation of superoxide. More recent studies have depended on the demonstration of a respiratory burst during neutrophil-mediated killing, the inhibition of killing by inhibitors of oxidative metabolism, and the lack of killing observed in cells from patients with genetic defects in oxidative pathways.

For example, Kazura et al. (1981) have attributed not only the eosinophil-mediated but also the neutrophil-mediated killing of schistosomula to oxidative mechanisms. As with eosinophils, they find that neutrophils, interacting with schistosomula in the presence of antibody, with or without complement, show a marked increase in production of hydrogen peroxide. Killing is inhibited by catalase at high concentrations, and also by the inhibitors of myeloperoxidase, azide, cyanide and aminotriazole. Furthermore, killing by neutrophils from patients with chronic granulomatous disease is less than that observed with neutrophils from normal individuals.

Oxygen-dependent mechanisms are also important in the killing of newborn larvae of *Trichinella spiralis* by mixed normal human leukocytes (Bass and Szejda, 1979a,b), in which the killing effect is largely attributable to neutrophils (Grover et al., 1983). Such killing can be inhibited by catalase, but is unchanged by superoxide dismutase and is enhanced by azide or cyanide, suggesting a direct effect of hydrogen peroxide rather than a peroxidase-dependent effect. Similar killing could be obtained in cell-free systems in which glucose/glucose oxidase and xanthine/xanthine oxidase were used to generate hydrogen peroxide and superoxide respectively: these effects were also inhibited by catalase, but unchanged by superoxide dismutase (Bass and Szejda, 1979b). The finding that leukocytes from patients with chronic granulomatous disease show a partial but incomplete reduction in their capacity to kill newborn larvae suggests that non-oxidative mechanisms may also be involved: but these may have been attributable to eosinophil-mediated effects in the mixed leukocyte preparations.

(d) *Enhancement of neutrophil functional activity.* In the same way that the capacity of eosinophils to kill helminths can be enhanced by exposure to a variety of mediators, so it is found that exogenous stimulation of neutrophils increases their killing capacity. Exposure of neutrophils to synthetic formyl methionyl peptides increases the expression of C3 receptors (Anwar and Kay, 1978; Kay, 1982). This is associated with an increased complement-dependent neutrophil-mediated killing of schistosomula (Anwar *et al.*, 1979) in a fashion analogous to the enhancement of eosinophil-mediated killing by the chemo-tactic tetrapeptides of ECF-A, described above. In addition, Hopper *et al.* (1981) have shown that lymphokine-containing culture supernatants prepared by stimulation of rat lymph node cells with concanavalin A or phytohaemagglutinin enhance the antibody- and complement-dependent killing of *Litomosoides carinii* microfilariae by normal rat neutrophils. Stimulation was associated with an increased capacity to phagocytose latex particles and opsonized erythrocytes, but with no increase in candidacidal activity. The activity differed from ESP (Greene and Colley, 1974), and from a monocyte-derived eosinophil stimulating activity (Veith and Butterworth, 1983), in being sensitive to heating at 56°C.

(e) *Species differences.* Neutrophils from different species differ less in both morphology and functional activity than eosinophils, but it may be assumed that species differences may account at least partially for some of the differences between different laboratories on the killing efficiency of neutrophils.

## 5. Conclusions

The emphasis in this section has been on the observed association between helminth infections, increased levels of eosinophils in the peripheral blood and IgE responses, and it is suggested that the eosinophil/IgE/mast cell axis may represent a specialized immune effector system that has evolved selectively to protect the host against large tissue-stage parasites. Helminth infections elicit an increase in both the numbers and the functional activity of circulating eosinophils: such cells then enter the site of invasion of tissue helminths, as a result of IgE-dependent mast cell reactions: and, having entered, they may then show a selective capacity to damage such helminths, by releasing their granule contents onto the surface of the organism. The proposed existence of this specialized system, however, does not preclude the involvement of more conventional granulocyte-mediated mechanisms, active against a wide range of target organisms: and, in some models, neutrophil-mediated killing through oxidative pathways has been shown to be markedly effective. The relative roles of eosinophil- and neutrophil-mediated mechanisms has to be tested for each host and parasite combination, and there is no good justification for extrapolation from one model to another.

### C. MACROPHAGE-MEDIATED DAMAGE

#### 1. *Introduction*

Although neutrophils and eosinophils differ markedly in ontogeny, morphology and biochemical and functional properties, they do show certain similarities: both are end cells, incapable of further division, that rapidly enter lesions to exert their functional effect, and that die as a result. In contrast, the macrophage, although ontogenetically related to the neutrophil, differs fundamentally in its functional properties. It is not an end cell: it is capable of further division within the tissues: it undergoes reversible changes in its functional state: and having achieved its phagocytic or cytotoxic effect, it does not necessarily die. Conventionally, the macrophage is regarded as a second line of defense against invading pathogens. Granulocytes rapidly enter the site of an invading pathogen, and exert a powerful but short-lived effect. Macrophages enter more slowly, and are relatively inactive in their unstimulated condition; but, after interaction with various mediators, they slowly acquire more potent functional properties, which can be expressed for long periods of time, thereby helping to eliminate pathogens that are resistant to the short-lived effects of granulocytes.

Macrophages bear cell-surface receptors for both Fc and C3, which can serve as ligands to mediate their interaction with a variety of targets. In the case of helminths, however—unlike the situation with neutrophils and eosinophils—there is not always an absolute requirement for such a ligand. Instead, a more important event is the change in the functional properties of resting macrophages that is induced by various mediators. This event, referred to as macrophage activation, and originally described as an essential component of the resistance of mice to challenge infection with *Listeria monocytogenes* (Mackaness, 1969), has now been demonstrated to be important in the destruction of a wide variety of intracellular parasites that resist killing both by granulocytes and by unstimulated macrophages, within which such parasites may indeed grow. Macrophage activation has been implicated as a major effector mechanism against some viruses, against many slow-growing intracellular bacteria, such as *Mycobacterium tuberculosis* (Patterson and Youmans, 1970) and against some intracellular protozoa, including various *Leishmania* species (Mauel *et al*., 1978), *Toxoplasma gondii* (Anderson *et al*., 1976) and *Trypanosoma cruzi* (Nogueira and Cohn, 1978). In addition to this intracellular killing, however, activated macrophages are also effective against a variety of extracellular targets, especially tumour cells, a process that depends on the generation and release of active oxygen metabolites (Nathan et al., 1979a,b).

It is perhaps not surprising, therefore, that such activated macrophages are also effective in killing a variety of non-phagocytosable helminths in an antibody-independent fashion. An unexpected finding, however, has been that macrophage activation, or a process strongly resembling activation, can be induced not only by T cell mediators but also by polymeric IgE, thus providing a further explanation for the selective advantage of the high IgE responses that are observed in helminth infections. In this particular case, the subsequent interaction of the macrophage with the helminth target *is* antibody-dependent, and more specifically IgE-dependent.

2. *Antibody-dependent macrophage-mediated damage*

(*a*) *Demonstration of IgE-dependent damage.* Early studies by Capron and his colleagues (1975a,b, 1976, 1977a,b, 1980, 1982; Joseph et al., 1977, 1978) showed that adherent peritoneal exudate cells from normal rats, or monocytes from human or baboon peripheral blood, in the presence of unheated immune serum, would adhere to schistosomula of *Schistosoma mansoni* and cause both release of $^{51}$Cr and microscopically detectable damage. The activity in immune serum was ablated by heating at 56°C, but was not restored by addition of fresh normal serum. This strongly suggested that complement was not involved in the reaction, and instead that the heat-labile moiety might be IgE. In confirmation of this, Capron *et al.* (1977a) showed that the activity in immune serum could be depleted by passage over an anti-IgE immunoadsorbent column, but not by absorption of $IgG_1$, $IgG_{2a}$, $IgG_{2b}$, IgM or IgA. The promoting activity in immune serum was also removed by passage over an anti-*S. mansoni* immunoadsorbent, suggesting the involvement of immune complexes: and, in confirmation of this, it was found that the immune complex-containing pellets formed after ultracentrifugation of immune serum were more active in inducing macrophage cytotoxicity than the complex-free supernatants.

In later studies, it was found that this IgE-dependent macrophage-mediated effect was not confined to an action on schistosomula: similar effects have now been demonstrated for *Dipetalonema viteae* (Haque et al., 1980) and *Litomosoides carinii* (Mehta et al., 1980). These findings suggest yet a further role for IgE in immunity against helminths, in addition to its involvement in mast cell degranulation and in eosinophil-mediated killing, as described in previous sections.

(*b*) *Macrophage receptors for IgE.* The demonstration of an IgE-dependent killing by macrophages implied the presence on the macrophage surface of receptors for IgE. Although IgG receptors had been thoroughly studied, there was no evidence for similar receptors for IgE. Subsequent experiments, however, confirmed the existence of such receptors. The affinity of binding of monomeric IgE is much lower than that of the mast cell receptor, especially at 4°C, and monomeric IgE is rapidly eluted from the cell surface by washing (Dessaint et al., 1979a). Aggregated IgE, however, and immune complexes containing IgE, bind with much higher affinity. The presence of IgE on the surface of macrophages recovered from immune animals could be demonstrated directly, by the use of peroxidase-labelled anti-IgE (Dessaint et al., 1979a). In addition, it was found that normal macrophages form rosettes with erythrocytes coated with IgE and that, when such macrophages are incubated for 24 hours in the presence of IgE, they form rosettes with erythrocytes coated with anti-IgE but not with IgE, implying that their IgE receptors are saturated during this culture period. If such cells are now washed and incubated for a further period in serum-free medium, they lose their capacity to react with anti-IgE coated cells but regain their capacity to react with IgE-coated cells, implying the progressive loss of IgE from the receptor.

The implication of these findings is that immune complexes containing IgE bind first to the surface of the macrophage: thereafter, if the complexes are in slight antibody excess, the free antibody-binding sites permit a further interaction with antigens on the surface of the target organism. Since these studies, other workers have confirmed the presence of IgE receptors on various cells of the monocyte–macrophage lineage, including human peripheral blood monocytes (Melewicz and Spiegelberg, 1980), a human monocyte cell line (Anderson and Spiegelberg, 1981), human alveolar macrophages (Joseph et al., 1981) and rat alveolar and peritoneal macrophages (Boltz-Nitulescu and Spiegelberg, 1981), and the characteristics of such receptors have begun to be defined.

*(c) Stimulation of macrophages after binding of IgE–antigen complexes.* Damage to schistosomula can be observed when macrophages, serum containing IgE, and schistosomula are mixed simultaneously and allowed to interact. However, the damage is considerably amplified if the macrophages are preincubated for 3 to 6 hours with the serum, before addition of the target (Joseph et al., 1977, 1978). This implies a slow enhancement of the functional properties of these macrophages, as a result of IgE binding. Such changes can be induced by immune serum containing IgE, but not by normal serum or by heated or IgE-depleted immune serum (Capron et al., 1977a): they can also be induced by incubation of normal macrophages with rat myeloma IgE protein and anti-IgE, which may be either intact or as its F(ab')$_2$ fragment (Dessaint et al., 1979b). The changes are somewhat similar

to those observed during 'classical' T lymphocyte-dependent activation, and include an increased synthesis and release of lysosomal enzymes such as β-glucuronidase (Capron et al., 1977a), increased secretion of plasminogen activator, increased glucosamine incorporation, the generation of the superoxide anion (Joseph et al., 1980) and an increase in intracellular levels of macrophage cyclic GMP (Dessaint et al., 1979b, 1980). The release of β-glucuronidase and the increase in cyclic GMP is triggered by dimeric or polymeric, but not by monomeric, IgE, implying a requirement for cross-linking of the surface receptors.

This initial stimulation of macrophages is induced by any form of polymeric IgE, independent of its specificity. The subsequent interaction of such stimulated macrophages with the target, however, depends on the continued presence of specific IgE, acting as a ligand. For some reason that is not clear, IgG fails to replace IgE in this second stage. Since macrophages bear receptors for IgG as well as IgE, it might be expected that IgG could act as a ligand during this stage: but it is possible that killing by such macrophages depends on a continuation of the IgE-dependent stimulation process throughout the killing event.

(d) *Mechanisms of IgE-dependent macrophage-mediated killing.* Damage to schistosomula is associated with an initial flattening of the macrophage onto the parasite surface (Capron et al., 1977a; Joseph et al., 1977). At this stage, no lesions to the membrane are seen, but the parasite becomes immotile. After 18 hours incubation, breaches are observed in the tegumental membrane, with macrophage pseudopods penetrating the membrane and phagocytosing internal components (Capron et al., 1977a): but this probably represents a late stage in the killing event. The mechanisms of the earlier stages of damage are not understood: by analogy with the neutrophil killing mechanisms, it might be postulated that an increased generation of hydrogen peroxide by IgE-stimulated macrophages is involved in the killing process, and Joseph (1982) quotes unpublished observations that macrophage-mediated killing can be inhibited by both catalase and cytochrome c. The effect of such macrophages is also inhibited by peptides released as a result of proteolytic cleavage of IgG on the surface of schistosomula (Auriault et al., 1980, 1981).

(e) *IgG-dependent macrophage-mediated killing.* The work of Capron and his colleagues showed that IgE could both stimulate macrophage function and act as a ligand for macrophage-mediated killing, whereas IgG failed to do so. Other workers have reported that IgG can also mediate macrophage-dependent adherence or killing, but these studies have been less extensive. For example, Perez and Smithers (1977) have found that rat macrophages adhere to schistosomula in the presence of IgG from immune serum, while Perrudet-Badoux et al. (1978, 1981) have described an IgG-dependent, complement-independent macrophage-mediated killing of newborn larvae of *Trichinella spiralis*. The extent and effectiveness of these killing reactions remains to be demonstrated.

3. *Antibody-independent macrophage-mediated damage*

(a) *Demonstration of the effect*. The studies described above on IgE-dependent macrophage-mediated damage had indicated that macrophages could damage helminths, provided that their functional properties had been altered in an appropriate fashion. Subsequently, other workers investigated the effects of macrophages activated by conventional T lymphocyte-dependent mechanisms, and acting in the absence of ligands. Studies *in vivo* had indicated that procedures that induced non-specific immunity with high levels of macrophage activation, such as infection with *Toxoplasma gondii* (Mahmoud *et al.*, 1976), immunization with BCG or with natural cord factor or its analogues (Civil *et al.*, 1978; Olds *et al.*, 1980b), or treatment with amphotericin B (Olds *et al.*, 1981b), led to the development of non-specific resistance to challenge with *Schistosoma mansoni*. Similarly, infection with *Toxoplasma gondii* or *Listeria monocytogenes*, or immunization with BCG, elicited a non-specific resistance to *Trichinella spiralis* infection (Wing and Remington, 1978; Grove and Civil, 1978; Copeland and Grove, 1979). Thereafter, various workers investigated the capacity of macrophages recovered from such animals, or of normal macrophages activated *in vitro* by exposure to T-cell mediators to kill various helminths. For example, activated macrophages recovered from mice immunized with BCG or *Corynebacterium parvum* were cytotoxic for schistosomula of *Schistosoma mansoni* and for newborn larvae of *Trichinella spiralis*, whereas macrophages 'stimulated' with thioglycollate or proteose peptone were not (Mahmoud *et al.*, 1979b; Mahmoud, 1980). Similarly, exposure of normal mouse macrophages to supernatants of lymphocyte cultures that had been stimulated with concanavalin A led to the development of the capacity of such macrophages to damage schistosomula of *Schistosoma mansoni* (Bout *et al.*, 1981).

In such cases, macrophage activation is elicited by exposure of T cells to antigen or mitogen. These cells then release mediators, including macrophage migration inhibitory factor (MIF) and macrophage activating factor (MAF), which act non-specifically on macrophages, enhancing their functional properties. *In vivo*, therefore, the antigen that elicits the initial T cell response need not be the same as that which is expressed on the surface of the target parasite: in other words, the expression of the effect of the activated macrophage is non-specific. In normal circumstances, however, macrophage activation will occur during the course of an infection, and the effect is then expressed on the organism that elicited the response. This process, which accounts for the development of immunity against many intracellular microorganisms, may also occur in some helminth infections. For example, James *et al.* (1982a,b) demonstrated that peritoneal exudate macrophages from *Schistosoma mansoni*-infected mice, in contrast to those from normal animals, were able to kill schistosomula in suspension cultures. This effect was independent of ligands, although it could be enhanced by the addition of immune mouse serum or concanavalin A to promote cell adherence. Older schistosomula, in contrast, are not susceptible to killing by activated macrophages, even when adherence is promoted by addition of anti-mouse erythrocyte sera (Sher *et al.*, 1982). The extent of macrophage activation that occurs in schistosome-infected mice is sufficient that cells recovered from egg granulomas are able to kill tumour cells *in vitro*, in a conventional assay for macrophage activation (Loveless *et al.*, 1982).

The studies described above relate to macrophages classically activated as a result of exposure to T cell mediators. In other situations, macrophages may exert a slight, non-specific, antibody-independent effect on helminths in the absence of known T cell responses. Ellner and Mahmoud (1979) have shown that monolayers of adherent cells prepared from the peripheral blood of normal human subjects caused microscopically detectable damage to schistosomula *in vitro*, associated with a lowered viability *in vivo*. The effect was not marked and required a high effector to target ratio for its expression; it could be enhanced by the addition of fresh immune serum. Subsequently, Olds *et al.* (1981a) showed that cultured monocytes from the blood of individuals heavily infected with *Schistosoma mansoni* showed a lower capacity to mediate damage in both control and immune sera. In contrast, monocytes from patients with tuberculosis showed an enhanced killing capacity, suggesting an element of macrophage activation in these patients. Recently, Peck *et al.* (1983) have demonstrated a relationship between the innate resistance of different species and strains of host to schistosome infection, and the capacity of their unstimulated blood monocytes or peritoneal macrophages to kill schistosomula *in vitro*, and have suggested that the mononuclear phagocyte may be important in determining initial susceptibility or resistance.

(*b*) *Mechanisms of damage*. In bacterial and protozoal killing systems, the effects of macrophage activation have usually been ascribed to an alteration in oxidative metabolism, with an increase in respiratory burst activity, which may be of prolonged duration. This effect has been shown most clearly with *Toxoplasma gondii* (Murray and Cohn, 1979). Activated macrophages have also been found to kill extracellular targets, especially tumour cells, in an oxygen-dependent fashion, and this has been ascribed to the extracellular release of hydrogen peroxide (Nathan *et al.*, 1979a,b). Evidence from helminth systems is less extensive. Olds *et al.* (1980a) have investigated the killing of schistosomula by *Corynebacterium parvum*-activated macrophages from two mouse strains. Activated macrophages from C57Bl/6 mice showed marked killing of schistosomula, in contrast to cells from BALB/c mice. This difference between the two strains was unrelated to the capacity of their macrophages to produce hydrogen peroxide, and was instead related to the release of arginase into the culture supernatant. Killing by activated macrophages was exaggerated in arginine-poor media, could be blocked by addition of arginine, and could be mimicked in cell-free systems by the addition of exogenous arginase. These findings suggest that oxidative mechanisms are not responsible for killing by activated macrophages, which instead appear to act by depleting the medium of free arginine. The relevance of such a phenomenon to possible killing by activated macrophages *in vivo* is unclear, although it is possible that a similar local depletion of arginine might occur in a microenvironment in which invading helminths are surrounded by activated macrophages. More recently, Peck *et al.* (1983) have reported that both arginase-dependent and oxygen-dependent mechanisms may be involved in killing by normal macrophages, the mechanisms varying in importance between different species.

### D. LYMPHOCYTE-MEDIATED DAMAGE

1. *Introduction*

Lymphocytes with cytotoxic effector properties, including cytolytic T lymphocytes (CTL), killer (K) cells and natural killer (NK) cells, differ from granulocytes and macrophages in that they are active only against extracellular targets: they are not phagocytic, nor are they capable of damaging bacteria. Although some types of effector lymphocyte are slightly active in killing some extra-cellular protozoa, including *Trypanosoma cruzi* epimastigotes (Abrahamson and Dias da Silva, 1977), their main effect is against mammalian cells, which may be allogeneic, virus-infected or neoplastic. In non-allogeneic reactions, the recognition and killing of target cells by CTL is restricted by a requirement for the expression by the target not only of virus-coded or of tumour-specific antigens, but also of products of the host's own major histocompatibility complex (MHC), especially the class I molecules encoded by the K and D loci in the mouse and the A, B and C loci in man (reviewed by Zinkernagel and Rosenthal, 1981). Since there is no reason why MHC products should generally be expressed on the surface of a helminth parasite, it is not surprising that CTL have usually been found to be inactive. NK cells are not restricted in this way, nor does their capacity to interact with a target depend upon the recognition of conventional antigens, although there is an element of selectivity in the targets recognized: tumour cells and transformed cells, for example, are more susceptible than normal cells to NK cell attack (reviewed by Herberman *et al.*, 1979). The effect of

K cells is strictly antibody-dependent, involving attachment through an Fc receptor to bound antibody on the surface of the target. There is no intrinsic reason why NK cells and K cells should in principle be unable to damage helminths, but there are no reported cases of such an activity.

## 2. *Cytolytic T lymphocytes*

An inability to detect an effect of CTL after recovery of lymphocytes from immunized animals has generally been attributed to a lack of expression of host MHC products on the surface of the target parasite. However, in one instance at least, host MHC products are demonstrable. Schistosomula of *Schistosoma mansoni*, during maturation, acquire on or in their tegumental membrane a variety of host-derived molecules (Smithers *et al.*, 1969; Clegg *et al,*. 1971). Originally, it was thought that only glycolipids could be taken up by such schistosomula (Goldring *et al.*, 1976, 1977a,b), but more recent work has shown that glycoproteins can also be incorporated, including both K and D region products of the mouse MHC (Sher *et al.*, 1978). MHC products are also found on adult worms (Gitter *et al.*, 1982). It is theoretically possible that such MHC products might be associated with parasite antigens in a fashion appropriate for MHC-restricted recognition by CTL. Experiments were therefore carried out to test whether CTL could, under artificial circumstances, bind to and kill schistosomula (Butterworth *et al.*, 1979c). CTL were elicited in A/J mice by immunization with $H-2^k$ or $H-2^d$ tumour cells, and were allowed to interact with schistosomula, either recovered from the lungs of $H-2^k$, $H-2^d$ or $H-2^b$ animals and bearing on their surface the relevant class I molecules, or with freshly prepared skin-stage schistosomula in the presence or absence of concanavalin A. Antigen-specific or mitogen-dependent adherence of lymphocytes was observed, and the lymphocyte population could be directly demonstrated to contain CTL: but no damage to the target schistosomula was seen, even after long periods of incubation at high effector to target ratios. In further experiments (Brown *et al.*, 1980) it was found that the adherent T cells were of the Ly-2 phenotype and that they recognized products of the K and D regions of the MHC, rather than the I region, on the surface of the schistosomula. These observations supported the hypothesis that the adherent T cells, although not active against schistosomula, were cytotoxic cells. In separate experiments, CTL were raised against minor alloantigens or against trinitrophenyl (TNP), and such CTL showed an MHC restriction in their capacity to kill tumour target cells (Vadas *et al.*, 1979c). These CTL also showed a specific adherence to the appropriate TNP-coated or control lung-stage schistosomula: and, in the case of CTLs with specificity for minor alloantigens, it was found that the minor alloantigen and the restricting MHC product could be acquired separately, by sequential passage through appropriate congenic animals. Again, however, there was no evidence of damage to the target.

In contrast, Ellner et al. (1982) have recently reported that human peripheral blood T lymphocytes can kill skin-stage schistosomula after stimulation with phytohaemagglutinin, concanavalin A or irrelevant antigens, as reflected by an uptake of methylene blue after 24 hours of coculture and by a reduced infectivity for mice. Lectin-induced cytotoxicity was associated with adherence of the cells to the parasite, and damage was reduced if the lectin was removed: antigen-induced cytotoxicity was of a much lower order of magnitude. Fractionation of peripheral blood T cells indicated that cells bearing the OKT8 marker were more active than those bearing the OKT4 marker. Unstimulated cells were inactive, but the capacity to induce damage developed rapidly, within 2 hours of addition of mitogen. Cells stimulated with neuraminidase and galactose oxidase, or with neuraminidase and peanut agglutinin, were also active. Damage was detectable after 2 hours of culture, reaching a plateau at 18 hours: ratios of 5000:1 were required to produce the maximum levels of killing, 50% of the target organisms. This interesting report is the first recorded case of lymphocyte-mediated damage to a multicellular target: and although the effect is artificial, in that the continued presence of a lectin ligand is required for the expression of a marked effect, it deserves much more study.

3. *K cells and NK cells*

There have been few experiments to test directly for an effect of K or NK cells, by the use of cell preparations which have simultaneously been shown, in conventional assays, to be enriched in either cell type. However, when procedures are used which are known to enrich for such cells—including, for example, the use of defibrinated blood fractionated on Ficoll-hypaque or metrizamide—the resulting cell preparations have consistently failed to show any antibody-dependent (K cell) or antibody-independent (NK cell) activity against schistosomula (Vadas et al., 1979a) or other helminth targets (Mackenzie et al., 1980, 1981; Greene et al., 1981). Although Attallah et al. (1980) demonstrated an increase in both K and NK cell activity early after infection of mice with *Schistosoma mansoni*, there was no evidence to suggest that these cells could act on target schistosomula.

4. *Conclusions*

The relative inability of effector lymphocytes of any type to kill helminths *in vitro* is perhaps not surprising, in view of the nature of their killing mechanisms. No degranulation of such cells has been observed during killing reactions, nor has damage convincingly been attributed to the release of toxic mediators. Instead, in the case of CTL and K cells, damage to the target is associated with deep invaginations of the target cell membrane that are induced by large, microfilament-containing processes of the killer cell (Sanderson and Glauert, 1977, 1979; Glauert and Sanderson, 1979; Sanderson, 1982): these alterations may induce the target cell to undergo zeiosis. The surface structure of most helminths is such that they might be relatively unaffected by such processes, since they present either a rigid cuticle or a syncytial tegumental layer whose cell bodies are buried deep below other structures. An alternative explanation that has been put forward for the damage mediated by cytotoxic lymphocytes is that the binding of lymphocytes to the surface of the target leads to a distortion of integral membrane proteins within the target cell membrane, creating an instability on the membrane (Berke and Clark, 1982). This hypothetical distortion may either not occur in helminth outer membranes or, if it does, may not lead to instability.

### E. CONCLUSIONS

The general principles to emerge from this summary of effector mechanisms is that helminths differ fundamentally in their susceptibility to attack by killing mechanisms that have previously been tested in the context of mammalian cells, bacteria or protozoa. First, they present non-phagocytosable targets to the host's defenses, and can therefore be killed only by cells that are capable of mediating extracellular events. Secondly, and within this restriction, they are more susceptible to some extracellular killing mechanisms than to others. Although it is difficult to make broad generalizations, it would seem that they are highly susceptible to attack by toxic mediators, especially lysosomal cationic proteins, that are released from effector cells during extracellular degranulation. They are also susceptible, but possibly to a lesser extent, to conventional oxidative killing mechanisms, especially those dependent on production of hydrogen peroxide.

These restrictions on susceptibility have resulted in the identification of a series of unusual killing events, not previously demonstrated in other target systems. Two of these appear to be particularly strong killing mechanisms, both involving the IgE responses that are such a characteristic feature of helminth infections. First, eosinophils can cause damage in the presence of IgE, IgG, complement, or IgG and complement together. The main feature of this reaction is a marked degranulation of the cell, with the release of granule contents. Although normal eosinophils are frequently effective, their activity is markedly enhanced in conditions of eosinophilia and after stimulation with mast-cell mediators released as a result of IgE-dependent degranulation. Secondly, macrophages can cause damage both after conventional T lymphocyte-dependent activation and after stimulation with immune complexes containing IgE. Although extremely potent, the mechanism of macrophage-mediated killing is not yet understood. In addition, neutrophils can also kill certain helminths, acting by oxidative mechanisms rather than by degranulation. Lymphocytes, however, with one interesting exception, have not been found to be active.

## IV. Effector Mechanisms Active Against Particular Helminth Species

### A. INTRODUCTION

The effector mechanisms described in Section III are generally active against a wide variety of helminth targets. However, different groups of helminth parasites show marked differences in their surface structure, as do different stages of a single parasite species. Consequently, the relative efficiency of the various cellular effector mechanisms differs according to the species or stage of parasite under consideration. This section summarizes the findings for most of the parasitic helminths that have been studied, both *in vitro* and by histological observations *in vivo*.

### B. PLATYHELMINTHES: TREMATODA

#### 1. *Introduction*

Studies on cellular effector mechanisms active against trematodes are confined to the order Digenea, in particular *Schistosoma mansoni* and *Fasciola hepatica*. The larval and adult stages of such organisms are characterized by the possession of an outer syncytial tegument that is bounded by an exposed plasma membrane that might be predicted to be extremely sensitive to immune attack, although individual parasites have evolved sophisticated means of evading such attack. In the eggs, in contrast, the fragile larva is enclosed within a hard chitinous shell that partially protects the larva from attack. It might be expected, therefore, that different mechanisms are active against the migrating larvae and adults on the one hand, and the eggs on the other hand, and this has proved to be correct.

#### 2. Schistosoma *species*

The schistosomes have been the most widely used targets for studies on cellular effector mechanisms *in vitro*. Most studies have involved the use of *Schistosoma mansoni*, as the parasite most readily maintained in the laboratory: but there is no evidence that other species differ markedly in their susceptibility to attack, although *Schistosoma japonicum* eggs elicit a different category of granulomatous reaction (Hsü *et al.*, 1973).

(*a*) *Schistosomula*. Until recently, it was considered that the young schistosomulum, immediately after penetration of the skin, was the only stage of the developing parasite that was susceptible to specific immune attack, and this is therefore the stage that has been most widely used for studies *in vitro*. After penetration of the skin and loss of the glycocalyx the organism produces, over the course of about 3 hours, an additional lipid bilayer on its outer tegumental membrane (Hockley and McLaren, 1973), a process that is common to all blood-stream flukes (McLaren and Hockley, 1977). Similar events occur if cercariae are mechanically induced to shed their tails, and subsequently cultured *in vitro*: but in this case the process is protracted, taking about 18 hours (Ramalho-Pinto *et al.*, 1974; Brink *et al.*, 1977). At this stage, such organisms express on their surface proteins and glycoproteins, which are available for external radiolabelling (Ramasamy, 1979; Snary *et al.*, 1980; Brink *et al.*, 1980; Taylor *et al.*, 1981; Dissous and Capron, 1981; Shah and Ramasamy, 1982) and some of which are recognized by antibodies present in sera from chronically infected mice, rats or humans. Such antibodies can be shown by immunofluorescence to bind to the surface of live schistosomula: and, in more recent studies, monoclonal antibodies have been raised which also bind to schistosomula (Verwaerde *et al.*, 1979; Taylor and Butterworth, 1982; Grzych *et al.*, 1982; Dissous *et al.*, 1982; Smith *et al.*, 1982). Some of these monoclonal antibodies confer a partial degree of protection against cercarial challenge after transfer into an appropriate recipient (Smith *et al.*, 1982; Grzych *et al.*, 1982).Thus, the live schistosomulum expresses surface antigens, and antibodies against such antigens are protective: it is therefore reasonable to expect that this stage could be susceptible to immune attack, both *in vitro* and *in vivo*.

Such schistosomula have been reported by different groups to be susceptible to most of the cellular effector mechanisms described in Section III. Active cells include eosinophils, acting in the presence of antibody, complement or both together (Butterworth *et al.*, 1975, 1977a,b, 1979a, 1980, 1982; Hsü *et al.*, 1977; Mackenzie *et al.*, 1977; Glauert *et al.*, 1978; McLaren *et al.*, 1978a; Ramalho-Pinto *et al.*, 1978; Capron *et al.*, 1978a,b, 1979, 1981a,b; Anwar *et al.*, 1979, 1980; Vadas *et al.*, 1979a,b, 1980a,b; Kassis *et al.*, 1979; McKean *et al.*, 1981; Incani and McLaren, 1981), neutrophils, again acting in the presence of antibody, complement, or both together (Dean *et al.*, 1974, 1975; Anwar *et al.*, 1979, 1980; Incani and McLaren, 1981; Kazura *et al.*, 1981; Moser and Sher, 1981), macrophages stimulated by, and acting in the presence of, immune complexes containing IgE (Capron *et al.*, 1975a,b, 1976, 1977a,b; Joseph *et al.*, 1977, 1978), activated macrophages (Mahmoud *et al.*, 1979b; Mahmoud, 1980; Olds *et al.*, 1980a; Bout *et al.*, 1981; James *et al.*, 1982a,b), cultured monocytes, acting in the presence or absence of antibody (Ellner and Mahmoud, 1979; Olds *et al.*, 1981a), and lectin-stimulated T lymphocytes (Ellner et al., 1982). The relative effectiveness of such cells, however, is still controversial, and this topic is discussed in detail in Section III.

Young schistosomula both express surface antigens and are susceptible to immune attack. During maturation, either in vitro or in vivo, they rapidly lose both their expression of surface antigens and their susceptibility to attack (Hockley and McLaren, 1973; McLaren et al., 1975). Until recently, this loss of susceptibility was attributed to an uptake by the parasite of a coating of host molecules, masking the underlying parasite antigens and preventing recognition by components of the immune system (Smithers et al., 1969; Clegg et al., 1971). Uptake of a variety of host molecules can be demonstrated (Clegg et al., 1971; Golding et al., 1976, 1977a,b; Sher et al., 1978), and the rate of uptake parallels the loss by the schistosomulum of antibody-binding capacity and of susceptibility to attack (McLaren et al., 1975; McLaren, 1980b). Such uptake is associated in vivo with an increase in intramembranous particles in the outer of the two lipid bilayers (McLaren et al., 1978b; Torpier et al., 1977; Torpier and Capron, 1980; McLaren, 1980a) and may be associated with the capacity of the outer bilayer to fuse with host cell membranes (Caulfield et al., 1980a,b). More recent evidence, however, suggests that uptake of host molecules need not be a prerequisite for loss of antibody-binding capacity and susceptibility to attack, since several groups have shown that schistosomula cultured in vivo in defined media, in the absence of any source of host molecules, also lose susceptibility and antibody-binding capacity (Dean, 1977; Tavares et al., 1978, 1980; Samuelson et al., 1980; Dessein et al., 1981). An alternative hypothesis, therefore, is that since the outer membrane is continuously being formed and shed (Samuelson and Caulfield, 1982), there is a simple failure of expression of antigens in the outer membrane formed by older larvae, even though antigens are present in the tegument as a whole (Hayunga et al., 1979a,b; Snary et al., 1980; Taylor et al., 1981; Butterworth et al., 1982). Whatever the explanation for the loss of expression of surface antigens, a necessary corollary is that the older schistosomula should not be susceptible to cellular attack in the presence of antibodies with specificity for parasite antigens, and this has generally proved to be correct (Dean, 1977; Tavares et al., 1978, 1980; Novato-Silva et al., 1980; Dessein et al., 1981b). In contrast, if antibodies are used with specificity either for host-derived molecules or for haptens that have artificially been attached to the parasite surface, then the organisms may be killed, although to a lesser degree (Moser et al., 1980).

Much recent evidence suggests that description given above, of a single phase of susceptibility during the early stages of larval development, may be an oversimplification. Studies in vivo have indicated that attrition of schistosomula in immune animals occurs in two distinct stages: one early, before the migration of schistosomula through the lungs, and one later (Miller and

Smithers, 1980; Smithers and Miller, 1980; Smithers and Gammage, 1980; Miller et al., 1981; Blum and Cioli, 1981). In addition, antibodies in sera from animals showing high levels of immunity after exposure to irradiated larvae (Bickle et al., 1979), in contrast to those from animals immunized by chronic infections, can be shown to bind to lung-stage schistosomula, although no immune damage has yet been reported as a consequence of this binding (Bickle and Ford, 1982). However, there is little information on these older, migrating schistosomula, and this important stage of potential susceptibility to attack remains to be investigated in more detail.

(b) *Adult worms*. Although the events that occur in the older migrating schistosomula are still controversial, it is reasonably clear that, by the time that they have reached the adult stage, they fail to express parasite antigens on their surface in large amounts (McLaren et al., 1975; Snary et al., 1980), and can continue to survive for long periods even in hosts that are immune to reinfection (Smithers and Terry, 1969; reviewed by Smithers and Terry, 1976). Under artificial circumstances, however, it can be shown that such worms are still susceptible to cell-mediated attack. In the presence of antibodies with specificity for host molecules, adult worms can be killed not only by complement but also by mixed peritoneal cells (Perez and Terry, 1973) and eosinophils (McLaren and Terry, 1982).

(c) *Eggs*. Schistosome eggs express stage-specific antigens (Pelley et al., 1976; Dunne et al., 1981), and the translation *in vitro* of mRNA recovered from eggs and adult worms respectively has revealed that a major component of egg mRNA, coding for an antigenic polypeptide, is not present in significant quantities in adult worms (J. S. Cordingley, D. W. Taylor, D. W. Dunne and A. E. Butterworth, unpublished observations). It is not surprising, therefore, that the immune response to eggs is quite different from that expressed against other stages. Classical observations showed that eggs deposited in the tissues *in vivo* elicit a thymus-dependent granulomatous reaction (Warren, 1972, 1982). This is associated with an infiltration of lymphocytes, macrophages and eosinophils, followed eventually by destruction of the eggs and by fibrosis. The formation of such granulomata can also be achieved *in vitro* (Bentley et al., 1982). There is currently much interest in the cellular components of the granuloma, in the differences in responses to eggs of different schistosome species, and in the mechanisms of fibrosis, but these topics are outside the scope of this review. However, some studies have been carried out on the mechanisms whereby the eggs within the granuloma may be destroyed. James and Colley (1976, 1978a,b,c) have demonstrated that murine eosinophils damage *S. mansoni* eggs, subsequently entering the shell and killing the contained miracidium. Eosinophils from *S. mansoni*-infected mice are more active than those from controls: the effect is abolished by trypsinization of the eosinophils, and can be restored by passive sensitization of trypsinized cells with infection serum, suggesting a role for cytophilic antibody in this reaction (James and Colley, 1978a). In addition, normal eosinophils stimulated by culture supernatants of isolated egg granulomas, containing eosinophil stimulation promoter (James and Colley, 1975), show an enhanced capacity to kill eggs (James and Colley, 1978b). Similarly, stimulated eosinophils from mice with *Trichinella spiralis* infection are also active, although after more prolonged periods of incubation. Killing of eggs is inhibited by cytochalasin B and by inhibitors of glycolysis and of aerobic respiration, and depends on the presence of calcium ions. These findings suggest a role for the eosinophil in destroying schistosome eggs, an hypothesis that is supported by the observation that mice treated with an anti-eosinophil serum show a prolonged survival of tissue eggs, associated with a reduction in granuloma size (Olds and Mahmoud, 1980). Histopathological observations in man are also consistent with an eosinophil-mediated damage to eggs (Hsü *et al.*, 1980).

2. Fasciola *species*

Acquired immunity to infection with *Fasciola hepatica* can be demonstrated in cattle (Doyle, 1971) and rats (Doy *et al.*, 1981a,b; Doy and Hughes, 1982c), but is weakly expressed, if at all, in sheep (Sinclair, 1973, 1975) and mice (Masake *et al.*, 1978). Studies on effector mechanisms active against recently excysted flukes have therefore mainly involved the use of bovine or rat materials. Duffus and Franks (1980) showed that cattle infected with *Fasciola hepatica* metacercariae would produce antibodies that bound to the outer glycocalyx of recently excysted larvae. However, these antibodies failed to induce damage, in the presence of complement or of bovine eosinophils or neutrophils (Duffus and Franks, 1980), although a major basic protein isolated from bovine eosinophils caused direct damage to the flukes at concentrations down to 1 μM (Duffus *et al.*, 1980). The presence of excess antibody on the surface of the parasite caused the accumulation of large aggregates of antigen–antibody complexes, which were eventually shed from the surface: and, when eosinophils or neutrophils were present, these cells were also shed (Duffus and Franks, 1980). It was suggested that, even though the isolated major basic protein was effective in inducing damage, the rapid turnover and shedding of the glycocalyx prevented intimate attachment of the effector cells to the parasite's surface. Similarly, when rat peritoneal exudate cells were incubated with newly excysted larvae in the presence of immune serum, a selective adherence of eosinophils was observed (Doy *et al.*, 1980). The antibodies mediating such adherence were demonstrable in the serum of infected rats, but not in rats artificially immunized with dead fluke antigens: and cell adherence occurred independently of complement.

However, a precipitate of material, similar to that seen in the presence of bovine antisera, could be observed around the larvae (Doy and Hughes, 1982b): and, when flukes that had been incubated in the presence of serum and cells (added freshly every 4 hours) were transferred to the peritoneal cavities of normal animals, there was no reduction in recovery from the liver 3 weeks later. These findings again suggested that the larvae were able to withstand the effects of eosinophils or other cells when tested in assays *in vitro*.

However, when newly excysted larvae were injected intraperitoneally into previously sensitized rats, and subsequently recovered at various intervals for examination by light and electron microscopy, there was evidence of marked damage to the parasite (Davies and Goose, 1981). Damage was associated with adherence of eosinophils, which degranulated on the surface, releasing cytochemically detectable peroxidase. Vacuoles, progressively increasing in size, formed in the tegument underlying areas of eosinophil attachment. At a later stage, some flukes entirely lost their teguments, and became surrounded by a variety of phagocytic cells, especially neutrophils. Mast cell attachment and degranulation was also observed during the first 5 minutes after injection, and it is suggested that the release of mediators may have contributed to eosinophil localization. Similar damage may be seen after surgical transfer of adult flukes (Bennett *et al.*, 1980).

Two main explanations may be put forward for these differences between the results obtained *in vitro* and *in vivo*. First, the conditions of the assay *in vitro* may be insufficient to permit full expression of damage mediated by eosinophils or other effector cells, and the parasite may escape under such conditions by virtue of its capacity to turn over its surface antigens, even though it would be susceptible to damage *in vivo*. Secondly, the eosinophils present in the previously sensitized rats may have been in a state of functional activation, in comparison with those used in the assays *in vitro*, and further activation may have occurred as a result of the release of mast cell mediators. It would be interesting to know whether the addition of mast cell or other mediators promotes eosinophil-mediated damage in the assays *in vitro*.

C. PLATYHELMINTHES: CESTODA

1. *Introduction*

A few reports are available on cell-mediated damage to parasitic cestodes of the order Cyclophyllidea. Such organisms resemble the trematodes in being bounded by a simple plasma membrane, but differ in that the surface layer presents to the host an array of long villus-like processes, the microtriches. These may be interdigitate with projections from host effector cells, thereby protecting the deeper tegumental structures from attack (Williams *et al.*, 1980). In general, interactions have been observed between parasites and host cells *in vitro* and *in vivo*, but the mechanisms of damage, if any, have not been analysed in detail.

2. Echinococcus *species*

The establishment and growth of protoscolices of *Echinococcus multilocularis* is markedly suppressed both in cotton rats bearing a primary infection (Rau and Tanner, 1973) and in animals immunized with BCG (Rau and Tanner, 1975; Reuben *et al.*, 1978), and unpurified peritoneal exudate cells from heavily infected animals kill protoscolices *in vitro* (Rau and Tanner, 1976). In subsequent studies on mice infected with *E. multilocularis*, it was found that cells from immune animals were more active than those from normal mice (Baron and Tanner, 1977). The protoscolicidal effect was attributable to an adherent macrophage, and could be enhanced by the presence of serum from immune animals. Similar damage was observed with activated macrophages recovered from BCG-immunized animals, tested on *E. granulosus* protoscolices, and with activated macrophages from animals infected with *Taenia crassiceps* tested on *E. multilocularis*. Studies by scanning electron microscopy confirmed the macrophage nature of the effector cell, and demonstrated a close adhesion of the cell with the extension of finger-like processes. It is suggested that immunity to reinfection in animals bearing a primary hydatid infection may be attributable to activated macrophages, whose effects are enhanced by antibody.

3. Taenia *species*

*Taenia solium* larvae elicit chronic granulomatous reactions in pig muscle, with extensive eosinophil infiltration and degranulation onto the surface of the parasite, but without evidence of parasite damage (Willms and Merchant, 1980). In contrast, when bulls were infected twice with *Taenia saginata* oncospheres, first orally and later by subcutaneous inoculation, the extensive eosinophil infiltration that occurred around both eggs and developing cysticerci was associated with damage to the parasite (Sterba *et al.*, 1981). In neither case, however, was there evidence for immunologically mediated events, and Camp and Leid (1982) have shown that saline extracts of *Taenia taeniaeformis* metacestodes are directly chemotactic and chemokinetic for equine polymorphonuclear leukocytes. The best model of immunity to *Taenia* species is the use of *Taenia taeniaeformis* in the rat (Leid, 1977; Hammerberg and Williams, 1978; Williams *et al.*, 1980). In this model, it is found that the young developing cysticerci are susceptible to killing by antibody and complement, without a requirement for effector cells. Older organisms, by releasing anticomplementary activities, evade such damage. Cell-mediated mechanisms appear to play little part in immune rejection.

### 4. *Other cestodes*

Information on cellular effector mechanisms active against other cestode parasites is sparse. Although *Mesocestoides corti* is particularly effective in eliciting a local eosinophilia after intraperitoneal inoculation into mice, there are no reports of the consequences of this cell infiltrate. Voge *et al.* (1979) describe the structure of the tetrathyridial stages, and report that host peritoneal cells may be observed in contact with the blade-shaped microvilli, but without evidence of damage. Hindsbo *et al.* (1982) have described the tissue reactions of intact or T-depleted rats to infection with *Hymenolepis diminuta*, and report a reduced mast cell and eosinophil response in the small intestine in T-depleted animals, associated with a delay in destrobilization and expulsion of the worms, but without direct evidence for a causal association between the two events.

#### D. NEMATODA

### 1. *Introduction*

Parasitic nematodes differ markedly from trematodes and cestodes, in that they present to the host's defenses a tough, rigid cuticle composed of layers of collagen-like proteins, with or without an external sheath, and it might be expected that this cuticle and sheath would be intrinsically resistant to immune attack. In spite of this, however, a variety of immune effector mechanisms have been described that are active against several species of nematodes of the orders Rhabditida, Spirurida and Enoplida. The most extensively studied has been *Trichinella spiralis*, but a variety of others have been shown to be susceptible to attack, including several that are important pathogens of man. A striking feature of the nematodes is that, although the appearance of the surface may be similar from one stage to the next, relatively few surface antigens may be expressed, and these may show a high degree of stage specificity (Philipp *et al.*, 1980; Maizels *et al.*, 1982). Thus responses that are active against one stage may be ineffective against another, in spite of their morphological similarities.

In general, it may tentatively be suggested that, although eosinophils act against a wide range of nematodes *in vitro*, and appear from histological studies to be extensively involved in nematode destruction *in vivo*, neutrophils may be more active against ensheathed larvae. This may be because eosinophils generally act by degranulation and the release of toxic products which are inactive against larval sheaths, although highly active against exsheathed parasites. Macrophages, either conventionally activated or acting through IgE-dependent mechanisms, are highly effective against some nematodes.

## 2. Trichinella *species*

*Trichinella spiralis* is readily maintained in laboratory hosts, and the mechanisms of immunity against various stages have been extensively studied *in vivo* (Ruitenberg and Duyzings, 1972; Despommier *et al.*, 1974, 1977; Despommier, 1977; Grove *et al.*, 1977; Wing and Remington, 1978; Ruitenberg *et al.*, 1980; Ngwenya, 1980; Dessein *et al.*, 1981a; Jungery and Ogilvie, 1982) and *in vitro* (Stankiewiecz and Jeska, 1973; Vermes *et al.*, 1972, 1974; McLaren *et al.*, 1977; Mackenzie *et al.*, 1978, 1980, 1981; Kazura and Grove, 1978; Kazura and Aikawa, 1980; Kazura 1981; Ljungstrom and Sundquist, 1979; Bass and Szejda, 1979; Anteunis *et al.*, 1980; Buys *et al.*, 1981; Grover *et al.*, 1983). Experiments *in vivo* suggest a role for immune effector mechanisms active against infective larvae, adult worms and newborn larvae (Despommier *et al.*, 1977). The expression of antigens on the surface of the parasite is markedly stage-specific (Philipp *et al.*, 1980) and different strains of mice respond differently to different antigens (Jungery and Ogilvie, 1982). Different responses may therefore be active against different stages and at different times of infection.

Some observations have been made on the exsheathed infective larvae *in vitro* (Stankiewicz and Jeska, 1973; Vernes *et al.*, 1979; Mackenzie *et al.*, 1980; Jungery and Ogilvie, 1982). These larvae are highly susceptible to cell-mediated damage in the presence of antibody or antibody and complement. Rat eosinophils are selectively active against such larvae, and damage induced by such eosinophils is associated with cell degranulation, rupture of the parasite cuticle and extrusion of internal components (Mackenzie *et al.*, 1980, 1981). Damage can occur in the presence either of antibody or of complement, activated by the alternative pathway. In this reaction, eosinophils and macrophages may exert a synergistic effect, with macrophages attaching to released eosinophil products.

Rat eosinophils are also more effective than rat neutrophils in damaging newborn larvae in the presence of antibody and complement (Mackenzie *et al.*, 1978, 1980, 1981), and murine eosinophils are also selectively capable of mediating antibody-dependent damage to newborn larvae (Kazura and Grove, 1978). Such larvae fail to activate complement by the alternative pathway, and the reaction is strictly antibody-dependent (Mackenzie *et al.*, 1980). In contrast, although normal human eosinophils can kill newborn larvae, neutrophils are more effective (Bass and Szedja, 1979; Kazura, 1981). Eosinophils from eosinophilic individuals are more active than those from normal individuals, however, especially in the presence of early baboon infection sera (Grover *et al.*, 1983). The mechanisms of killing of newborn larvae by human granulocytes, described earlier (p. 170), depend primarily on a respiratory burst with the generation of hydrogen peroxide.

*Trichinella spiralis* infection is associated with a marked eosinophilia, and the T-dependence of this response has been examined in detail (p. 160). Some evidence suggests that these eosinophils may contribute to immunity *in vivo*. Grove *et al.* (1977), for example, have demonstrated that depletion of eosinophils by treatment with an anti-eosinophil serum leads to an increase in the numbers of muscle-stage larvae recovered after a primary infection, while Dessein *et al.* (1981a) have shown that depletion of IgE, by neonatal treatment of rats with an anti-epsilon serum, reduces the eosinophilic infiltrate around muscle-stage larvae of a primary infection and increases the number of larvae that can be recovered. However, other mechanisms may also be involved *in vivo*: induction of macrophage activation, for example, can protect against challenge (Wing and Remington, 1978; Grove and Civil, 1978).

3. Nippostrongylus *species*

*Nippostrongylus brasiliensis* has been extensively used as a model for the investigation of the control of IgE responses (Ishizaka and Ishizaka, 1978). In addition, the mechanisms of immunity to *Nippostrongylus brasiliensis* in the rat have been thoroughly studied by Ogilvie and colleagues (Ogilvie and Love, 1974; Ogilvie *et al.*, 1977a,b, 1980), the general conclusions being that a combination of direct antibody-mediated damage and a subsequent cell-mediated event is responsible for the expulsion of the adult worm, with a non-immunoglobulin-bearing lymphocyte being involved in the final stages. McLaren *et al.* (1977) and Mackenzie *et al.* (1980, 1981) have shown that, as with other helminths, eosinophils can attach to various stages of *Nippostrongylus brasiliensis* in the presence of antibody, with or without complement, discharging their granule contents but failing to damage the organism. Macrophages and neutrophils also adhere: the adherence of macrophages being long-lasting, whereas that of neutrophils is of shorter duration. Neutrophils adhere through small areas of membrane flattening, at which nitroblue tetrazolium reduction can be observed (Mackenzie *et al.*, 1981): the deposit of reduced dye remaining as a visible 'footprint' on the surface of the parasite after detachment of the cell. In the case of infective larvae, macrophage adherence leads to damage to the parasite, but only in the presence of complement-containing sera (Mackenzie *et al.*, 1980).

4. Ascaris *species*

*Ascaris* species have been particularly studied for their capacity to elicit local or systemic eosinophilia and to induce IgE responses (Nielsen *et al.*, 1974; Hirashima and Hiyashi, 1976; Tanaka *et al.*, 1979), and there have been few reports on their susceptibility to attack by cellular effector mechanisms. However, Ziprin and Jeska (1975) have reported adherence of murine peritoneal exudate cells to *Ascaris suum* larvae, which was both antibody and complement dependent, and Thompson *et al.* (1977) have described damage to such larvae associated with neutrophil infiltration detectable by electron microscopy.

5. Strongyloides *species*

Moqbel (1980) has studied the histological responses of rats to primary or multiple challenge with *Strongyloides ratti*. Primary infections are terminated by rejection of adult worms from the intestine; during secondary infections, the adult worms and pulmonary phases are the main targets for immune attack; while, in multiple infections, the early cutaneous phase is damaged. In all cases, rejection is associated with a tissue eosinophilia and, during rejection at the skin stage, eosinophilic granulomata are rapidly formed, with eosinophils in close contact with the invading larvae. These findings strongly suggest a role for the eosinophil in immune rejection.

6. Dictyocaulus *species*

*Dictyocaulus viviparus*, the cattle lungworm, is a particularly suitable parasite for investigation of immune effector mechanisms, since it is possible to protect cattle against a challenge infection by immunization with irradiated larvae (Jarrett *et al.*, 1960): and this forms the basis for a commercially useful vaccine. However, relatively little work on immune effector mechanisms has been carried out. Knapp and Oakley (1981) have demonstrated a preferential adherence of bovine eosinophils, in comparison with other cells present in peripheral blood, to *Dictyocaulus viviparus* larvae. This adherence is dependent on heat-stable factors, presumably IgG antibodies, in hyperimmune sera, acting in concert with heat-labile factors, presumably complement, present in normal sera, and is associated with damage to the larvae, as reflected by a loss of motility. More recently, D. A. Lammas and W. P. Duffus (personal communication) have confirmed these findings, and have developed a mode for investigating immune damage to *Dictyocaulus filaria* L3 larvae in the sheep. Larvae are introduced into the teat canals of control sheep, of sheep stimulated by inoculation of *Dictyocaulus* antigen into the teat canal 1 day previously, and of animals immunized by infection and stimulated in the same way. An eosinophil-rich exudate is observed in immune, but not in control, animals, and extensive destruction of the larvae occurs over a 1 to 4 hour period.

These various findings suggest a role for the eosinophil in conferring protection against *Dictyocaulus* species, an hypothesis that is supported by the observation, in immune cattle, of eosinophil-rich foci in the lungs at the site of migration of the larval stages of *Dictyocaulus viviparus* (Jarrett and Sharp, 1963).

## 7. Nematospiroides *species*

Mice immunized with *Nematospiroides dubius* by subcutaneous vaccination develop high levels of immunity against challenge with infective larvae (Ninnemann and Lueker, 1974), and in such animals damage to larvae implanted intraperitoneally in chambers is associated with macrophage adherence. Immunization with a single infection leads to the development of a partial resistance to superinfection that lasts for several weeks, while a second immunizing infection leads to a marked enhancement in resistance to challenge (Prowse *et al.*, 1978), with the possible involvement of two distinct mechanisms of immunity dependent respectively on eosinophils and macrophages. Peritoneal cells, possible activated macrophages, recovered from immune mice can damage infective larvae *in vitro* in the absence of added antibody (Chaicumpa and Jenkin, 1978), and this is associated with a loss of the capacity of the larvae to mature after injection into mice. This effect is enhanced by the presence of immune serum, or after alternative pathway activation of complement (Prowse *et al.*, 1979). This requirement for antibodies in mediating immunity is supported by the observation that Biozzi high-responder mice develop greater levels of immunity after a primary infection than low-responder mice, which show hyperactive macrophages but low antibody levels (Jenkins and Carrington, 1981). In this infection, therefore, more than one effector mechanism may be involved, and the relative contribution of each component remains to be determined.

## 8. Dirofilaria *species*

El-Sadr *et al.* (1983) have investigated possible immune effector mechanisms whereby microfilariae of *Dirofilaria immitis* are cleared from the bloodstream during canine infection. Both mononuclear cells and granulocytes adhere to microfilariae in the presence of immune serum, but only granulocytes are capable of damaging the larvae. Sera from animals with occult infection are more effective than those from animals with patent infection, the opsonizing activity being associated with an IgM fraction. Killing of microfilariae is markedly enhanced by the presence of fresh normal dog serum as a source of complement, and immunofluorescence studies revealed the presence of both IgM and C3 on the surface of opsonized larvae. Although both eosinophils and neutrophils could kill microfilariae, neutrophils were the more effective cell population in this case.

## 9. Dipetalonema *species*

Infection of hamsters or rats with the filarial nematode *Dipetalonema viteae* leads to the development of immunity against both the microfilariae and the infective larvae. The disappearance of microfilariae from the blood of infected hamsters is associated with the development of opsonizing 19S antibodies, presumably IgM, which promote the adherence of peritoneal cells from normal uninfected hamsters (Tanner and Weiss, 1978). In experiments *in vitro*, mononuclear cells are the predominant adherent cell type, although eosinophils are occasionally found. A different pattern was observed, however, when microfilariae were introduced in micropore chambers *in vivo* (Weiss and Tanner, 1979; Rudin *et al.*, 1980). In immune, amicrofilaraemic animals, microfilariae were rapidly destroyed in chambers with a membrane pore size of 3 or 5 μm, but not in chambers of 0.3 μm, implying a requirement for cell entry into the chamber. In non-immune animals, microfilariae were only destroyed if they had been preincubated with immune serum or with a mercaptoethanol-sensitive, 19S fraction. In the early stages of the reaction, eosinophils and neutrophils were the predominant adherent cell type, although lymphocytes and monocytes also adhered: in the later stages of destruction, the monocyte contribution was more pronounced, and it was suggested that more than one cell type contributed to the damaging process, which was associated with disintegration of the cuticle and lysis of the hypodermis.

In the rat model, as with other helminth infections, a greater importance has been ascribed to IgE-dependent mechanisms, in association with either macrophages or eosinophils. Sera from infected but amicrofilaraemic rats confer on normal rat macrophages the capacity to adhere to and kill microfilariae *in vitro* over a period of 16 to 24 hours (Haque *et al.*, 1980). The activity in immune serum is heat-labile and depleted on anti-IgE immunoadsorbents: and, as in the schistosomulum system, is cytophilic for the macrophage rather than opsonic for the parasite. Damage is associated with cell adherence and spreading, followed by degranulation of the cell and the release of granule contents onto the surface, with breaking of the microfilarial cuticle (Ouassi *et al.*, 1981). Similar IgE-dependent damage can be observed with eosinophils (Haque *et al.*, 1981) associated with eosinophil degranulation and immobilization of the parasite. In this reaction, eosinophils and macrophages may be acting synergistically, the initial eosinophil-mediated immobilization being followed by macrophage adherence and disruption of the organism.

Eosinophils and macrophages may also act in concert to damage the infective larvae of *Dipetalonema viteae*, in a complement-dependent antibody-independent fashion (Haque *et al.*, 1982). Infective larvae cause alternative but not classical pathway activation of rat complement in the absence of antibody, and C3 can subsequently be demonstrated on the surface of the parasite. Both eosinophils and macrophages adhere to C3-coated larvae, but macrophage adherence cannot occur without simultaneous adherence of eosinophils. These degranulate onto the surface of the organism, causing extensive damage which can be markedly enhanced by replacing spent eosinophils with fresh populations. Eosinophil adherence can occur in the absence of mast cells, but is markedly enhanced when mast cells are present. It is possible that this killing reaction may contribute to non-specific resistance to infection in the non-immune host.

Adult *Dipetalonema* may also be eliminated by cellular mechanisms. Worms and McLaren (1982) have examined the reaction to an adult female *Dipetalonema setariosum* recovered from the pleural cavity of *Meriones libycus*. Damage to the worm was observed, with local breaching of the cuticle and infiltration by macrophages: macrophages were also present and degranulating on the surface of the organism, and there was no evidence for the involvement of other cell types.

10. Litomosoides *species*

Albino rats infected with *Litomosoides carinii* develop a patent infection but subsequently clear the microfilariae from the peripheral blood about 6 months after the initial infection, and histological examination of pleural exudates *in vivo* have demonstrated the adherence of lymphocytes, macrophages and eosinophils to dead or dying microfilariae (Bagai and Subrahmanyam, 1970). Subsequent experiments demonstrated an antibody-dependent adherence of rat spleen or peritoneal exudate cells to microfilariae *in vitro*, associated with release of $^{51}$Cr (Subrahmanyam et al., 1976) and uptake of trypan blue into immotile organisms (Mehta et al., 1980). Antibodies mediating adherence in latent rat serum (Mehta et al., 1980) and in serum from hyperimmunized animals (Mehta et al., 1982) were of the IgE class, in that they were sensitive to heating at 56°C and would be absorbed and eluted from anti-IgE immunoadsorbents. IgG antibodies were ineffective on their own, but addition of fresh normal rat serum to heated latent serum partially restored the adherence and killing (Mehta et al., 1980), suggesting that IgG and complement might also be involved. However, no experiments were reported in which purified IgG was tested in the presence of fresh normal serum. Although both human (Mehta et al., 1981b) and rat (Mehta et al., 1982) eosinophils can adhere to microfilariae in the presence of immune serum, they fail to cause damage. Instead, damage by fractionated rat cells is seen with macrophages and neutrophils (Mehta et al., 1982), and the effect of neutrophils can be enhanced by supernatants from cultures of lymph node lymphocytes stimulated with concanavalin A or phytohaemagglutinin (Hopper et al., 1981).

The lack of effect of eosinophils in this model may be attributable to the fact that the microfilariae are ensheathed, and the sheath may prevent the released granule contents from attacking the underlying organism. In addition, it may be noted that these eosinophils were recovered from oil-elicited peritoneal exudates, and may have been functionally abnormal. The fascinating suggestion that neutrophils may have Fc receptors for IgE needs further investigation.

11. Brugia *species*

Immune responses to *Brugia pahangi* and *Brugia malayi* have been studied in several experimental hosts, including rhesus monkeys (Wong *et al.*, 1977), cats (Johnson *et al.*, 1981), jirds (Vincent *et al.*, 1980; Karavodin and Ash, 1982), ferrets (Crandall *et al.*, 1982) and rats (Gusmao *et al.*, 1981), as well as in man (Sim, 1981; Sim *et al.*, 1982; Piessens and Dias da Silva, 1982). Histological responses to microfilariae and adult worms involve an extensive cellular infiltrate containing eosinophils, macrophages and giant cells, with or without damage to the contained parasite (Vincent *et al.*, 1980; Crandall *et al.*, 1982). Observations *in vitro* have shown that infective larvae of *Brugia malayi* can be damaged by mixed human peripheral blood leukocytes in the presence of serum from patients with tropical pulmonary eosinophilia, elephantiasis, or amicrofilaraemic symptomatic filariasis (Sim, 1981; Sim *et al.*, 1982). Microfilariae of *Brugia malayi* can similarly be damaged by human leukocytes in an antibody- and complement-dependent fashion (Piessens and Dias da Silva, 1982). Peritoneal exudate cells from normal jirds kill *Brugia pahangi* microfilariae in the presence of sera from infected animals (Karavodin and Ash, 1982). The activity in such sera was sensitive to heat inactivation, and was only partially restored by the addition of fresh normal jird serum as a source of complement, possibly suggesting an IgE-dependent effect. The effect was associated with macrophage adherence, and could be inhibited by immune complexes precipitated from the sera of chronically infected animals. Feline eosinophils and neutrophils have also been shown to adhere to *Brugia malayi* microfilariae (Johnson *et al.*, 1981). In this case, adherence could be mediated either by heat-labile factors present in fresh sera of several species, an effect associated with the deposition of C3 on the surface of the larvae, or by IgG antibodies present in the sera of a small proportion of infected cats. Complement-dependent adherence was observed only with blood microfilariae recovered from animals whose infections had been patent for more than 3 weeks. In contrast, antibody-dependent adherence could be observed with all microfilariae, including those recovered from animals with recently patent infections or those produced *in vitro*. This suggests that the capacity to fix complement directly is acquired only during the maturation of microfilariae into the older forms.

12. Wuchereria *species*

Higashi and Chowdhury (1970) described a selective antibody-dependent adherence of eosinophils to infective larvae of *Wuchereria bancrofti*. Subsequently, Mehta *et al.* (1981a) have studied the nature of the immunoglobulin and effector cells involved in antibody-dependent cell-mediated damage to microfilariae of *Wuchereria bancrofti*. As with their findings in *Litomosoides carinii* infections of rodents, they report that although both human eosinophils and neutrophils adhere to microfilariae in the presence of sera or IgG fractions from amicrofilaraemic individuals, only the neutrophils are capable of inducing damage. This finding held true irrespective of whether the cells were obtained from normal individuals or from infected individuals with or without microfilaraemia.

13. Onchocerca *species*

Histological studies have been made of the reactions to adult worms of *Onchocerca gibsoni* (Nitisuwirjo and Ladds, 1980) and *Onchocerca volvulus* (Burchard *et al.*, 1979), with no clear evidence of immune damage to the parasite. Similarly, in untreated onchocercal dermatitis there is no evidence of cellular reactions around, or damage to, the microfilariae, although there are extensive perivascular infiltrates (Burchard and Beirther, 1978). However, after treatment with diethylcarbamazine, extensive destruction of microfilariae in the skin is observed, with a predominantly eosinophilic infiltrate (Gibson *et al.*, 1976). Furthermore, *Onchocerca volvulus* microfilariae can be damaged *in vitro* by antibody-dependent cell-mediated mechanisms (Greene *et al.*, 1981). In the presence of inactivated immune serum, both human eosinophils and human neutrophils kill microfilariae, the effect of eosinophils being preferentially enhanced by the addition of fresh normal serum as a source of complement. Some eosinophil-mediated damage is also seen when fresh microfilariae are tested in the presence of complement alone, without immune serum.

14. *Other nematodes*

Histological observations on tissue reactions to other helminths, including *Trichostrongylus colubriformis* (Rothwell and Dineen, 1972; Rothwell, 1975; Rothwell and Love, 1975), *Capillaria hepatica* (Solomon and Soulsby, 1973) and *Toxocara canis* (Kayes and Oaks, 1978; Sugane and Oshima, 1982) strongly suggest the involvement of cellular effector mechanisms in parasite damage. In *T. colubriformis* infection, for example, expulsion of intestinal parasites is associated with a local accumulation of basophils and eosinophils (Rothwell and Dineen, 1972), while the granulomatous reaction in murine toxocariasis is associated with eosinophil infiltration and degranulation, with a later macrophage accumulation (Kayes and Oaks, 1978). Athymic nude mice fail to show an eosinophil infiltration, and no larval destruction occurs in these animals (Sugane and Oshima, 1982). However, although these findings implicate eosinophil-mediated reactions, frequently T lymphocyte-dependent, in immune rejection, there is no direct evidence for an effector function for these cells. Along different lines, Tanaka and Torisu (1978) have demonstrated a marked local eosinophil accumulation around *Anisakis* larvae, and have shown that this may be attributable to the release from the larvae of eosinophil-selective chemotactic factors. Again, however, the role of the eosinophil in mediating damage to the larvae remains to be determined.

## V. THE ROLE OF CELLULAR EFFECTOR MECHANISMS IN IMMUNITY *In vivo*

### A. INTRODUCTION

Assays for studying cell-mediated effector mechanisms *in vitro* are of great value in testing which effector mechanisms are or are not effective against a given parasite, and in analysing in detail the mode of action of those mechanisms that are shown to be effective. In addition, apart from rare genetic defects in particular effector systems, they offer the only approach to the study of those immune effector mechanisms that are active in man. However, the simple demonstration that a host can mount a particular response to a given parasite does not necessarily imply that that response is active in mediating immunity *in vivo*. Evidence for a role for a given response must instead come from one of several different types of observation. First, histological observations on tissues in which immune damage to a parasite is occurring allow the identification of those cells that are associated with parasite rejection. The problem about such studies is that, except in very clearcut cases, it is usually difficult to be certain whether a given cell type has initiated parasite damage, or whether it is present in the lesion as a secondary consequence of damage that has already been initiated by some other event. Secondly, experimental animal models may be manipulated in various ways, and the effect of such manipulation on the induction or expression of immunity against a given parasite may then be tested. Such manipulations may include the use of animals that differ genetically in their capacity to produce or express a given mechanism; the depletion of components of the immune system; and the transfer of such components from an immune to a normal animal. The problem inherent in such studies is the question of the specificity of the manipulation used. For studies on the role of different categories of antibodies and of lymphocytes, highly specific and refined methods are now available: but, unfortunately, this is not yet true for studies on the role of accessory cells, including eosinophils, neutrophils and macrophages. Thirdly, correlations may be made between the presence or absence of a particular mechanism and the presence or absence of immunity, under different conditions of infection or immunization. This type of approach is the only one that is applicable on a large scale to man. The problem is that an observed correlation may be fortuitous: a given response may be associated with immunity only because it is also associated with another, unidentified response which is actually involved in mediating immunity. There is no easy solution to this problem, and the only possible approach is to test for the presence or absence of as wide a range as possible of candidate effector mechanisms.

With all of these approaches, a further difficulty is that a host may have more than one effective mechanism of immunity, and that different mechanisms may be operative at different times or under different conditions of infection or immunization.

Because of these various problems, studies on the role of different cellular effector mechanisms *in vivo* have so far proved somewhat inconclusive. An enormous amount of work has been carried out, especially in experimental animal models, and it is not possible in this review to be in any way comprehensive: the selected examples that follow serve to illustrate the types of approach that may be used, and the difficulties that may be encountered.

### B. HISTOLOGICAL STUDIES

Details of histological findings in various individual helminth infections have been described in Section IV. In this section, the skin response to schistosomula of *Schistosoma* species is considered as an example of some of the problems in interpretation that may be encountered. Such studies have been carried out in a variety of host species, including mice (von Lichtenberg et al., 1976; Colley et al., 1977; Savage and Colley, 1980; Maeda et al., 1982), rats (Bentley et al., 1981a,b), rhesus monkeys (von Lichtenberg and Ritchie, 1961; Hsü et al., 1974, 1975, 1979, 1981) and baboons (Seitz et al., 1980, and unpublished observations). In mice, an anamnaestic skin reaction is observed in animals immune to *Schistosoma mansoni* with a marked eosinophil infiltrate and with demonstrable interactions between the eosinophils and the schistosomulum tegument (von Lichtenberg et al., 1976). However, similar infiltrates can be seen in the skin of mice sensitized with schistosome antigens, but not immune to challenge, indicating that the observed reaction, by itself, is not sufficient to account for immunity (Savage and Colley, 1980). In murine *Schistosoma japonicum* infection, neutrophil infiltrates are more pronounced (Maeda et al., 1982). In the rat, challenge of hosts previously infected with *S. mansoni* is associated with a more intense exudative inflammatory reaction than is seen in normal hosts, and with a more marked contribution of eosinophils to the granulocytic infiltrate (Bentley et al., 1981b). In rhesus monkeys immunized with irradiated cercariae, challenge is associated with localization of immunoglobulins and C3 (Hsü et al., 1981), degranulation of mast cells, emigration of eosinophils, deposition of IgE on the surface of invading schistosomula and, eventually, death of the schistosomula within eosinophilic abscesses (Hsü et al., 1979).

Similarly, in chronically infected baboons, the death of schistosomula in the skin is associated with marked eosinophil adherence and degranulation on the tegument and with infiltration of eosinophils into the interior of the dead organism (Seitz et al., 1980, and unpublished observations). Live schistosomula in the same animals, or in control uninfected animals, show no eosinophil adherence.

These findings strongly suggest a role for eosinophil-mediated killing, more particularly in the primate than the rodent models. The difficulty, however, is in distinguishing between cause and effect. Although it would seem plausible—especially in the baboon model, in which a heterogeneity of responses to different schistosomula within an individual animal is observed— to conclude that eosinophils are inducing death of the schistosomula, this need not be the case: the schistosomula may be killed by some other process, not identified histologically, and the eosinophils may only subsequently interact with the killed organism. Although supporting evidence from experimental systems *in vitro* and *in vivo* makes this second possiblity unlikely in this case, the 'chicken-and-egg' problem has to be remembered in the interpretation of all such histological findings, in the absence of support from other experimental studies.

C. MANIPULATION OF EXPERIMENTAL ANIMAL MODELS

These studies are too extensive for detailed descriptions, but some general features may be noted about the approaches that may be used, and the information that they may yield about the role of cell-mediated effector mechanisms in immunity to helminths.

1. *Genetic analysis of immunity*

Such studies may involve either conventional genetic selection and backcross experiments (Wakelin, 1975a,b, 1978, 1980) or the use of inbred strains of animals that have previously been shown to differ in specific genetic traits. The first approach has particularly been applied to the study of intestinal nematodes, especially *Trichuris muris* and *Trichinella spiralis*; and, for example, Wakelin and Donachie (1980) have shown that the rate of expulsion of *Trichinella spiralis* in B10 congenic mice is controlled by genes unlinked to the MHC, and is related to the generation of intestinal inflammatory reactions. Similarly, the susceptibility of different strains of mice to *Taenia taeniaeformis* infection is related to the rate at which such mice can generate an antibody response, capable of damaging the parasite before it can escape from antibody-mediated attack (Mitchell et al., 1977, 1980). Such experiments, although useful, do not give detailed information about the nature of the immune effector mechanisms involved in worm rejection, beyond saying that they are antibody-dependent. Similarly, the demonstration of MHC-linked gene involvement in resistance to *Trichinella spiralis* infection (Wassom *et al.*, 1979) and in *Schistosoma mansoni* pathology (Claas and Deelder, 1979, 1980) provides information more about the control of the induction of the immune response than about its expression: and this is equally true of the use of athymic (nude) mice, or of mice showing high or low antibody responses.

Selective genetic traits that would be more use for the analysis of immune effector mechanisms as such would include abnormalities in various complement components, in IgE responses (Ovary *et al.*, 1978), in macrophage responses to T-dependent activation (Meltzer *et al.*, 1980), in the induction of blood eosinophilia (Vadas, 1982), and in the functional properties of accessory cells, including eosinophils or neutrophils. Few such studies have yet been performed in the context of helminth immunity, but this is a rapidly expanding field, and the increasing availability of strains with highly selective defects should soon allow a convincing demonstration of the role of individual mechanisms.

## 2. *Transfer and depletion experiments*

Classically, the demonstration of a role for a specific acquired immune response has involved either the passive transfer of serum or purified immunoglobulins, or the adoptive transfer of T or B lymphocytes to syngeneic recipients. In some cases, such as the granulomatous reaction to *Schistosoma mansoni* eggs, the transfer of T cells has been found to be a necessary and sufficient condition for the transfer of the reaction (reviewed by Warren, 1982), implying the involvement of T lymphocyte-dependent mechanisms not involving antibodies. A striking feature of helminth infections in general, however, is the importance of antibodies in mediating immunity to challenge. For example, the transfer of serum or of purified immunoglobulins has been shown to confer protection against challenge with *Schistosoma mansoni* in both the mouse (Sher *et al.*, 1975) and more strikingly the rat (Bazin *et al.* 1980). In the rat, absorption of either IgE or $IgG_{2a}$ antibodies from immune sera lower the capacity of such sera to transfer immunity (Capron *et al.*, 1980), and suppression of antibody formation by treatment of neonatal rats with anti-μ serum prevents the establishment of immunity (Bazin *et al.*, 1980). Furthermore, monoclonal antibodies against schistosomulum surface antigens can also transfer protection (Grzych *et al.*, 1982).

However, the demonstration of a role for antibody in mediating protection *in vivo* does not necessarily imply that antibody-dependent cellular effector mechanisms are involved since, depending on the isotype of the transferred antibody, antibody-dependent complement-mediated killing could be equally important. Tests for a role for complement can involve either the use of animals congenitally deficient in various complement components, or the depletion of complement, for example by treatment of the recipient with cobra venom factor. There are relatively few examples of the use of this approach: in those cases in which it has been carried out, especially with *Schistosoma mansoni* infections, the evidence suggests that complement plays little, if any, major role (Santoro et al., 1982).

Even when a role of complement can be demonstrated, this does not necessarily imply that immunity depends on the lytic activity of complement, with activation of the complete pathway, since activation of earlier components, especially C3, may enhance the antibody-dependent effect of accessory cells, as described in detail in Section III. A better approach to testing for a role for such accessory cells, therefore, is to deplete the recipient of immune serum of these cells or their precursors, and this may be achieved either by whole-body irradiation or by the use of antisera with specificity for individual cell types. For example, Sher (1977) has reported that whole-body irradiation prevents the expression of early phase immunity to *Schistosoma mansoni*, as reflected by a decrease in the recovery of challenge schistosomula during their migration through the lungs, in mice passively immunized with serum. Immunity can be restored by reconstitution with bone marrow cells. Mahmoud and his colleagues (Mahmoud et al., 1974, 1975a; Mahmoud, 1977) have produced rabbit anti-mouse eosinophil sera, and have used such sera selectively to deplete eosinophils *in vivo*. Such depletion is associated with a loss of immunity to *Schistosoma mansoni* challenge, both after primary infection and in the recipients of immune serum (Mahmoud et al., 1975b), with a prolonged survival of schistosome eggs *in vivo* (Olds and Mahmoud, 1980), and with a reduction in immunity to *Trichinella spiralis* (Grove et al., 1977). These were heterologous antisera, and there must be some doubts about their exact specificity, although their functional activity in reducing eosinophils *in vivo* seems quite clearcut: it would be extremely worthwhile if these studies were repeated with more specific reagents, especially monoclonal antibodies.

A further, and very conclusive, approach would be to reconstitute depleted animals with specific cell types. This approach has generally proved useful in the demonstration of a role for lymphocytes of particular groups, since such cells can be obtained at high levels of purity, by positive or negative selection on the basis of particular markers, and since they then show the capacity for extensive replication in the recipient. This approach, however, has not proved profitable in the demonstration of a role for accessory effector cells. Such cells are either incapable or poorly capable of further division in the recipient, and large numbers would have to be administered. Alternatively, reconstitution with precursor cells could be attempted, but such cells usually have the capacity for differentiation into a variety of end cells: it is thus not possible to interpret the results of such reconstitution in terms of a particular cell.

At present, therefore, the situation in most helminth infections is that a role for antibody can be demonstrated by passive transfer experiments: that an effect of accessory cells can frequently be inferred by depletion experiments: but that the exact nature of such accessory cells cannot usually be determined. Further progress in this area must depend on the development of more specific techniques for depletion of accessory cells, such as monoclonal antibodies, and of new approaches to cell transfer.

Along these general lines, however, some individual experimental systems have yielded some striking and important results. For example, Dessein *et al.* (1981a) have examined the effects of IgE depletion on the response of rats to *Trichinella spiralis* infection, and have found that treatment of neonatal rats with anti-epsilon serum both suppresses the subsequnt production of IgE and is associated with an increase in the number of recoverable muscle-stage larvae. This effect is associated with a reduced infiltration of eosinophils around the larvae, suggesting a protective role for the eosinophil-IgE axis.

### D. CORRELATIVE STUDIES IN EXPERIMENTAL ANIMALS AND MAN

Some indication of immunological events occurring in man can be obtained by histological examination of specimens obtained at autopsy or by biopsy, for example of tissues in which helminth migration or helminth-associated damage is occurring. Such investigations have, in general, supported the findings from similar studies in experimental animal models, and have been useful, for example, in establishing the pathogenesis and modulation of the schistosome egg granuloma. However, most specimens can only be taken in an uncontrolled fashion on single or infrequent occasions, and it is difficult to establish the aetiology of the observed reactions or their effects on the parasite. Manipulative procedures of the types that are commonly used in experimental animal systems are inapplicable in man, except when they involve some therapeutic procedure such as passive immunization or immunosuppression. 'Experiments of nature', in which there are selective deficiencies of particular effector systems, are useful, but usually so rare that they can only be investigated in the context of the most abundant of potential parasites: an example being the increased susceptibility of patients with chronic granulomatous disease to infection with a variety of opportunist bacteria. Other genetic characteristics may also be useful, such as the association of HLA with helminth infection or disease (Salem *et al.*, 1979; Sasazuki *et al.*, 1980), but provide general, rather than detailed, information on immunity or pathogenesis. The only method of analysing immune mechanisms in man that is generally applicable on a wide scale is therefore to carry out correlative studies on the relationship between immunity *in vivo* and a variety of immune responses that can be detected either *in vitro* or by simple, non-invasive tests *in vivo*, such as skin-testing for immediate or delayed hypersensitivity reactions. Such correlative studies must be treated with caution, since it is clearly possible to detect false correlations: but no other methods are suitable.

Before coming to the limited data available from human studies, it should be pointed out that such correlative studies may also be useful in experimental animal models. For example, Capron *et al.* (1977b, 1980) have, in addition to the transfer and depletion experiments described above, reported a close correlation between the development of immunity to *Schistosoma mansoni* in the rat and the presence of antibodies of either anaphylactic isotype (IgE or IgG$_{2a}$). During the early stages of immunity, up to 6 weeks after infection, IgG$_{2a}$ antibody, capable of mediating eosinophil-dependent killing as well as mast cell degranulation, is the predominant isotype. Thereafter, during the later stages of infection when immunity is still marked but IgG$_{2a}$ antibodies decline, IgE responses, capable of mediating both eosinophil and macrophage-dependent damage, become more pronounced. This and similar observations suggest that more than one antibody-dependent cell-mediated effector mechanism may be operative at different stages of infection.

Correlative studies in man are difficult and expensive to perform, requiring careful epidemiological design and a close coordination of parasitological, clinical, behavioural and immunological observations. Perhaps the most successful recent example, in infectious diseases, has been the demonstration of an immunologically determined spectrum of disease in leprosy (reviewed by Bullock, 1981). No such example yet exists for helminth infections, although the same approach is valid: many problems remain to be overcome. In schistosomiasis, for example, the difficulty is in determining the extent to which immunity may limit superinfection or of reinfection after treatment, before one can start to ask questions about the mechanism of such immunity. For many years, it has been recognized that the distribution of prevalence or intensity of infection in a population living in an endemic area shows a characteristic pattern, with a rise in prevalence or intensity during the first 10 or 20 years of life, followed by a decline. Originally, it was assumed that this decline was attributable to a slow spontaneous death of adult worms from early infections, together with a slowly acquired resistance to superinfection. More recently, however, it has been argued that this decline could equally well be attributable to the same slow death of adult worms, together with a reduced water contact in the older age groups (Warren, 1973). Any study on immunity, therefore, must take into account the possibility of changes in exposure, and various attempts at this have been made, with more or less success (Kloetzel and Da Silva, 1967; Bradley and McCullough,

1973; McCullough and Bradley, 1973). One promising approach comes from a recent study by Sturrock et al. (1983), who have shown that reinfection after treatment of children with Schistosoma mansoni infection occurs more slowly in those who have both raised eosinophil counts and anti-schistosomular antibodies detectable in a $^{51}$Cr release assay than in those who have neither. This difference was unrelated either to previous intensity of infection or to where the children lived in relation to the contaminated waterbodies, and was therefore probably unrelated to their levels of exposure. This preliminary observation is now being followed by a more detailed study, in which reinfection rates after treatment are being related not only to immune responses detectable in vitro but also to observed water contact habits (Butterworth et al., 1984). The extent of information available from studies in vitro now justifies the adoption of this sort of large-scale field study on immunity, not only in schistosomiasis but also in other helminth infections.

E. CONCLUSIONS

The purpose of this section has been to emphasize the difficulty involved in going from the demonstration of a particular effector mechanism in vitro to the demonstration that the mechanism is indeed involved in mediating immunity in vivo, either in experimental animals or in man. However, methods for making this extrapolation do exist, and the data available so far strongly suggest that antibody-dependent cell-mediated effector mechanisms may be of great importance in mediating immunity to helminths, and thus justify further studies on the nature and mode of action of such mechanisms by experiments in vitro. The importance of such studies from the point of view of human disease lies in rational approaches to vaccination, through the identification of relevant antigens, their production in large amounts by recombinant DNA techniques, and their use in the induction of appropriate immune responses (Butterworth et al., 1982).

VI. Summary and Conclusions

Although it is difficult to draw any sweeping conclusions that would be applicable to all helminth infections, the main features that are emphasized in this review may be summarized briefly.

(1) Pathogenic helminths, although extremely diverse in structure and behaviour, have one common feature, namely that they present to the host's defenses large, non-phagocytosable surfaces.

(2) Because of this, they are susceptible to a range of effector mechanisms differing either quantitatively or qualitatively from those that are active against other parasites or against normal or abnormal host cells. As an extreme example, the various types of cytotoxic lymphocyte, with one interesting exception, are inactive against helminths.

(3) Instead, helminth infections are characterized by high IgE responses and increased numbers of circulating eosinophils. Such eosinophils are activated, and show a marked capacity to kill a variety of target helminths *in vitro*. Further activation may occur in response to mast cell mediators released as a result of IgE-dependent degranulation; and IgE, as well as IgG and complement, can mediate eosinophil attachment and killing. It may therefore be suggested that the eosinophil/IgE/mast cell axis represents a powerful host defense against helminth infections.

(4) IgE can also mediate macrophage-dependent killing of several helminths, a process which involves a functional change in the macrophage, resembling activation.

(5) Although eosinophil-mediated and IgE-dependent macrophage-mediated effects are particularly potent, other effector cells are not excluded: in certain circumstances, neutrophils and conventionally activated macrophages may be equally or more effective. Neutrophils appear to act solely by oxidative killing mechanisms, whereas degranulation and the release of toxic granule contents is equally or more important in eosinophil-mediated damage.

(6) Different stages of different helminths vary in their degree of susceptibility to different mechanisms. Eosinophils appear to be somewhat less active than neutrophils against ensheathed nematodes, whereas trematodes and exsheathed nematodes are highly susceptible to eosinophil attack.

(7) In many experimental helminth infections, studies *in vivo* suggest a role for antibody-dependent cell-mediated immune effector mechanisms. The identity of the effector cell is difficult to establish because of a lack of techniques for specific manipulation of individual cell types, but histological studies frequently point to a strong eosinophil or macrophage involvement.

(8) The development and analysis of *in vitro* assays allows the study of immune effector mechanisms in man. Such studies are difficult to perform, but preliminary work in human schistosomiasis again suggests a role for antibody-dependent cell-mediated effector mechanisms. Further investigations of a more extensive nature both in schistosomiasis and in other human helminth infections, may permit the identification both of relevant target and of relevant immune responses, and hence provide a logical approach towards the development of vaccines for use in man.

ACKNOWLEDGEMENTS

Studies by the author and colleagues referred to in this review were carried out with the financial support of the Wellcome Trust, the Medical Research Council, the Edna McConnell Clark Foundation, and the World Health Organization.

REFERENCES

Abrahamson, I. A. and Dias Da Silva, W. (1977). Antibody-dependent cell-mediated cytotoxicity against *Trypanosoma cruzi*. *Parasitology* **75**, 317–323.

Ackerman, S. J., Gleich, G. J., Weller, P. F. and Ottesen, E. A. (1981). Eosinophilia and elevated serum levels of eosinophil major basic protein and Charcot–Leyden crystal protein (lysophospholipase) after treatment of patients with Bancroft's filariasis. *Journal of Immunology* **127**, 1093–1098.

Ackerman, S. J., Durack, D. T. and Gleich, G. J. (1983). The Gordon phenomenon revisited: purification, localization and physiochemical properties of the human eosinophil-derived neurotoxin. *In* "Immunobiology of the Eosinophil" (T. Yoshida and M. Torisu, eds), pp. 181–200. Elsevier Biomedical, New York.

Anderson, C. L. and Spiegelberg, H. L. (1981). Macrophage receptors for IgE: binding of IgE to specific IgE Fc receptors on a human macrophage cell line, U 937. *Journal of Immunology* **126**, 2470–2473.

Anderson, S. E., Bautista, S. and Remington, J. S. (1976). Induction of resistance to *Toxoplasma gondii* in human macrophages by soluble lymphocyte products. *Journal of Immunology* **117**, 381–387.

Anteunis, A., Perrudet-Badoux, A., Astesano, A., Boussac-Aron, Y. and Binaghi, R. (1980). Étude ultrastructurale de la destruction immune de larves nouveau-nées de *Trichinella spiralis* par des cellules péritonéales. *Comptes rendus, Academie des sciences (Paris)* D **290**, 979–981.

Anwar, A. R. E. and Kay, A. B. (1977). Membrane receptors for IgG and complement (C4, C3b, and C3d) on human eosinophils and neutrophils and their relation to eosinophilia. *Journal of Immunology* **119**, 976–982.

Anwar, A. R. E. and Kay, A. B. (1978). Enhancement of human eosinophil complement receptors by pharmacologic mediators. *Journal of Immunology* **121**, 1245–1250.

Anwar, A. R. E., Smithers, S. R. and Kay, A. B. (1979). Killing of schistosomula of *Schistosoma mansoni* coated with antibody and/or complement by human leukocytes *in vitro*: requirement for complement in preferential killing by eosinophils. *Journal of Immunology* **122**, 628–637.

Anwar, A. R. E., McKean, J. R., Smithers, S. R. and Kay, A. B. (1980). Human eosinophil- and neutrophil-mediated killing of schistosomula of *Schistosoma mansoni in vitro*. I. Enhancement of complement-dependent damage by mast cell-derived mediators and formyl methionyl peptides. *Journal of Immunology* **124**, 1122–1129.

Archer, G. T. and Hirsch, J. G. (1963). Motion picture studies on degranulation of horse eosinophils during phagocytosis. *Journal of Experimental Medicine* **118**, 287–294.

Archer, G. T., Robson, J. E. and Thompson, A. R. (1977). Eosinophilia and mast cell hyperplasia induced by parasite phospholipid. *Pathology* **9**, 137–153.

Attallah, A. M., Lewis, F. A., Urritia-Shaw, A., Folks, T. and Yeatman, T. J. (1980). Natural killer cells (NK) and antibody-dependent cell-mediated cytotoxicity (ADCC) components of *Schistosoma mansoni* infection. *International Archives of Allergy and Applied Immunology* **63**, 351–354.

Auriault, C., Joseph, M., Dessaint, J.-P. and Capron, A. (1980). Inactivation of rat macrophages by peptides resulting from cleavage of IgG by *Schistosoma* larvae proteases. *Immunology Letters* **2**, 135–139.

Auriault, C., Pestel, J., Joseph, M., Dessaint, J. P. and Capron, A. (1981). Interaction between macrophages and *Schistosoma mansoni* schistosomula: role of IgG peptides and aggregates on the modulation of beta-glucuronidase release and the cytotoxicity against schistosomula. *Cellular Immunology* **62**, 15–27.

Auriault, C., Capron, M. and Capron, A. (1982). Activation of rat and human eosinophils by soluble factor(s) released by *Schistosoma mansoni* schistosomula. *Cellular Immunology* **66**, 59–69.

Babior, B. M. (1978). Oxygen-dependent microbial killing by phagocytes. *New England Journal of Medicine* **298**, 659–668.

Baehner, R. and Johnston, R. B. (1971). Metabolic and bactericidal activities of human eosinophils. *British Journal of Haematology* **20**, 277–285.

Bagai, R. C. and Subrahmanyam, D. (1970). Studies on the host parasite relationship in albino rats infected with *Litomosoides carinii*. *Nature* **228**, 682–684.

Baron, R. W. and Tanner, C. E. (1977). *Echinococcus multilocularis* in the mouse: the *in vitro* protoscolicidal activity of peritoneal macrophages. *International Journal for Parasitology* **7**, 489–495.

Bass, D. A. and Szejda, P. (1979a). Eosinophils versus neutrophils in host defense. Killing of newborn larvae of *Trichinella spiralis* by human granulocytes. *Journal of Clinical Investigation* **64**, 1415–1422.

Bass, D. A. and Szejda, P. (1979b). Mechanisms of killing of newborn larvae of *Trichinella spiralis* by neutrophils and eosinophils. Killing by generators of hydrogen peroxide *in vitro*. *Journal of Clinical Investigation* **64**, 1558–1564.

Bass, D. A., Grover, W. H., Lewis, J., Szejda, P., DeChatelet, L. R. and McCall, C. E. (1980). Comparison of human eosinophils from normals and patients with eosinophilia. *Journal of Clinical Investigation* **66**, 1265–1273.

Basten, A. and Beeson, P. B. (1970). Mechanism of eosinophilia. II. Role of the lymphocyte. *Journal of Experimental Medicine* **131**, 1288–1305.

Basten, A., Boyer, M. H. and Beeson, P. B. (1970). Mechanism of eosinophilia. I. Factors affecting the eosinophil response of rats to *Trichinella spiralis*. *Journal of Experimental Medicine* **131**, 1271–1287.

Bazin, H. and Pauwels, R. (1982). IgE and Ig2a isotypes in the rat. *Progress in Allergy* **32**, 52–104.

Bazin, H. Capron, A., Capron, M., Joseph, M., Dessaint, J. P. and Pauwels, R. (1980). Effect of neonatal injection of anti-$\mu$ antibodies on immunity to schistosomes (*Schistosoma mansoni*) in the rat. *Journal of Immunology* **124**, 2373–2377.

Bennett, C. E., Hughes, D. L. and Harness, E. (1980). *Fasciola hepatica*: changes in tegument during killing of adult flukes surgically transferred to sensitized rats. *Parasite Immunology* **2**, 39–55.

Bentley, A. G., Carlisle, A. S. and Phillips, S. M. (1981a). Ultrastructural analysis of the cellular response to *Schistosoma mansoni*: initial and challenge infections in the rat. *American Journal of Tropical Medicine and Hygiene* **30**, 102–112.

Bentley, A. G., Carlisle, A. S. and Phillips, S. M. (1981b). Ultrastructural analysis of the cellular response to *Schistosoma mansoni*. II. Inflammatory response in rodent skin. *American Journal of Tropical Medicine and Hygiene* **30,** 815–824.

Bentley, A. G., Doughty, B. L. and Phillips, S. M. (1982). Ultrastructural analysis of the cellular response to *Schistosoma mansoni*. III. The *in vitro* granuloma. *American Journal of Tropical Medicine and Hygiene* **31,** 1168–1180.

Benveniste, J. (1974). Platelet-activating factor, a new mediator of anaphylaxis and immune complex deposition from rabbit and human basophils. *Nature* **249,** 581–582.

Berke, G. and Clark, W. R. (1982). T lymphocyte-mediated cytolysis—a comprehensive theory. I. The mechanism of CTL-mediated cytolysis. *Advances in Experimental Medicine and Biology* **146,** 57–68.

Bickle, Q. D. and Ford, M. J. (1982). Studies on the surface antigenicity and susceptibility to antibody-dependent killing of developing schistosomula using sera from chronically infected mice and mice vaccinated with irradiated cercariae. *Journal of Immunology* **128,** 101–106.

Bickle, Q. D., Taylor, M. G., Doenhoff, M. J. and Nelson, G. S. (1979). Immunization of mice with gamma-irradiated intramuscularly injected schistosomula of *Schistosoma mansoni*. *Parasitology* **79,** 209–222.

Blanden, R. V. (1974). T cell response to viral and bacterial infection. *Transplantation Reviews* **19,** 56–88.

Blum, K. and Cioli, D. (1981). *Schistosoma mansoni*: age-dependent susceptibility to immune elimination of schistosomula artificially introduced into preinfected mice. *Parasite Immunology* **3,** 13–24.

Boltz-Nitulescu, G. and Spiegelberg, H. L. (1981). Receptors specific for IgE on rat alveolar and peritoneal macrophages. *Cellular Immunology* **59,** 106–114.

Boswell, R. N., Austen, K. F. and Goetzl, E. J. (1978). Intermediate molecular weight eosinophil chemotactic factors in rat peritoneal mast cells: immunologic release, granule association, and demonstration of structural heterogeneity. *Journal of Immunology* **120,** 15–20.

Bout, D. T., Joseph, M., David, J. R. and Capron, A. R. (1981). *In vitro* killing of *S. mansoni* schistosomula by lymphokine-activated mouse macrophages. *Journal of Immunology* **127,** 1–5.

Bradley, D. J. and McCullough, F. S. (1973). Egg output stability and the epidemiology of *Schistosoma haematobium*. Part II. An analysis of the epidemiology of endemic *S. haematobium*. *Transactions of the Royal Society of Tropical Medicine and Hygiene* **67,** 491–500.

Brink, L. H., McLaren, D. J. and Smithers, S. R. (1977). *Schistosoma mansoni*: a comparative study of artificially-transformed schistosomula and schistosomula recovered after cercarial penetration of isolated skin. *Parasitology* **74,** 73–86.

Brink, L. H., Krueger, K. L. and Harris, C. (1980). Stage-specific antigens of Schistosoma mansoni. *In* "The Host Invader Interplay" (H. Van den Bossche, ed.), pp. 393–404. Elsevier/North Holland Biomedical Press, Amsterdam.

Brown, A. P., Burakoff, S. J. and Sher, A. (1980). Specificity of alloreactive T lymphocytes that adhere to lung stage schistosomula of *Schistosoma mansoni*. *Journal of Immunology* **124,** 2516–2518.

Bryceson, A. D., Warrell, D. A. and Pope, H. M. (1977). Dangerous reactions to treatment of onchocerciasis with carbamazine. *British Medical Journal* **i,** 742–744.

Bullock, W. E. (1981). Immunobiology of leprosy. *In* "Immunology of human infection. Part I: Bacteria, Mycoplasmae, Chlamydiae and Fungi" (A. J. Nahmias and R. J. O'Reilly, eds), pp. 369–390. Plenum, New York.

Burchard, G. D. and Bierther, M. (1978). Electron microscopical studies on onchocerciasis. I. Mesenchyme reaction in untreated onchocercal dermatitis and ultrastructure of the microfilariae. *Tropenmedizin und Parasitologie* **29**, 451–461.
Burchard, G. D., Buttner, D. W. and Bierther, M. (1979). Electron microscopical studies on onchocerciasis. III. The onchocerca-nodule. *Tropenmedizin und Parasitologie* **30**, 103–112.
Butterworth, A. E. (1977). The eosinophil and its role in immunity to helminth infection. *Current Topics in Microbiology and Immunology* **77**, 127–168.
Butterworth, A. E. and Vadas, M. A. (1979). Immunological studies on schistosomes cultured *in vitro*. In "Practical Tissue Culture Applications" (K. Maramorosch and H. Hirumi, eds), pp. 287–307. Academic Press, New York.
Butterworth, A. E., Sturrock, R. F., Houba, V. and Rees, P. H. (1974). Antibody-dependent cell-mediated damage to schistosomula *in vitro*. *Nature* **252**, 503–505.
Butterworth, A. E., Sturrock, R. F., Houba, V., Mahmoud, A. A. F., Sher, A. and Rees, P. H. (1975). Eosinophils as mediators of antibody-dependent damage to schistosomula. *Nature* **256**, 727–729.
Butterworth, A. E., Sturrock, R. F., Houba, V. and Taylor, R. (1976a). *Schistosoma mansoni* in baboons. Antibody-dependent cell-mediated damage to $^{51}$Cr-labelled schistosomula. *Clinical and Experimental Immunology* **25**, 95–102.
Butterworth, A. E., Coombs, R. R. A., Gurner, B. W. and Wilson, A. B. (1976b). Receptors for antibody-opsonic adherence on the eosinophils of guinea-pigs. *International Archives of Allergy and Applied Immunology* **51**, 368–377.
Butterworth, A. E., David, J. R., Franks, D., Mahmoud, A. A. F., David, P. H., Sturrock, R. F. and Houba, V. (1977a). Antibody-dependent eosinophil-mediated damage to $^{51}$Cr-labeled schistosomula of *Schistosoma mansoni*: damage by purified eosinophils. *Journal of Experimental Medicine* **145**, 136–150.
Butterworth, A. E., Remold, H. G., Houba, V., David, J. R., Franks, D., David, P. H. and Sturrock, R. F. (1977b). Antibody-dependent eosinophil-mediated damage to $^{51}$Cr-labeled schistosomula of *Schistosoma mansoni*: mediation by IgG, and inhibition by antigen-antibody complexes. *Journal of Immunology* **118**, 2230–2236.
Butterworth, A. E., Vadas, M. A., Wassom, D. L., Dessein, A., Hogan, M., Sherry, B., Gleich, G. J. and David, J. R. (1979a). Interactions between human eosinophils and schistosomula of *Schistosoma mansoni*. II. The mechanism of irreversible eosinophil adherence. *Journal of Experimental Medicine* **150**, 1456–1471.
Butterworth, A. E., Wassom, D. L., Gleich, G. J., Loegering, D. A. and David, J. R. (1979b). Damage to schistosomula of *Schistosoma mansoni* induced directly by eosinophil major basic protein. *Journal of Immunology* **122**, 221–229.
Butterworth, A. E., Vadas, M. A., Martz, E. and Sher, A. (1979c). Cytolytic T lymphocytes recognize alloantigens on schistosomula of *Schistosoma mansoni*, but fail to induce damage. *Journal of Immunology* **122**, 1314–1321.
Butterworth, A. E., Vadas, M. A. and David, J. R. (1980). Mechanisms of eosinophil-mediated helminthotoxicity. In "The Eosinophil in Health and Disease" (A. A. F. Mahmoud and K. F. Austen, eds), pp. 253–273. Grune and Stratton, New York.
Butterworth, A. E., Taylor, D. W., Veith, M. C., Vadas, M. A., Dessein, A., Sturrock, R. F. and Wells, E. (1982). Studies on the mechanisms of immunity in human schistosomiasis. *Immunological Reviews* **61**, 5–39.
Butterworth, A. E., Dalton, P. R., Dunne, D. W., Mugambi, M., Ouma, J. H., Richardson, B. A., Arap Siongok, T. K. and Sturrock, R. F. (1984). Immunity after treatment of human schistosomiasis mansoni. I. Study design, pretreatment observations and the results of treatment. *Transactions of the Royal Society of Tropical Medicine and Hygiene*, in press.

Buys, J., Wever, R., van Stigt, R. and Ruitenberg, E. J. (1981). The killing of newborn larvae of *Trichinella spiralis* by eosinophil peroxidase *in vitro*. *European Journal of Immunology* **11**, 843–845.

Camp, C. J. and Leid, R. W. (1982). Chemokinetic factors obtained from the larval stage of the cestode, *Taenia taeniaeformis*. *Parasite Immunology* **4**, 373–381.

Capron, A., Dessaint, J. P., Capron, M. and Bazin, H. (1975a). Specific IgE antibodies in immune adherence of normal macrophages to *Schistosoma mansoni* schistosomules. *Nature* **253**, 474–475.

Capron, A., Bazin, H., Dessaint, J. P. and Capron, M. (1975b). Role des anticorps IgE specifiques dans l'adherence immune de macrophages normaux aux schistosomules de *Schistosoma mansoni*. *Comptes rendus, Académie des sciences (Paris) D* **280**, 927–930.

Capron, A., Dessaint, J. P., Joseph, M., Rousseaux, R. and Bazin, H. (1976). Macrophage cytotoxicity mediated by IgE antibody. *In* "Molecular and Biological Aspects of the Acute Allergic Reaction" (S. G. O. Johansson, K., Strandberg and B. Urnäs, eds), pp. 153–171. Plenum Publishing Corporation, New York.

Capron, A., Dessaint, J. P., Joseph, M., Rousseaux, R., Capron, M. and Bazin, H. (1977a). Interaction between IgE complexes and macrophages in the rat: a new mechanism of macrophage activation. *European Journal of Immunology* **7**, 315–322.

Capron, A., Dessaint, J. P., Joseph, M., Torpier, G., Capron, M., Rousseaux, R., Santoro, F. and Bazin, H. (1977b). IgE and cells in schistosomiasis. *American Journal of Tropical Medicine and Hygiene* **26**, Supplement, 39–47.

Capron, A., Dessaint, J. P., Capron, M., Joseph, M. and Pestel, J. (1980). Role of anaphylactic antibodies in immunity to schistosomes. *American Journal of Tropical Medicine and Hygiene* **29**, 849–857.

Capron, A., Dessaint, J. P., Capron, M., Joseph, M. and Torpier, G. (1982). Effector mechanisms of immunity to schistosomes and their regulation. *Immunological Reviews* **61**, 41–66.

Capron, M., Capron, A., Torpier, G., Bazin, H., Bout, D. and Joseph, M. (1978a). Eosinophil-dependent cytotoxicity in rat schistosomiasis. Involvement of IgG2a antibody and role of mast cells. *European Journal of Immunology* **8**, 127–133.

Capron, M., Rousseaux, J., Mazingue, C., Bazin, H. and Capron, A. (1978b). Rat mast cell-eosinophil interaction in antibody-dependent eosinophil cytotoxicity to *Schistosoma mansoni* schistosomula. *Journal of Immunology* **121**, 2518–2525.

Capron, M., Torpier, G. and Capron, A. (1979). *In vitro* killing of *S. mansoni* schistosomula by eosinophils from infected rats: role of cytophilic antibodies. *Journal of Immunology* **123**, 2220–2230.

Capron, M., Capron, A., Dessaint, J. P., Torpier, G., Johansson, S. G. O. and Prin, L. (1981a). Fc receptors for IgE on human and rat eosinophils. *Journal of Immunology* **126**, 2087–2092.

Capron, M., Bazin, H., Joseph, M. and Capron, A. (1981b). Evidence for IgE-dependent cytotoxicity by rat eosinophils. *Journal of Immunology* **126**, 1764–1768.

Capron, M., Capron, A., Goetzl, E. J. and Austen, K. F. (1981c). Tetrapeptides of the eosinophil chemotactic factor of anaphylaxis (ECF-A) enhance eosinophil Fc receptor. *Nature* **289**, 71–73.

Caulfield, J. P., Korman, G., Butterworth, A. E., Hogan, M. and David, J. R. (1980a). The adherence of human neutrophils and eosinophils to schistosomula: evidence for membrane fusion between cells and parasites. *Journal of Cell Biology* **86**, 46–63.

Caulfield, J. P., Korman, G., Butterworth, A. E., Hogan, M. and David, J. R. (1980b). Partial and complete detachment of neutrophils and eosinophils from schistosomula: evidence for the establishment of continuity between a fused and normal parasite membrane. *Journal of Cell Biology* **86**, 64–76.

Caulfield, J. P., Hein, A., Moser, G. and Sher, A. (1981). Light and electron microscopic appearance of rat peritoneal mast cells adhering to schistosomula of *Schistosoma mansoni* by means of complement or antibody. *Journal of Parasitology* **67**, 776–783.

Caulfield, J. P., Korman, G. and Samuelson, J. C. (1982). Human neutrophils endocytose multivalent ligands from the surface of schistosomula of *Schistosoma mansoni* before membrane fusion. *Journal of Cell Biology* **94**, 370–378.

Chaicumpa, V. and Jenkin, C. R. (1978). Studies *in vitro* on the reaction of peritoneal exudate cells from mice immune to infection with *Nematospiroides dubius* with the infective third stage larvae of this parasite. *Australian Journal of Experimental Biology and Medical Science* **56**, 61–68.

Civil, R. H., Warren, K. S. and Mahmoud, A. A. F. (1978). Conditions of bacille Calmette-Guerin induced resistance to *Schistosoma mansoni* in mice. *Journal of Infectious Diseases* **137**, 550–558.

Claas, F. H. J. and Deelder, A. M. (1979). H-2 linked immune response to murine experimental *Schistosoma mansoni* infections. *Journal of Immunogenetics* **6**, 167–175.

Claas, F. H. J. and Deelder, A. M. (1980). Influence of the I region of the H-2 complex on the immune response to *Schistosoma mansoni* infection. *Acta Leidensia* **48**, 23–27.

Clark, R. A. F., Gallin, J. I. and Kaplan, A. P. (1975). The selective eosinophil chemotactic activity of histamine. *Journal of Experimental Medicine* **142**, 1462–1476.

Clegg, J. A. and Smithers, S. R. (1972). The effects of immune rhesus monkey serum on schistosomula of *Schistosoma mansoni* during cultivation *in vitro*. *International Journal of Parasitology* **2**, 79–98.

Clegg, J. A., Smithers, S. R. and Terry, R. J. (1971). Acquisition of human antigens by *Schistosoma mansoni* during cultivation *in vitro*. *Nature* **232**, 653–654.

Cline, M. J., Hanifin, J. and Lehrer, R. I. (1968). Phagocytosis by human eosinophils. *Blood* **32**, 922–934.

Cohen, S. G. and Sapp, T. M. (1969). Phagocytosis of bacteria by eosinophils in infections-related asthma. *Journal of Allergy* **44**, 113–117.

Colley, D. G. (1972). *Schistosoma mansoni*: eosinophilia and the development of lymphocyte blastogenesis in response to soluble egg antigen in inbred mice. *Experimental Parasitology* **32**, 520–526.

Colley, D. G. (1973). Eosinophils and immune mechanisms. Eosinophil stimulation promoter (ESP): a lymphokine induced by specific antigen or phytohaemagglutinin. *Journal of Immunology*, **110**, 1419–1423.

Colley, D. G. (1974). Variations in peripheral blood eosinophil levels in normal and *Schistosoma mansoni*-infected mice. *Journal of Laboratory and Clinical Medicine* **83**, 871–876.

Colley, D. G. (1976). Eosinophils and immune mechanisms. IV. Culture conditions, antigen requirements, production kinetics and immunologic specificity of the lymphokine eosinophil stimulation promoter. *Cellular Immunology* **24**, 328–335.

Colley, D. G. (1980). Lymphokine-related eosinophil responses. *Lymphokine Reports* **1**, 133–155.

Colley, D. G., Savage, A. M. and Lewis, F. A. (1977). Host responses induced and elicited by cercariae, schistosomula, and cercarial antigenic preparations. *American Journal of Tropical Medicine and Hygiene* **26**, 88–95.

Connell, J. T. (1968). Morphological changes in eosinophils in allergic disease. *Journal of Allergy* **41**, 1–9.

Copeland, D. and Grove, D. I. (1979). Effects of *Toxoplasma gondii* (Gleadle strain) on the host–parasite relationship in trichinosis. *International Journal of Parasitology*, **9**, 205–211.

Crandall, R. B., McGreevy, P. B., Connor, D. H., Crandall, C. A., Neilson, J. T. and McCall, J. W. (1982). The ferret (*Mustela putorius furo*) as an experimental host for *Brugia malayi* and *Brugia pahangi*. *American Journal of Tropical Medicine and Hygiene* **31**, 752–759.

Czarnetzki, B. M. (1978). Eosinophil chemotactic factor release from neutrophils by *Nippostrongylus brasiliensis* larvae. *Nature* **271**, 553–554.

David, J. R., Butterworth, A. E., Remold, H. G., David, P. H., Houba, V. and Sturrock, R. F. (1977). Antibody-dependent, eosinophil-mediated damage to $^{51}$Cr-labeled schistosomula of *Schistosoma mansoni*: effect of metabolic inhibitors and other agents which alter cell function. *Journal of Immunology* **118**, 2221–2229.

David, J. R., Vadas, M. A., Butterworth, A. E., de Brito, P. A., Carvalho, E. M., David, R. A., Bina, J. C. and Andrade, Z. A. (1980). Enhanced helminthotoxic capacity of eosinophils from patients with eosinophilia. *New England Journal of Medicine* **303**, 1147–1152.

Davies, C. and Goose, J. (1981). Killing of newly excysted juveniles of *Fasciola hepatica* in sensitized rats. *Parasite Immunology* **3**, 81–96.

Day, R. P. (1970). Eosinophil cell separation from human peripheral blood. *Immunology* **18**, 955–959.

Dean, D. A. (1977). Decreased binding of cytotoxic antibody by developing *Schistosoma mansoni*. Evidence for a surface change independent of host antigen adsorption and membrane turnover. *Journal of Parasitology* **63**, 418–426.

Dean, D. A., Wistar, R. and Murrell, K. D. (1974). Combined *in vitro* effects of rat antibody and neutrophilic leukocytes on schistosomula of *Schistosoma mansoni*. *American Journal of Tropical Medicine and Hygiene* **23**, 420–428.

Dean, D. A., Wistar, R. and Chen, P. (1975). Immune response of guinea pigs to *Schistosoma mansoni*. I. *In vitro* effects of antibody and neutrophils, eosinophils and macrophages on schistosomula. *American Journal of Tropical Medicine and Hygiene* **24**, 74–82.

DeChatelet, L. R., Shirley, P. S., McPhail, L. C., Huntley, C. C., Mus, H. B. and Bass, D. A. (1977). Oxidative metabolism of the human eosinophil. *Blood* **50**, 525–534.

Densen, P., Mahmoud, A. A., Sullivan, J., Warren, K. S. and Mandell, G. L. (1978). Demonstration of eosinophil degranulation on the surface of opsonized schistosomules by phase-contrast cinemicrography. *Infection and Immunity* **22**, 282–285.

Despommier, D. (1977). Immunity to *Trichinella spiralis*. *American Journal of Tropical Medicine and Hygiene* **26**, 68–75.

Despommier, D., Weisbroth, S. and Fass, C. (1974). Circulating eosinophils and trichinosis in the rat: the parasitic stage responsible for induction during infection. *Journal of Parasitology* **60**, 280–284.

Despommier, D. D., Campbell, W. C. and Blair, L. S. (1977). The *in vivo* and *in vitro* analysis of immunity to *Trichinella spiralis* in mice and rats. *Parasitology* **74**, 109–119.

Dessaint, J.-P., Torpier, G., Capron, M., Bazin, H. and Capron, A. (1979a). Cytophilic binding of IgE to the macrophage. I. Binding characteristics of IgE on the surface of macrophages in the rat. *Cellular Immunology* **46**, 12–23.

Dessaint, J.-P., Capron, A., Joseph, M. and Bazin, H. (1979b). Cytophilic binding of IgE to the macrophage. II. Immunologic release of lysosomal enzyme from macrophages by IgE and anti-IgE in the rat: a new mechanism of macrophage activation. *Cellular Immunology* **46**, 24–34.

Dessaint, J.-P., Waksman, B. H., Metzger, H. and Capron, A. (1980). Cytophilic binding of IgE to the macrophage. III. Involvement of cyclic GMP and calcium in macrophage activation by dimeric or aggregated rat myeloma IgE. *Cellular Immunology* **51**, 280–292.

Dessein, A. J., Parker, W. L., James, S. L. and David, J. R. (1981a). IgE antibody and resistance to infection. I. Selective suppression of the IgE antibody response in rats diminishes the resistance and the eosinophil response to *Trichinella spiralis* infection. *Journal of Experimental Medicine* **153**, 423–436.

Dessein, A. J., Samuelson, J. F., Butterworth, A. E., Hogan, M., Sherry, B. A., Vadas, M. A. and David, J. R. (1981b). Immune evasion by *Schistosoma mansoni*: loss of susceptibility to antibody or complement-dependent eosinophil attack by schistosomula cultured in medium free of macromolecules. *Parasitology* **82**, 357–374.

Dessein, A. J., Vadas, M. A., Nicola, N. A., Metcalf, D. and David, J. R. (1982). Enhancement of human blood eosinophil cytotoxicity by semi-purified eosinophil colony-stimulating factor(s). *Journal of Experimental Medicine* **156**, 90–103.

Dessein, A. J., Lenzi, H. L., Vadas, M. A. and David, J. R. (1983a). A new class of eosinophil activators that enhance eosinophil helminthotoxicity. *In* "Immunobiology of the Eosinophil" (T. Yoshida and M. Torisu, eds), pp. 369–382. Elsevier Biomedical, New York.

Dessein, A. J., Butterworth, A. E., Vadas, M. A. and David, J. R. (1983b). Maturation *in vivo* of *Schistosoma mansoni* schistosomula after culture *in vitro* with granulocytes and antibody. *Infection and Immunity* **39**, 225–232.

Dissous, C. and Capron, A. (1981). Isolation and characterisation of surface antigens from *Schistosoma mansoni* schistosomula. *Molecular and Biochemical Parasitology* **3**, 215–225.

Dissous, C., Grzych, J.-M. and Capron, A. (1982). *Schistosoma mansoni* surface antigen defined by a rat monoclonal IgG2a. *Journal of Immunology* **129**, 2232–2234.

Dobson, N. J., Lambris, J. D. and Ross, G. D. (1981). Characteristics of isolated erythrocyte complement receptor type one (CR$_1$, C4b-C3b receptor) and CR$_1$-specific antibodies. *Journal of Immunology* **126**, 693–698.

Doy, T. G. and Hughes, D. L. (1982a). The role of the thymus in the eosinophil response of rats infected with *Fasciola hepatica*. *Clinical and Experimental Immunology* **47**, 74–76.

Doy, T. G. and Hughes, D. L. (1982b). *In vitro* cell adherence to newly excysted *Fasciola hepatica*: failure to affect their subsequent development in rats. *Research in Veterinary Science* **32**, 118–120.

Doy, T. G. and Hughes, D. L. (1982c). Evidence for two distinct mechanisms of resistance in the rat to reinfection with *Fasciola hepatica*. *International Journal of Parasitology* **12**, 357–361.

Doy, T. G., Hughes, D. L. and Harness, E. (1980). The selective adherence of rat eosinophils to newly excysted *Fasciola hepatica in vitro*. *Research in Veterinary Science* **29**, 98–101.

Doy, T. G., Hughes, D. L. and Harness, E. (1981a). Hypersensitivity in rats infected with *Fasciola hepatica*: possible role in protection against a challenge infection. *Research in Veterinary Science* **30**, 360–363.

Doy, T. G., Hughes, D. L. and Harness, E. (1981b). The heterologous protection of rats against a challenge with *Fasciola hepatica* by prior infection with the nematode *Nippostrongylus brasiliensis*. *Parasite Immunology* **3**, 171–180.

Doyle, J. J. (1971). Acquired immunity to experimental infection with *Fasciola hepatica* in cattle. *Research in Veterinary Science* **12**, 527–534.

Duffus, W. P. and Franks, D. (1980). *In vitro* effect of immune serum and bovine granulocytes on juvenile *Fasciola hepatica*. *Clinical and Experimental Immunology* **41**, 430–440.

Duffus, W. P., Thorne, K. and Oliver, R. (1980). Killing of juvenile *Fasciola hepatica* by purified bovine eosinophil proteins. *Clinical and Experimental Immunology* **40**, 336–344.

Dunne, D. W., Lucas, S., Bickle, Q., Pearson, S., Madgwick, L. Bain, J. and Doenhoff, M. J. (1981). Identification and partial purification of an antigen ($\omega_1$) from *Schistosoma mansoni* eggs which is putatively hepatotoxic in T-cell deprived mice. *Transactions of the Royal Society of Tropical Medicine and Hygiene* **75**, 54–71.

Ehlenberger, A. G. and Nussenzweig, V. (1977). The role of membrane receptors for C3b and C3d in phagocytosis. *Journal of Experimental Medicine* **145**, 357–371.

Ellner, J. J. and Mahmoud, A. A. F. (1979). Killing of schistosomula of *Schistosoma mansoni* by normal human monocytes. *Journal of Immunology* **123**, 949–951.

Ellner, J. J., Olds, G. R., Lee, C. W., Kleinhenz, M. E. and Edwards, K. L. (1982). Destruction of the multicellular parasite *Schistosoma mansoni* by T lymphocytes. *Journal of Clinical Investigation* **70**, 369–378.

El-Sadr, W. M., Aikawa, M. and Greene, B. M. (1983). *In vitro* immune mechanisms associated with clearance of microfilariae of *Dirofilaria immitis*. *Journal of Immunology* **130**, 428–434.

Enzenauer, R. W. and Yamaoka, R. M. (1982). Eosinophilic meningitis and hydrocephalus in an infant. *Archives of Neurology* **39**, 380–381.

Fearon, D. T. (1980). Identification of the membrane glycoprotein that is the C3b receptor of the human erythrocyte, polymorphonuclear leukocyte, B lymphocyte and monocyte. *Journal of Experimental Medicine* **152**, 20–30.

Fine, D. P., Buchanan, R. D. and Colley, D. G. (1973). *Schistosoma mansoni* infection in mice depleted of thymus-dependent lymphocytes. I. Eosinophilia and immunologic responses to a schistosomal egg preparation. *American Journal of Pathology* **71**, 193–206.

Gärtner, I. (1980). Separation of human eosinophils in density gradients of polyvinylpyrrolidine-coated silica gel (Percoll). *Immunology* **40**, 133–136.

Gentilini, M., Pinon, J. M., Sulahian, A., Nosny, Y. and Maillard, F. (1975). Poumon eosinophile filarien, interêt de l'immunologie (à propos de 9 cas). *Bulletin de la Societé de Pathologie Exotique* **68**, 496–503.

Gibson, D. W., Connor, D. H., Brown, H. L., Fuglsang, H., Anderson, J., Duke, B. O. L. and Buck, A. A. (1976). Onchocercal dermatitis: ultrastructural studies of microfilariae and host tissues, before and after treatment with diethylcarbamazine (Hetrazan[R]). *American Journal of Tropical Medicine and Hygiene* **25**, 74–80.

Gitter, B. D., McCormick, S. L. and Damian, R. T. (1982). Murine alloantigen acquisition by *Schistosoma mansoni*: presence of H-2K determinants on adult worms and failure of allogeneic lymphocytes to recognize acquired MHC gene products on schistosomula. *Journal of Parasitology* **68**, 513–518.

Glauert, A. M. and Butterworth, A. E. (1977). Morphological evidence for the ability of eosinophils to damage antibody-coated schistosomula. *Transactions of the Royal Society of Tropical Medicine and Hygiene* **71**, 392–395.

Glauert, A. M. and Sanderson, C. J. (1979). The mechanism of K cell (antibody-dependent) mediated cytotoxicity. III. The ultrastructure of K cell projections and their possible role in target cell killing. *Journal of Cell Science*, **35**, 355–366.

Glauert, A. M., Butterworth, A. E., Sturrock, R. F. and Houba, V. (1978). The mechanism of antibody-dependent, eosinophil-mediated damage to schistosomula of *Schistosoma mansoni in vitro*: a study by phase-contrast and electron microscopy. *Journal of Cell Science* **34**, 173–192.

Glauert, A. M., Oliver, R. C. and Thorne, K. J. (1980). The interaction of human eosinophils and neutrophils with non-phagocytosable surfaces: a model for studying cell-mediated immunity in schistosomiasis. *Parasitology* **80**, 525–537.

Gleich, G. J., Loegering, D. A. and Maldonado, J. E. (1973). Identification of a major basic protein in guinea pig eosinophil granules. *Journal of Experimental Medicine* **137**, 1459–1471.

Gleich, G. J., Loegering, D. A., Kueppers, F., Bajaj, S. P. and Mann, K. G. (1974). Physicochemical and biological properties of the major basic protein from guinea pig eosinophil granules. *Journal of Experimental Medicine* **140**, 313–332.

Gleich, G. J., Loegering, D. A., Mann, K. G. and Maldonado, J. E. (1976). Comparative properties of the Charcot-Leyden crystal protein and the major basic protein from human eosinophils. *Journal of Clinical Investigation* **57**, 633–640.

Goetzl, E. J. and Austen, K. F. (1975). Purification and synthesis of eosinophilotactic tetrapeptides of human lung tissue: identification as eosinophil chemotactic factor of anaphylaxis. *Proceedings of the National Academy of Sciences of the United States of America* **72**, 4123–4127.

Goetzl, E. J. and Austen, K. F. (1976). Structural determinants of the eosinophil chemotactic activity of the acidic tetrapeptides of ECF-A. *Journal of Experimental Medicine* **144**, 1424–1437.

Goetzl, E. J. and Gorman, R. R. (1978). Chemotactic and chemokinetic stimulation of human eosinophil and neutrophil polymorphonuclear leukocytes by 12-L-hydroxy-5,8,10-heptadecatrienoic acid (HHT). *Journal of Immunology* **120**, 526–531.

Goetzl, E. J., Wasserman, S. I. and Austen, K. F. (1975). Eosinophil polymorphonuclear leukocyte function in immediate hypersensitivity. *Archives of Pathology* **99**, 1–4.

Goetzl, E. J., Woods, J. M. and Gorman, R. R. (1977). Stimulation of human eosinophil and neutrophil polymorphonuclear leukocyte chemotaxis and random migration by 12-L-hydroxy-5,8,10,14-eicosatetraenoic acid (HETE). *Journal of Clinical Investigation* **59**, 179–183.

Goldring, O. L., Clegg, J. A., Smithers, S. R. and Terry, R. J. (1976). Acquisition of human blood group antigens by *Schistosoma mansoni*. *Clinical and Experimental Immunology* **26**, 181–187.

Goldring, O. L., Kusel, J. R. and Smithers, S. R. (1977a). *Schistosoma mansoni*: origin *in vitro* of host-like surface antigens. *Experimental Parasitology* **43**, 82–93.

Goldring, O. L., Sher, A., Smithers, S. R. and McLaren, D. J. (1977b). Host antigens and parasite antigens of murine *Schistosoma mansoni*. *Transactions of the Royal Society of Tropical Medicine and Hygiene* **71**, 144–148.

Greene, B. M. and Colley, D. G. (1974). Eosinophils and immune mechanisms. II. Partial characterization of the lymphokine eosinophil stimulation promoter. *Journal of Immunology* **113**, 910–917.

Greene, B. M. and Colley, D. G. (1976). Eosinophils and immune mechanisms. III. Production of the lymphokine eosinophil stimulation promoter by mouse T lymphocytes. *Journal of Immunology* **116**, 1078–1083.

Greene, B. M., Taylor, H. R. and Aikawa, M. (1981). Cellular killing of microfilariae of *Onchocerca volvulus*: eosinophil and neutrophil-mediated immune serum-dependent destruction. *Journal of Immunology* **127**, 1611–1618.

Grove, D. I. and Civil, R. H. (1978). *Trichinella spiralis*. Effects on the host–parasite relationship in mice of BCG (attenuated *Mycobacterium bovis*). *Experimental Parasitology* **44**, 181–189.

Grove, D. I. Mahmoud, A. A. F. and Warren, K. S. (1977). Eosinophils and resistance to *Trichinella spiralis*. *Journal of Experimental Medicine* **145**, 755–759.

Grover, W. H., Winkler, H. H. and Normansell, D. E. (1978). Phagocytic properties of isolated human eosinophils. *Journal of Immunology* **121**, 718–725.

Grover, W. H., Butterworth, A. E., Sturrock, R. F. and Bass, D. A. (1983). Killing of *Trichinella* larvae by granulocytes: activation of eosinophils during eosinophilia and role of granulocyte-specific opsonius. *In* "Immunobiology of the Eosinophil" (T. Yoshida and M. Torisu, eds), pp. 327–342. Elsevier Biomedical, New York.

Grzych, J. M., Capron, M., Bazin, H. and Capron, A. (1982). *In vitro* and *in vivo* effector function of rat IgG2a monoclonal anti-*S. mansoni* antibodies. *Journal of Immunology* **129**, 2739–2743.

Gupta, S., Ross, G. D. and Good, R. A. (1976). Surface characteristics of human eosinophils. *Journal of Allergy and Clinical Immunology* **57**, 189. (Abstract.)

Gusmao, R. D., Stanley, A. M. and Ottesen, E. A. (1981). *Brugia pahangi*: immunologic evaluation of the differential susceptibility of filarial infection in inbred Lewis rats. *Experimental Parasitology* **52**, 147–159.

Hammerberg, B. and Williams, J. F. (1978). Interaction between *Taenia taeniaeformis* and the complement system. *Journal of Immunology* **120**, 1039–1045.

Haque, A., Joseph, M., Ouaissi, M. A., Capron, M. and Capron, A. (1980). IgE antibody-mediated cytotoxicity of rat macrophages against microfilariae of *Dipetalonema viteae in vitro*. *Clinical and Experimental Immunology* **40**, 487–495.

Haque, A., Ouaissi, A., Joseph, M., Capron, M. and Capron, A. (1981). IgE antibody in eosinophil- and macrophage-mediated *in vitro* killing of *Dipetalonema viteae* microfilariae. *Journal of Immunology* **127**, 716–725.

Haque, A., Ouaissi, A., Santoro, F., des Moutis, I. and Capron, A. (1982). Complement-mediated leukocyte adherence to infective larvae of *Dipetalonema viteae* (Filarioidea): requirement for eosinophils or eosinophil products in effecting macrophage adherence. *Journal of Immunology* **129**, 2219–2225.

Hayunga, E. G., Murrell, K. D., Taylor, D. W. and Vannier, W. E. (1979a). Isolation and characterisation of surface antigens from *Schistosoma mansoni*. Evaluation of techniques for radioisotope labelling of surface proteins from adult worms. *Journal of Parasitology* **65**, 488–496.

Hayunga, E. G., Murrell, K. D., Taylor, D. W. and Vannier, W. E. (1979b). Isolation and characterisation of surface antigens from *Schistosoma mansoni*. II. Antigenicity of radiolabelled proteins from adult worms. *Journal of Parasitology* **65**, 497–506.

Henson, P. M. (1969). The adherence of leukocytes and platelets induced by fixed IgG antibody or complement. *Immunology* **16**, 107–121.

Herberman, R. B., Djeu, J. Y., Kay, H. D., Ortaldo, J. R., Riccardi, C., Bonnard, G. D., Holden, H. T., Fagnani, R., Santoni, A. and Puccetti, P. (1979). Natural killer cells: characteristics and regulation of activity. *Immunological Reviews* **44**, 43–70.

Higashi, G. I. and Chowdhury, A. B. (1970). In vitro adhesion of eosinophils to infective larvae of *Wucheria bancrofti*. *Immunology* **19**, 65–83.

Hindsbo, O., Andreassen, J. and Ruitenberg, J. (1982). Immunological and histopathological reactions of the rat against the tapeworm *Hymenolepis diminuta* and the effects of anti-thymocyte serum. *Parasite Immunology* **4**, 59–76.

Hirashima, M. and Hayashi, H. (1976). The mediation of tissue eosinophilia in hypersensitivity reaction. I. Isolation of two different chemotactic factors from DNP-Ascaris extract-induced skin lesion in guinea-pig. *Immunology* **30**, 203–212.

Hirashima, M., Yodoi, J. and Izhizaka, K. (1980). Regulatory role of IgE binding factors from rat T lymphocytes. III. IgE-specific suppressive factor with IgE-binding activity. *Journal of Immunology* **125**, 1442–1448.

Hockley, D. J. and McLaren, D. J. (1973). *Schistosoma mansoni*: changes in the outer membrane of the tegument during development from cercariae to adult worm. *International Journal of Parasitology* **3**, 13–25.

Hopper, K. E., Mehta, K., Subrahmanyam, D. and Nelson, D. S. (1981). Enhanced adhesion of rat neutrophils to *Litomosoides carinii* microfilariae in the presence of culture supernatants from mitogen-stimulated lymph node cells. *Clinical and Experimental Immunology* **45**, 633–641.

Hsü, S. Y., Hsü, H. F., Lust, G. L., Davis, J. R. and Eveland, L. K. (1973). Comparative studies on the lesions caused by eggs of *Schistosoma japonicum* and *Schistosoma mansoni* in the liver of hamsters, guinea pigs and albino rats. *Annals of Tropical Medicine and Parasitology* **67**, 349–356.

Hsü, S. Y. L., Hsü, H. F., Penick, G. D., Lust, G. L. and Osborne, J. W. (1974). Dermal hypersensitivity to schistosome cercariae in rhesus monkeys during immunization and challenge. I. Complex hypersensitivity reactions of a well-immunized monkey during the challenge. *Journal of Allergy and Clinical Immunology* **54**, 339–349.

Hsü, S. Y., Hsü, H. F., Penick, G. D., Hanson, H. O., Schiller, H. J. and Cheng, (1975). Mechanism of immunity of schistosomiasis: histopathologic study of lesions elicited in rhesus monkeys during immunizations and challenge with cercariae of *Schistosoma japonicum*. *Journal of the Reticuloendothelial Society* **18**, 167–185.

Hsü, S. Y., Hsü, H. F., Isacson, P. and Cheng, H. F. (1977). In vitro schistosomulicidal effect of immune serum and eosinophils, neutrophils and lymphocytes. *Journal of the Reticuloendothelial Society* **21**, 153–162.

Hsü, S. Y., Hsu, H. F., Penick, G. D., Hanson, H. O., Schiller, H. J. and Cheng, H. F. (1979). Immunoglobulin E, mast cells, and eosinophils in the skin of rhesus monkeys immunized with X-irradiated cercariae of *Schistosoma japonicum*. *International Archives of Allergy and Applied Immunology* **59**, 383–393.

Hsü, S. Y., Hsü, H. F., Mitros, F. A., Helms, C. M. and Solomon, R. I. (1980). Eosinophils as effector cells in the destruction of *Schistosoma mansoni* eggs in granulomas. *Annals of Tropical Medicine and Parasitology* **74**, 179–183.

Hsü, S. Y., Hsü, H. F. and Hanson, H. O. (1981). Immunoglobulins and complement in the skin of rhesus monkeys immunized with X-irradiated cercariae of *Schistosoma japonicum*. *Zeitschrift für Parasitenkunde* **66**, 133–143.

Hubscher, T. (1975). Role of the eosinophil in the allergic reactions. I. EDI—an eosinophil-derived inhibitor of histamine release. *Journal of Immunology* **114**, 1379–1388.

Incani, R. N. and McLaren, D. J. (1981). Neutrophil-mediated cytotoxicity to schistosomula of *Schistosoma mansoni in vitro*: studies on the kinetics of complement and/or antibody-dependent adherence and killing. *Parasite Immunology* **3**, 107–126.

Ishikawa, T., Dalton, A. C. and Arbesman, C. E. (1972). Phagocytosis of *Candida albicans* by eosinophilic leukocytes. *Journal of Allergy and Clinical Immunology* **49** 311–315.

Ishikawa, T., Wicher, K. and Arbesman, C. E. (1974). *In vitro* and *in vivo* studies on uptake of antigen–antibody complexes by eosinophils. *International Archives of Allergy and Applied Immunology* **46**, 230–248.

Ishizaka, K. and Ishizaka, T. (1978). Mechanisms of reaginic hypersensitivity and IgE antibody response. *Immunological Reviews* **41**, 109–148.

Jakubowski, M. S. and Barnard, D. E. (1971). Anaphylactic shock during operation for hydatid disease. *Anaesthesiology* **34**, 197–199.

James, S. L. and Colley, D. G. (1975). Eosinophils and immune mechanisms: production of the lymphokine eosinophil stimulation promoter (ESP) *in vitro* by isolated intact granulomas. *Journal of the Reticuloendothelial Society* **18**, 283–293.

James, S. L. and Colley, D. G. (1976). Eosinophil-mediated destruction of *Schistosoma mansoni* eggs. *Journal of the Reticuloendothelial Society* **20**, 359–374.

James, S. L. and Colley, D. G. (1978a). Eosinophil-mediated destruction of *Schistosoma mansoni* eggs *in vitro*. II. The role of cytophilic antibody. *Cellular Immunology* **38**, 35–47.

James, S. L. and Colley, D. G. (1978b). Eosinophil-mediated destruction of *Schistosoma mansoni* eggs. III. Lymphokine involvement in the induction of eosinophil functional abilities. *Cellular Immunology* **38**, 48–58.

James, S. L. and Colley, D. G. (1978c). Eosinophil-mediated destruction of *S. mansoni* eggs. IV. Effects of several inhibitory substances on eosinophil function. *Cellular Immunology* **38**, 59–67.

James, S. L., Lazdins, J. K., Meltzer, M. S. and Sher, A. (1982a). Macrophages as effector cells of protective immunity in murine schistosomiasis. *Cellular Immunology* **67**, 255–266.

James, S. L., Sher, A., Lazdins, J. K. and Meltzer, M. S. (1982b). Macrophages as effector cells of protective immunity in murine schistosomiasis. II. Killing of newly transformed schistosomula *in vitro* by macrophages activated as a consequence of *Schistosoma mansoni* infection. *Journal of Immunology* **128**, 1535–1540.

Jarrett, E. E. E. (1973). Reaginic antibodies and helminth infection. *Veterinary Record* **93**, 480–483.

Jarrett, E. E. E. and Miller, H. R. (1982). Production and activities of IgE in helminth infection. *Progress in Allergy* **31**, 178–233.

Jarrett, W. F. H. and Sharp, N. C. G. (1963). Vaccination against parasitic disease: reactions in vaccinated and immune hosts in *Dictyocaulus viviparus* infection. *Journal of Parasitology* **49**, 177–189.

Jarrett, W. F. H., Jennings, M. W., McIntyre, W. I. M., Mulligan, W. and Urquhart, G. M. (1960). Immunological studies on *Dictyocaulus viviparus* infection: active immunisation with whole worm vaccine. *Immunology* **3**, 135–143.

Jenkins, D. C. and Carrington, T. S. (1981). *Nematospiroides dubius*: the course of primary, secondary and tertiary infections in high and low responder Biozzi mice. *Parasitology* **82**, 311–318.

Johansson, S. G. O., Mellbin, T. and Vahlquist, B. (1968). Immunoglobulin levels in Ethiopian preschool children with special reference to high concentrations of immunoglobulin E (IgND). *Lancet* **i**, 1118–1121.

John, D. T. and Martinez, A. J. (1975). Animal model of human disease. Central nervous system infection with the nematode *Angiostrongylus cantonensis*. Animal model: eosinophilic meningoencephalitis in mice infected with *Angiostrongylus cantonensis*. *American Journal of Pathology* **80**, 345–348.

Johnson, G. R. and Metcalf, D. (1978). The nature of cells forming erythroid colonies in agar after stimulation by spleen conditioned medium. *Journal of Cellular Physiology* **94**, 243–252.

Johnson, G. R., Dresch, C. and Metcalf, D. (1977). Heterogeneity in human neutrophil macrophage, and eosinophil progenitor cells demonstrated by velocity sedimentation preparation. *Blood* **50**, 823–831.

Johnson, P., Mackenzie, C. D., Suswillo, R. R. and Denham, D. A. (1981). Serum-mediated adherence of feline granulocytes to microfilariae of *Brugia pahangi in vitro*: variations with parasite maturation. *Parasite Immunology* **3**, 69–80.

Jong, E. C., Mahmoud, A. A. F. and Klebanoff, S. J. (1981). Peroxidase-mediated toxicity to schistosomula of *Schistosoma mansoni. Journal of Immunology* **126**, 468–471.

Joseph, M. (1982). Effector functions of phagocytic cells against helminths. *Clinics in Allergy and Immunology* **2**, 567–596.

Joseph, M., Dessaint, J. P. and Capron, A. (1977). Characteristics of macrophage cytotoxicity induced by IgE immune complexes. *Cellular Immunology* **34**, 247–258.

Joseph, M., Capron, A., Butterworth, A. E., Sturrock, R. F. and Houba, V. (1978). Cytotoxicity of human and baboon mononuclear phagocytes against schistosomula *in vitro*: induction by immune complexes containing IgE and *Schistosoma mansoni* antigens. *Clinical and Experimental Immunology* **33**, 48–56.

Joseph, M., Tonnel, A. B., Capron, A. and Voisin, C. (1980). Enzyme release and superoxide anion production by human alveolar macrophages stimulated with immunoglobulin E. *Clinical and Experimental Immunology* **40**, 416–422.

Joseph, M., Tonnel, A. B., Capron, A. and Dessaint, J. P. (1981). The interaction of IgE antibody with human alveolar macrophages and its participation in the inflammatory processes of lung allergy. *Agents and Actions* **11**, 619–622.

Jungery, M. and Ogilvie, B. M. (1982). Antibody response to stage-specific *Trichinella spiralis* surface antigens in strong and weak responder mouse strains. *Journal of Immunology* **129**, 839–843.

Karavodin, L. M. and Ash, L. R. (1982). Inhibition of adherence and cytotoxicity by circulating immune complexes formed in experimental filariasis. *Parasite Immunology* **4**, 1–12.

Kassis, A. I., Aikawa, M. and Mahmoud, A. F. (1979). Mouse antibody-dependent eosinophil and macrophage adherence and damage to schistosomula of *Schistosoma mansoni. Journal of Immunology* **122**, 398–405.

Kater, L. A., Goetzl, E. J. and Austen, K. F. (1976). Isolation of human eosinophil phospholipase D. *Journal of Clinical Investigation* **57**, 1173–1180.

Katz, D. H. (1982). IgE antibody responses *in vitro*: from rodents to man. *Progress in Allergy* **32**, 105–160.

Kay, A. B. (1982). Complement receptor enhancement by chemotactic factors. *Molecular Immunology* **19**, 1307–1311.

Kay, A. B. and Austen, K. F. (1971). The IgE-mediated release of an eosinophil leukocyte chemotactic factor from human lung. *Journal of Immunology* **107**, 899–902.

Kay, A. B., Stechschulte, D. J. and Austen, K. F. (1971). An eosinophil leukocyte chemotactic factor of anaphylaxis. *Journal of Experimental Medicine* **133**, 602–619.

Kayes, S. G. and Oaks, J. A. (1978). Development of the granulomatous response in murine toxocariasis. Initial events. *American Journal of Pathology* **93**, 277–294.

Kazura, J. W. (1981). Host defense mechanisms against nematode parasites: destruction of newborn *Trichinella spiralis* larvae by human antibodies and granulocytes. *Journal of Infectious Diseases* **143**, 712–718.

Kazura, J. W. and Grove, D. I. (1978). Stage-specific antibody-dependent eosinophil-mediated destruction of *Trichinella spiralis*. *Nature* **274**, 588–589.

Kazura, J. W. and Aikawa, M. (1980). Host defense mechanisms against *Trichinella spiralis* infection in the mouse: eosinophil-mediated destruction of newborn larvae *in vitro*. *Journal of Immunology* **124**, 355–361.

Kazura, J. W., Mahmoud, A. A., Karb, K. S. and Warren, K. S. (1975). The lymphokine eosinophil stimulation promoter and human schistosomiasis mansoni. *Journal of Infectious Diseases* **132**, 702–706.

Kazura, J. W., Fanning, M. M., Blumer, J. L. and Mahmoud. A. A. (1981). Role of cell-generated hydrogen peroxide in granulocyte-mediated killing of schistosomula of *Schistosoma mansoni in vitro*. *Journal of Clinical Investigation* **67**, 93–102.

Kishimoto, T. (1982). IgE class-specific suppressor T cells and regulation of the IgE response. *Progress in Allergy* **32**, 265–317.

Klebanoff, S. J. (1975). Antimicrobial mechanisms in neutrophilic polymorphonuclear leukocytes. *Seminars in Hematology* **12**, 117–142.

Klebanoff, S. J., Durack, D. T., Rosen, H. and Clark, R. A. (1977). Functional studies on human peritoneal eosinophils. *Infection and Immunity* **17**, 167–173.

Klebanoff, S. J., Jong, E. C. and Henderson, W. R. (1980). The eosinophil peroxidase: purification and biological properties. *In* "The Eosinophil in Health and Disease" (A. A. F. Mahmoud and K. F. Austen, eds), pp. 99–114. Grune and Stratton, New York.

Kloetzel, K. and Da Silva, J. R. (1967). Schistosomiasis mansoni acquired in adulthood: behavior of egg counts and the intradermal test. *American Journal of Tropical Medicine and Hygiene* **16**, 167–169.

Knapp, N. H. and Oakley, G. A. (1981). Cell adherence to larvae of *Dictyocaulus viviparus in vitro*. *Research in Veterinary Science* **31**, 389–391.

Lehrer, S. B. and Bozelka, B. E. (1982). Mouse IgE. *Progress in Allergy* **32**, 8–51.

Leid, R. W. (1977). Immunity to the metacestode of *Taenia taeniaeformis* in the laboratory rat. *American Journal of Tropical Medicine and Hygiene* **26**, 54–60.

Lewis, D. M., Lewis, J. C., Loegering, D. A. and Gleich, G. J. (1978). Localization of the guinea pig eosinophil major basic protein to the core of the granule. *Journal of Cell Biology* **77**, 702–713.

Litt, M. (1964). Studies in experimental eosinophilia. VI. Uptake of immune complexes by eosinophils. *Journal of Cell Biology* **23**, 355–364.

Ljungstrom, I. and Sundquist, K. G. (1979). Lymphocyte activation induced by *Trichinella spiralis* infection reflected as spontaneous DNA synthesis *in vitro*. *Clinical and Experimental Immunology* **38**, 381–388.

Loveless, S. E., Wellhausen, S. R., Boros, D. L. and Heppner, G. H. (1982). Tumoricidal macrophages isolated from liver granulomas of *Schistosoma mansoni*-infected mice. *Journal of Immunology* **128**, 284–288.

Mackaness, G. B. (1969). The influence of immunologically committed lymphoid cells on macrophage activity *in vivo*. *Journal of Experimental Medicine* **129**, 973–992.

Mackenzie, C. D., Ramalho-Pinto, F. J., McLaren, D. J. and Smithers, S. R. (1977). Antibody-mediated adherence of rat eosinophils to schistosomula of *Schistosoma mansoni in vitro*. *Clinical and Experimental Immunology* **30**, 97–104.

Mackenzie, C. D., Preston, P. M. and Ogilvie, B. M. (1978). Immunological properties of the surface of parasitic nematodes. *Nature* **276**, 826–828.

Mackenzie, C. D., Jungery, M., Taylor, P. M. and Ogilvie, B. M. (1980). Activation of complement, the induction of antibodies to the surface of nematodes and the effect of these factors and cells on worm survival *in vitro*. *European Journal of Immunology* **10**, 594–601.

Mackenzie, C. D., Jungery, M., Taylor, P. M. and Ogilvie, B. M. (1981). The *in-vitro* interaction of eosinophils, neutrophils, macrophages and mast cells with nematode surfaces in the presence of complement or antibodies. *Journal of Pathology* **133**, 161–175.

Maeda, S., Irie, Y. and Yasuraoka, K. (1982). Resistance of mice to secondary infection with *Schistosoma japonicum*, with special reference to neutrophil enriched response to schistosomula in the skin of immune mice. *Japanese Journal of Experimental Medicine* **52**, 111–118.

Mahmoud, A. A. F. (1977). Antieosinophil serum. *American Journal of Tropical Medicine and Hygiene* **26**, 151–158.

Mahmoud, A. A. F. (1980). Nonspecific resistance to schistosomiasis. In "The Host-Invader Interplay" (H. Van den Bossche, ed.), pp. 417–426. Elsevier/North Holland Biomedical Press, Amsterdam.

Mahmoud, A. A. F., Kellermeyer, R. W. and Warren, K. S. (1974). Production of monospecific rabbit antihuman eosinophil serums and demonstration of a blocking phenomenon. *New England Journal of Medicine* **290**, 417–420.

Mahmoud, A. A. F., Warren, K. S. and Graham, R. C. (1975a). Antieosinophil serum and the kinetics of eosinophilia in Schistosomiasis mansoni. *Journal of Experimental Medicine*, **142**, 560–574.

Mahmoud, A. A. F., Warren, K. S. and Peters, P. A. (1975b). A role for the eosinophil in acquired resistance to *Schistosoma mansoni* infection as determined by antieosinophil serum. *Journal of Experimental Medicine*, **142**, 805–813.

Mahmoud, A. A. F., Warren, K. S. and Strickland, G. T. (1976). Acquired resistance to infection with *Schistosoma mansoni* induced by *Toxoplasma gondii*. *Nature* **263**, 56–57.

Mahmoud, A. A. F., Stone, M. K. and Kellermeyer, R. W. (1977). Eosinophilopoietin: a circulating low molecular weight peptide-like substance which stimulates the production of eosinophils in mice. *Journal of Clinical Investigation* **60**, 675–682.

Mahmoud, A. A. F., Stone, M. K. and Tracy, J. W. (1979a). Eosinophilopoietin production: relationship to eosinophilia of infection and thymus function. *Transactions of the Association of American Physicians* **355–359**.

Mahmoud, A. A. F., Peters, P. A., Civil, R. H. and Remington, J. S. (1979b). *In vitro* killing of schistosomula of *Schistosoma mansoni* by BCG and *C. parvum*-activated macrophages. *Journal of Immunology* **122**, 1655–1657.

Maizels, R. M., Philipp, M. and Ogilvie, B. M. (1982). Molecules on the surface of parasite nematodes as probes of the immune response in infection. *Immunological Reviews* **61**, 109–136.

Mantovani, B. (1975). Different roles of IgG and complement receptors in phagocytosis by polymorphonuclear leukocytes. *Journal of Immunology* **115**, 15–17.

Martz, E. (1977). Mechanism of specific tumor cell lysis by alloimmune T lymphocytes: resolution and characterization of discrete steps in the cellular interaction. *Contemporary Topics in Immunobiology* **7**, 301–361.

Masake, R. A., Wescott, R. B., Spencer, G. R. and Lang, B. Z. (1978). The pathogenesis of primary and secondary infection with *Fasciola hepatica* in mice. *Veterinary Pathology* **15**, 763–769.

Mauel, J., Buchmuller, Y. and Behin, R. (1978). Studies on the mechanism of macrophage activation. I. Destruction of intracellular *Leishmania enriettii* in macrophages activated by coculture with stimulated lymphocytes. *Journal of Experimental Medicine* **148**, 393–407.

McCullough, F. S. and Bradley, D. J. (1973). Egg output stability and the epidemiology of *Schistosoma haematobium*. Part I. Variation and stability in *Schistosoma haematobium* egg counts. *Transactions of the Royal Society of Tropical Medicine and Hygiene* **67**, 475–490.

McGarry, M. P. and Miller, A. M. (1974). Evidence for the humoral stimulation of eosinophil granulocytopoiesis in *in vivo* diffusion chambers. *Experimental Haematology* **2**, 372–379.

McKean, J. R., Anwar, A. R. and Kay, A. B. (1981). *Schistosoma mansoni*: complement and antibody damage, mediated by human eosinophils and neutrophils, in killing schistosomula *in vitro*. *Experimental Parasitology* **51**, 307–317.

McLaren, D. J. (1980a). "*Schistosoma mansoni*: the parasite surface in relation to host immunity". Research Studies Press, John Wiley and Sons, Chichester.

McLaren, D. J. (1980b). Ultrastructural observations on the interaction between host cells and parasitic helminths. *In* "The Host Invader Interplay" (H. Van den Bossche, ed.), pp. 85–98. Elsevier/North-Holland Biomedical Press, Amsterdam.

McLaren, D. J. and Hockley, D. J. (1977). Blood flukes have a double outer membrane. *Nature* **269**, 147–149.

McLaren, D. J. and Ramalho-Pinto, F. J. (1979). Eosinophil-mediated killing of schistosomula of *Schistosoma mansoni in vitro*: synergistic effect of antibody and complement. *Journal of Immunology* **123**, 1431–1438.

McLaren, D. J. and Terry, R. J. (1982). The protective role of acquired host antigens during schistosome maturation. *Parasite Immunology* **4**, 129–148.

McLaren, D. J., Clegg, J. A. and Smithers, S. R. (1975). Acquisition of host antigens by young *Schistosoma mansoni* in mice: correlation with failure to bind antibody *in vivo*. *Parasitology* **70**, 67–75.

McLaren, D. J., Mackenzie, C. D. and Ramalho-Pinto, F. J. (1977). Ultrastructural observations on the *in vitro* interaction between rat eosinophils and some parasitic helminths (*Schistosoma mansoni*, *Trichinella spiralis* and *Nippostrongylus brasiliensis*). *Clinical and Experimental Immunology* **30**, 105–118.

McLaren, D. J., Ramalho-Pinto, F. J. and Smithers, S. R. (1978a). Ultrastructural evidence for complement and antibody-dependent damage to schistosomula of *Schistosoma mansoni* by rat eosinophils *in vitro*. *Parasitology* **77**, 313–324.

McLaren, D. J., Hockley, D. J., Goldring, O. L. and Hammond, B. J. (1978b). A freeze-fracture study of the developing tegumental outer membrane of *Schistosoma mansoni*. *Parasitology* **76**, 327–348.

McLaren, D. J., McKean, J. R., Olsson, I., Venge, P. and Kay, A. B. (1981). Morphological studies on the killing of schistosomula of *Schistosoma mansoni* by human eosinophil and neutrophil cationic proteins *in vitro*. *Parasite Immunology* **3**, 359–373.

Mehta, K., Sindhu, R. K., Subramanyam, D. and Nelson, D. S. (1980). IgE-dependent adherence and cytotoxicity of rat spleen and peritoneal cells to *Litomosoides carinii* microfilariae. *Clinical and Experimental Immunology* **41**, 107–114.

Mehta, K., Sindhu, R. K., Subrahmanyam, D., Hopper, K., Nelson, D. S. and Rao, C. K. (1981a). Antibody-dependent cell-mediated effects in bancroftian filariasis. *Immunology* **43**, 117–123.

Mehta, K., Subrahmanyam, D., Hopper, K., Nelson, D. S. and Rao, C. K. (1981b). IgG-dependent human eosinophil-mediated adhesion and cytotoxicity of *Litomosoides carinii* larvae. *Indian Journal of Medical Research* **74**, 226–230.

Mehta, K., Sindhu, R. K., Subrahmanyam, D., Hopper, K. and Nelson, D. S. (1982). IgE-dependent cellular adhesion and cytotoxicity to *Litomosoides carinii* microfilariae—nature of effector cells. *Clinical and Experimental Immunology* **48**, 477–484.

Melewicz, F. M. and Spiegelberg, H. L. (1980). Fc receptors for IgE on a subpopulation of human peripheral blood monocytes. *Journal of Immunology* **125**, 1026–1031.

Meltzer, M. S., Ruco, L. P., Boraschi, D., Mannel, D. and Edelstein, M. C. (1980). Genetics of macrophage tumor cytotoxicity. *In:* "Genetic Control of Natural Resistance to Infection and Malignancy" (E. Skamene, P. Kongshavn and M. Landy, eds), pp. 537–554. Academic Press, New York.

Metcalf, D., Parker, J., Chester, H. M. and Kincade, P. W. (1974). Formation of eosinophilic-like granulocytic colonies by mouse bone marrow cells *in vitro*. *Journal of Cellular Physiology* **84**, 275–289.

Mickenberg, I. D., Root, R. K. and Wolff, S. M. (1972). Bactericidal and metabolic properties of human eosinophils. *Blood* **39**, 67–80.

Miller, A. M. and McGarry, M. P. (1976). A diffusible stimulator of eosinopoiesis produced by lymphoid cells as demonstrated with diffusion chambers. *Blood* **48**, 293–300.

Miller, A. M., Colley, D. G. and McGarry, M. P. (1976). Spleen cells from *Schistosoma mansoni*-infected mice produce diffusible stimulator of eosinophilopoiesis *in vivo*. *Nature* **262**, 586–587.

Miller, K. L. and Smithers, S. R. (1980). *Schistosoma mansoni*: the attrition of a challenge infection in mice immunized with highly irradiated live cercariae. *Experimental Parasitology* **50**, 212–221.

Miller, K. L., Smithers, S. R. and Sher, A. (1981). The response of mice immune to *Schistosoma mansoni* to a challenge infection which bypasses the skin: evidence for two mechanisms of immunity. *Parasite Immunology* **3**, 25–31.

Mitchell, G. F., Goding, J. W. and Richard, M. D. (1977). Increased susceptibility of certain mouse strains and hypothymic mice to *Taenia taeniaeformis* and analysis of passive transfer with serum. *Australian Journal of Experimental Biology and Medical Science* **55**, 165–175.

Mitchell, G. F., Rajasekariah, G. R. and Richard, M. D. (1980). A mechanism to account for mouse strain variation in resistance to the larval cestode, *Taenia taeniaeformis*. *Immunology* **39**, 481–495.

Moqbel, R. (1980). Histopathological changes following primary, secondary and repeated infections of rats with *Strongyloides ratti*, with special reference to tissue eosinophils. *Parasite Immunology* **2**, 11–27.

Moser, G. and Sher, A. (1981). Studies of the antibody-dependent killing of schistosomula of *Schistosoma mansoni* employing haptenic target antigens. II. *In vitro* killing of TNP-schistosomula by human eosinophils and neutrophils. *Journal of Immunology* **126**, 1025–1029.

Moser, G., Wassom, D. L. and Sher, A. (1980). Studies of the antibody-dependent killing of schistosomula of *Schistosoma mansoni* employing haptenic target antigens. I. Evidence that the loss in susceptibility to immune damage undergone by developing schistosomula involves a change unrelated to the masking of parasite antigens by host molecules. *Journal of Experimental Medicine* **152**, 41–53.

Murray, H. W. and Cohn, Z. A. (1979). Macrophage oxygen-dependent antimicrobial activity. I. Susceptibility of *Toxoplasma gondii* to oxygen intermediates. *Journal of Experimental Medicine* **150**, 938–949.

Nathan, C. F., Brokner, L. H., Silverstein, S. C. and Cohn, Z. A. (1979a). Extracellular cytolysis by activated macrophages and granulocytes. I. Pharmacologic triggering of effector cells and the release of hydrogen peroxide. *Journal of Experimental Medicine* **149**, 84–99.

Nathan, C. F., Silverstein, S. C., Brokner, L. H. and Cohn, Z. A. (1979b). Extracellular cytolysis by activated macrophages and granulocytes. II. Hydrogen peroxide as a mediator of cytotoxicity. *Journal of Experimental Medicine* **149**, 100–113.

Neva, F. A., Kaplan, A. P., Pacheco, G., Gray, L. and Danaraj, T. J. (1975). Tropical eosinophilia. A human model of parasitic immunopathology, with observations on serum IgE levels before and after treatment. *Journal of Allergy and Clinical Immunology* **55**, 422–429.

Newman, S. L. and Johnston, R. B. (1979). Role of binding through C3B and IgG in polymorphonuclear neutrophil function: studies with trypsin-generated C3b. *Journal of Immunology* **123**, 1839–1846.

Ngwenya, B. Z. (1980). Altered lysophospholipase B responsiveness in lactating mice infected with intestinal nematode parasites. *Parasitology* **81**, 17–26.

Nicola, N. A., Metcalf, D., Johnson, G. R. and Burgess, A. W. (1978). Preparation of colony stimulating factors from human placental conditioned medium. *Leukemia Research* **2**, 313–322.

Nicola, N. A., Metcalf, D., Johnson, G. R. and Burgess, A. W. (1979). Separation of functionally distinct human granulocyte-macrophage colony stimulating factors. *Blood* **54**, 614–623.

Nielsen, K., Fogh, L. and Andersen, S. (1974). Eosinophil response to migrating *Ascaris suum* larvae in normal and congenitally thymus-less mice. *Acta Pathologica et Microbiologica Scandinavica (B)* **82**, 919–920.

Ninnemann, J. L. and Lueker, D. C. (1974). Mechanisms of murine immunity to *Nematospiroides dubius* after subcutaneous vaccination. *Journal of Parasitology* **60**, 980–984.

Nitisuwirjo, S. and Ladds, P. W. (1980). A quantitative histopathological study of *Onchocerca gibsoni* nodules in cattle. *Tropenmedizin und Parasitologie* **31**, 467–474.

Nogueira, N. and Cohn, Z. A. (1978). *Trypanosoma cruzi*: in vitro induction of macrophage microbicidal activity. *Journal of Experimental Medicine* **148**, 288–300.

North, R. J. (1981). Immunity of *Listeria monocytogenes*. In "Immunology of Human Infection. Part I: Bacteria, Mycoplasmae, Chlamydiae, and Fungi" (A. J. Nahmias and R. J. O'Reilly, eds), pp. 201–220. Plenum Press, New York.

Novato-Silva, E., Nogueira-Machado, J. A. and Gazzinelli, G. (1980). *Schistosoma mansoni*: comparison of the killing effect of granulocytes and complement with or without antibody on fresh and cultured schistosomula *in vitro*. *American Journal of Tropical Medicine and Hygiene* **29**, 1263–1267.

Ogilvie, B. M. and Love, R. J. (1974). Co-operation between antibodies and cells in immunity to a nematode parasite. *Transplantation Reviews* **19**, 147–169.

Ogilvie, B. M., Mackenzie, C. D. and Love, R. J. (1977a). Lymphocytes and eosinophils in the immune response of rats to initial and subsequent infections with *Nippostrongylus brasiliensis*. *American Journal of Tropical Medicine and Hygiene* **26**, 61–67.

Ogilvie, B. M., Love, R. J., Jarra, W. and Brown, K. N. (1977b). *Nippostrongylus brasiliensis* infection in rats. The cellular requirement for worm expulsion. *Immunology* **32**, 521–528.

Ogilvie, B. M., Askenase, P. W. and Rose, M. E. (1980). Basophils and eosinophils in three strains of rats and in athymic (nude) rats following infection with the nematodes *Nippostrongylus brasiliensis* or *Trichinella spiralis*. *Immunology* **39**, 385–389.

Olds, G. R. and Mahmoud, A. A. F. (1980). Role of host granulomatous response in murine schistosomiasis mansoni: eosinophil-mediated destruction of eggs. *Journal of Clinical Investigation* **66**, 1191–1199.

Olds, G. R., Ellner, J. J., Kearse, L. A., Kazura, J. W. and Mahmoud, A. A. F. (1980a). Role of arginase in killing of schistosomula of *Schistosoma mansoni*. *Journal of Experimental Medicine* **151**, 1557–1562.

Olds, G. R., Chedid, L., Lederer, E. and Mahmoud, A. A. F. (1980b). Induction of resistance to *Schistosoma mansoni* by natural cord factor and synthetic lower homologues. *Journal of Infectious Diseases* **141**, 473–478.

Olds, G. R., Ellner, J. J., el-Kholy, A. and Mahmoud, A. A. F. (1981a). Monocyte-mediated killing of schistosomula of *Schistosoma mansoni*: alterations in human Schistosomiasis mansoni and tuberculosis. *Journal of Immunology* **127**, 1538–1542.

Olds, G. R., Stewart, S. J. and Ellner, J. J. (1981b). Amphotericin B-induced resistance to *Schistosoma mansoni*. *Journal of Immunology* **126**, 1667–1670.

Olsson, I. and Venge, P. (1974). Cationic proteins of human granulocytes. II. Separation of the cationic proteins of the granules of leukemic myeloid cells. *Blood* **44**, 235–246.

Olsson, I., Venge, P., Spitznagel, J. K. and Lehrer, R. I. (1977). Arginine-rich cationic proteins of human eosinophil granules. Comparison of the constituents of eosinophilic and neutrophilic leukocytes. *Laboratory Investigation* **36**, 493–500.

Orange, R. P., Murphy, R. C. and Austen, K. F. (1974). Inactivation of slow reacting substance of anaphylaxis (SRS-A) by arylsulfatases. *Journal of Immunology* **113**, 316–322.

Ottesen, E. A., Stanley, A. M., Gelfand, J. A., Gadek, J. E., Frank, M. M., Nash, T. E. and Cheever, A. W. (1977). Immunoglobulin and complement receptors on human eosinophils and their role in cellular adherence to schistosomules. *American Journal of Tropical Medicine and Hygiene* **26**, 134–141.

Ottesen, E. A., Neva, F. A., Paranjape, R. S., Tripathy, S. P., Thiruvengadam, K. V. and Beaven, M. A. (1979). Specific allergic sensitisation to filarial antigens in tropical eosinophilia syndrome. *Lancet* **i**, 1158–1161.

Ottolenghi, A. (1969). The relationship between eosinophilic leukocytes and phospholipase B activity in some rat tissues. *Lipids* **5**, 531–538.

Ottolenghi, A., Weatherly, N. F., Kocan, A. A. and Larsh, J. E. (1977). *Angiostrongylus cantonensis*: phospholipase in nonsensitized and sensitized rats after challenge. *Infection and Immunity* **15**, 13–18.

Ouaissi, M. A., Hague, A. and Capron, A. (1981). *Dipetalonema viteae*: ultrastructural study on the *in vitro* interaction between rat macrophages and microfilariae in the presence of IgE antibody. *Parasitology* **82**, 55–62.

Ovary, Z., Itaya, T., Watanabe, N. and Kojima, S. (1978). Regulation of IgE in mice. *Immunological Reviews* **41**, 26–51.

Owhashi, M. and Ishii, A. (1982). Purification and characterization of a high molecular weight eosinophil chemotactic factor from *Schistosoma japonicum* eggs. *Journal of Immunology* **129**, 2226–2231.

Parillo, J. E. and Fauci, A. S. (1978). Human eosinophils: purification and cytotoxic capability of eosinophils from patients with the hypereosinophilic syndrome. *Blood* **51**, 457–466.

Pascual, J. E., Bouli, R. P. and Aguiar, H. (1981). Eosinophilic meningoencephalitis in Cuba, caused by *Angiostrongylus cantonensis*. *American Journal of Tropical Medicine and Hygiene* **30**, 960–962.

Patterson, R. J. and Youmans, G. P. (1970). Demonstration in tissue culture of lymphocyte-mediated immunity to tuberculosis. *Infection and Immunity* **1**, 600–603.

Peck, C. A., Carpenter, M. D. and Mahmoud, A. A. F. (1983). Species-related innate resistance to *Schistosoma mansoni*. Role of mononuclear phagocytes in schistosomula killing *in vitro*. *Journal of Clinical Investigation* **71**, 66–72.

Pelley, R. P., Pelley, R. J., Hamburger, J., Peters, P. A. and Warren, K. S. (1976). *Schistosoma mansoni* soluble egg antigens: I. Identification and purification of three major antigens and the employment of radioimmunoassay for their further characterization. *Journal of Immunology* **117**, 1553–1560.

Perez, H. and Terry, R. J. (1973). The killing of adult *Schistosoma mansoni* in vitro in the presence of antisera to host antigenic determinants and peritoneal cells. *International Journal of Parasitology* **3**, 499–503.

Perez, H. A. and Smithers, S. R. (1977). *Schistosoma mansoni*: in the rat: the adherence of macrophages to schistosomula *in vitro* after sensitization with immune serum. *International Journal of Parasitology* **7**, 315–320.

Perrudet-Badoux, A., Anteunis, A., Dumitrescu, S. M. and Binaghi, R. A. (1978). Ultrastructural study of the immune interaction between peritoneal cells and larvae of *Trichinella spiralis*. *Journal of the Reticuloendothelial Society* **24**, 311–314.

Perrudet-Badoux, A., Binaghi, R. A., Boussac-Aron, Y. and Ruitenberg, E. J. (1981). Antibody-dependent mechanisms of immunity against migrating larvae of *Trichinella spiralis*. *Veterinary Parasitology* **8**, 89–94.

Philipp, M., Parkhouse, R. M. and Ogilvie, B. M. (1980). Changing proteins on the surface of a parasitic nematode. *Nature* **287**, 538–540.

Piessens, W. F. and Dias da Silva, W. (1982). Complement-mediated adherence of cells to microfilariae of *Brugia malayi*. *American Journal of Tropical Medicine and Hygiene* **31**, 297–301.

Pincus, S. H., Butterworth, A. E., David, J. R., Robbins, M. and Vadas, M. A. (1981). Antibody-dependent eosinophil-mediated damage to schistosomula of *Schistosoma mansoni*: lack of requirement for oxidative metabolism. *Journal of Immunology* **126**, 1794–1799.

Presentey, B. and Szapiro, L. (1969). Hereditary deficiency in peroxidase and phospholipids in eosinophilic granulocytes. *Acta Haematologica* **41**, 359 (1969).

Pritchard, D. I. and Eady, R. P. (1981). Eosinophilia in athymic nude (rnu/rnu) rats—thymus-independent eosinophilia? *Immunology* **43**, 409–416.

Prowse, S. J., Ey, P. L. and Jenkin, C. R. (1978). Immunity to *Nematospiroides dubius*: cell and immunoglobulin changes associated with the onset of immunity in mice. *Australian Journal of Experimental Biology and Medical Science* **56**, 237–246.

Prowse, S. J., Ey, P. L. and Jenkin, C. R. (1979). Alternative pathway activation of complement by a murine parasitic nematode (*Nematospiroides dubius*). *Australian Journal of Experimental Biology and Medical Science* **57**, 459–466.

Rabellino, E. M. and Metcalf, D. (1975). Receptors for C3 and IgG on macrophage, neutrophil and eosinophil colony cells grown *in vitro*. *Journal of Immunology* **115**, 688–692.

Ramalho-Pinto, F. J., Gazzinelli, G., Howells, R. E., Mota-Santos, T. A., Figueiredo, E. A. and Pellegrino, J. (1974). *Schistosoma mansoni*: defined system for stepwise transformation of cercaria to schistosomule *in vitro*. *Experimental Parasitology* **36**, 360–372.

Ramalho-Pinto, F. J., McLaren, D. J. and Smithers, S. R. (1978). Complement-mediated killing of schistosomula of *Schistosoma mansoni* by rat eosinophils *in vitro*. *Journal of Experimental Medicine* **147**, 147–156.

Ramalho-Pinto, F. J., De Rossi, R. and Smithers, S. R. (1979). Murine *Schistosomiasis mansoni*: anti-schistosomula antibodies and the IgG subclasses involved in the complement- and eosinophil-mediated killing of schistosomula *in vitro*. *Parasite Immunology* **1**, 295–308.

Ramasamy, R. (1979). Surface proteins on schistosomula and cercariae of *Schistosoma mansoni*. *International Journal of Parasitology* **9**, 491–493.

Rand, T. H. and Colley, D. G. (1982). Influence of a lymphokine fraction containing eosinophil stimulation promoter (ESP) on oxidative and degranulation responses of murine eosinophils. *Cellular Immunology* **71**, 334–345.

Rand, T. H., Turk, J., Maas, R. L. and Colley, D. G. (1982). Arachidonic and metabolism of the murine eosinophil. II. Involvement of the lipoxygenase pathway in the response to the lymphokine eosinophil stimulation promoter. *Journal of Immunology* **129**, 1239–1244.

Rau, M. E. and Tanner, C. E. (1973). *Echinococcus multilocularis* in the cotton rat. The effect of pre-existing subcutaneous cysts on the development of a subsequent intraperitoneal inoculum of protoscolices. *Canadian Journal of Zoology* **51**, 55–59.

Rau, M. E. and Tanner, C. E. (1975). BCG suppresses growth and metastases of hydatid infections. *Nature* **256**, 318–319.

Rau, M. E. and Tanner, C. E. (1976). *Echinococcus multilocularis* in the cotton rat. The *in vitro* protoscolicidal activity of peritoneal cells. *International Journal of Parasitology* **6**, 195–198.

Reuben, J. M., Tanner, C. E. and Rau, M. E. (1978). Immunoprophylaxis with BCG of experimental *Echinococcus multilocularis* infections. *Infection and Immunity* **21**, 135–139.

Rosen, L. R., Chappell, G. L., Laquer, G. D., Wallace, G. D. and Weinstein, P. P. (1962). Eosinophilic meningoencephalitis caused by a metastrongylid lung worm of rats. *Journal of the American Medical Association* **179**, 620–624.

Rothwell, T. L. W. (1975). Studies of the responses of basophil and eosinophil leucocytes and mast cells to the nematode *Trichostrongylus colubriformis*. I. Observations during the expulsion of first and second infections by guinea-pigs. *Journal of Pathology* **116**, 51–60.

Rothwell, T. L. W. and Dineen, J. K. (1972). Cellular reactions in guinea-pigs following primary and challenge infection with *Trichostrongylus colubriformis* with special reference to the roles played by eosinophils and basophils in rejection of the parasite. *Immunology* **22**, 733–745.

Rothwell, T. L. W. and Love, R. J. (1975). Studies of the responses of basophil and eosinophil leucocytes and mast cells to the nematode *Trichostrongylus colubriformis*. II. Changes in cell numbers following infection of thymectomised and adoptively or passively immunised guinea-pigs. *Journal of Pathology* **116**, 183–194.

Rudin, W., Tanner, M., Bauer, P. and Weiss, N. (1980). Studies on *Dipetalonema viteae* (Filarioidea). 5. Ultrastructural aspects of the antibody-dependent cell-mediated destruction of microfilariae. *Tropenmedizin und Parasitologie* **31**, 194–200.

Ruitenberg, E. J. and Duyzings, M. J. (1972). An immunohistological study of the immunological response of the rat to infection with *Trichinella spiralis*. *Journal of Comparative Pathology* **82**, 401–407.

Ruitenberg, E. J., Perrudet-Badoux, A., Boussac-Aron, Y. and Elgersma, A. (1980). *Trichinella spiralis* infection in animals genetically selected for high and low antibody production. Studies on intestinal pathology. *International Archives of Allergy and Applied Immunology* **62**, 104–110.

Ruscetti, F. W., Cypess, R. H. and Chervenick, P. A. (1976). Specific release of neutrophilic- and eosinophilic-stimulating factors from sensitized lymphocytes. *Blood* **47**, 757–765.
Sabesin, S. M. (1963). A function of the eosinophil: phagocytosis of antigen–antibody complexes. *Proceedings of the Society for Experimental Biology and Medicine* **113**, 667–670.
Sakai, N., Johnstone, C. and Weiss, L. (1981). Bone marrow cells associated with heightened eosinophilopoiesis: an electron microscope study of murine bone marrow stimulated by *Ascaris suum*. *American Journal of Anatomy* **161**, 11–32.
Salem, E. A., Ishaac, S. and Mahmoud, A. A. F. (1979). Histocompatibility-linked susceptibility for hepatosplenomegaly in human *Schistosoma mansoni*. *Journal of Immunology* **123**, 1829–1831.
Salmon, S. E., Cline, M. J., Schultz, J. and Lehrer, R. I. (1970). Myeloperoxidase deficiency. Immunologic study of a genetic leukocyte defect. *New England Journal of Medicine* **282**, 250–253.
Samuelson, J. C. and Caulfield, J. P. (1982). Loss of covalently labeled glycoproteins and glycolipids from the surface of newly transformed schistosomula of *Schistosoma mansoni*. *Journal of Cell Biology* **94**, 363–369.
Samuelson, J. C., Sher, A. and Caulfield, J. P. (1980). Newly transformed schistosomula spontaneously lose surface antigens and C3 acceptor sites during culture. *Journal of Immunology* **124**, 2055–2057.
Sanderson, C. J. (1982). Morphological aspects of lymphocyte mediated cytotoxicity. *Advances in Experimental Biology and Medicine* **146**, 3–21.
Sanderson, C. J. and Glauert, A. M. (1977). The mechanism of T cell mediated cytotoxicity. V. Morphological studies by electron microscopy. *Proceedings of the Royal Society of London, B*, **198**, 315–323.
Sanderson, C. J. and Glauert, A. M. (1979). The mechanism of T cell mediated cytotoxicity. VI. T cell projections and their role in target cell killing. *Immunology* **36**, 119–129.
Santoro, F., Lachmann, P. J., Capron, A. and Capron, M. (1979). Activation of complement by *Schistosoma mansoni* schistosomula: killing of parasites by the alternative pathway and requirement of IgG for classical pathway activation. *Journal of Immunology* **123**, 1551–1557.
Santoro, F., Vandemeulebroucke, B., Liebart, M. C. and Capron, A. (1982). *Schistosoma mansoni*: role *in vivo* of complement in primary infection in mice. *Experimental Parasitology* **54**, 40–46.
Saran, R. (1973). Cytoplasmic vacuoles of eosinophils in tropical pulmonary eosinophilia. *American Review of Respiratory Disease* **108**, 1283–1284.
Sasazuki, T., Ohta, N., Kaneoka, R. and Kojima, S. (1980). Association between an HLA haplotype and low responsiveness to schistosomal worm antigen in man. *Journal of Experimental Medicine* **152**, 314–318.
Savage, A. M. and Colley, D. G. (1980). The eosinophil in the inflammatory response to cercarial challenge of sensitized and chronically infected CBA/J mice. *American Journal of Tropical Medicine and Hygiene*, **29**, 1268–1278.
Schriber, R. A. and Zucker-Franklin, D. (1975). Induction of blood eosinophilia by pulmonary embolization of antigen-coated particles: the relationship to cell-mediated immunity. *Journal of Immunology* **114**, 1348–1353.
Seitz, H. M., Cottrell, B. and Sturrock, R. F. (1980). The penetration of *Schistosoma mansoni* cercariae, a histological study on baboon skin. *Zentralblatt für Bakteriologie, 1. Abt. Ref.* **267**, 305.

Shah, J. and Ramasamy, R. (1982). Surface antigens on cercariae, schistosomula and adult worms of *Schistosoma mansoni*. *International Journal of Parasitology* **12**, 451–461.

Sher, A. (1976). Complement-dependent adherence of schistosomula to mast cells. *Nature* **263**, 334–336.

Sher, A. (1977). Immunity against *Schistosoma mansoni* in the mouse. *American Journal of Tropical Medicine and Hygiene* **26** Supplement, 20–28.

Sher, A., Smithers, S. R. and Mackenzie, P. (1975). Passive transfer of acquired resistance to *Schistosoma mansoni* in laboratory mice. *Parasitology* **70**, 347–357.

Sher, A., Hall, B. F. and Vadas, M. A. (1978). Acquisition of murine major histocompatibility complex gene products by schistosomula of *Schistosoma mansoni*. *Journal of Experimental Medicine* **148**, 46–56.

Sher, A., James, S. L., Simpson, A. J., Lazdins, J. K. and Meltzer, M. S. (1982). Macrophages as effector cells of protective immunity in murine schistosomiasis. III. Loss of susceptibility to macrophage-mediated killing during maturation of *S. mansoni* schistosomula from the skin to the lung stage. *Journal of Immunology* **128**, 1876–1879.

Sher, R. and Glover, A. (1976). Isolation of human eosinophils and their lymphocyte-like rosetting properties. *Immunology* **31**, 337–341.

Sim, B. K. (1981). Ultrastructure of antibody-dependent cell-mediated destruction of *Brugia malayi* infective larvae *in vitro*. *Southeast Asian Journal of Tropical Medicine and Public Health* **12**, 618–619.

Sim, B. K., Kwa, B. H. and Mak, J. W. (1982). Immune responses in human *Brugia malayi* infections: serum-dependent cell-mediated destruction of infective larvae *in vitro*. *Transactions of the Royal Society of Tropical Medicine and Hygiene* **76**, 362–370.

Sinclair, K. B. (1973). The resistance of sheep to *Fasciola hepatica*: studies on the development and pathogenicity of challenge infections. *British Veterinary Journal* **129**, 236–250.

Sinclair, K. B. (1975). The resistance of sheep to *Fasciola hepatica*: studies on the pathophysiology of challenge infections. *Research in Veterinary Science* **19**, 296–303.

Smith, M. A., Clegg, J. A., Snary, D. and Trejdosiewicz, A. J. (1982). Passive immunization of mice against *Schistosoma mansoni* with an IgM monoclonal antibody. *Parasitology* **84**, 83–89.

Smithers, S. R. and Gammage, K. (1980). The recovery of *Schistosoma mansoni* from the skin, lungs and hepatic portal system of naive mice and mice previously exposed to *S. mansoni*. Evidence for two phases of parasite attrition in immune mice. *Parasitology* **80**, 289–298.

Smithers, S. R. and Miller, K. L. (1980). Protective immunity in murine schistosomiasis mansoni: evidence for two distinct mechanisms. *American Journal of Tropical Medicine and Hygiene* **29**, 832–841.

Smithers, S. R. and Terry, R. J. (1969). Immunity in schistosomiasis. *Annals of the New York Academy of Science* **160**, 826–840.

Smithers, S. R. and Terry, R. J. (1976). The immunology of schistosomiasis. *Advances in Parasitology* **14**, 399–423.

Smithers, S. R., Terry, R. J. and Hockley, D. J. (1969). Host antigens in schistosomiasis. *Proceedings of the Royal Society of London, B*, **171**, 483–494.

Snary, D., Smith, M. A. and Clegg, J. A. (1980). Surface proteins of *Schistosoma mansoni* and their expression during morphogenesis. *European Journal of Immunology*, **10**, 573–575.

Solomon, G. B. and Soulsby, E. J. (1973). Granuloma formation to *Capillaria hepatica* eggs. I. Descriptive definition. *Experimental Parasitology* **33**, 458–467.
Spry, C. J. F. and Tai, P. C. (1976). Studies on blood eosinophils. II. Patients with Löffler's cardiomyopathy. *Clinical and Experimental Immunology* **24**, 423–434.
Stankiewicz, M. and Jeska, E. L. (1973). Leucocytes and *Trichinella spiralis*. 3. Importance of heat labile and heat stable substances in peritoneal exudate fluid for cell adherence reactions to infective larvae. *Immunology* **25**, 827–834.
Sterba, J., Slais, J., Machnicka, B. and Schandl, V. (1981). Development of oncospheres of *Taenia saginata* after a concomitant infection by oral and subcutaneous routes. *Folia Parasitologica* **28**, 353–358.
Stirewalt, M. A. and Uy, A. (1969). *Schistosoma mansoni*: cercarial penetration and schistosomule collection in an *in vitro* system. *Experimental Parasitology* **26**, 17–28.
Sturrock, R. F., Butterworth, A. E., Houba, V., Karamsadkar, S. D. and Kimani, R. (1978). *Schistosoma mansoni* in the Kenyan baboon (*Papio anubis*): the development and predictability of resistance to homologous challenge. *Transactions of the Royal Society of Tropical Medicine and Hygiene* **72**, 251–261.
Sturrock, R. F., Kimani, R., Joseph, M., Butterworth, A. E., David, J. R., Capron, A. and Houba, V. (1981). Heat-labile and heat-stable anti-schistosomular antibodies in Kenyan schoolchildren infected with *Schistosoma mansoni*. *Transactions of the Royal Society of Tropical Medicine and Hygiene* **75**, 219–227.
Sturrock, R. F., Kimani, R., Cottrell, B. J., Butterworth, A. E., Seitz, H. M., Siongok, T. A. and Houba, V. (1983). Observations on possible immunity to reinfection among Kenyan school children after treatment for *Schistosoma mansoni*. *Transactions of the Royal Society of Tropical Medicine and Hygiene* **77**, 363–371.
Subrahmanyam, D., Rao, Y. V., Mehta, K. and Nelson, D. S. (1976). Serum-dependent adhesion and cytotoxicity of cells to *Litomosoides carinii* microfilariae. *Nature* **260**, 529–530.
Suemara, M., Yodoi, J., Hirashima, M. and Ishizaka, K. (1980). Regulatory role of IgE-binding factors from rat T lymphocytes. I. Mechanism of enhancement of IgE response by IgE-potentiating factor. *Journal of Immunology* **125**, 148–154.
Sugane, K. and Oshima, T. (1982). Eosinophilia, granuloma formation and migratory behaviour of larvae in the congenitally athymic mouse infected with *Toxocara canis*. *Parasite Immunology* **4**, 307–318.
Swisher, S. N. (1956). Non-specific adherence of platelets and leucocytes to antibody-sensitized red cells. A mechanism producing thrombocytopenia and leucopenia during incompatible transfusions. *Journal of Clinical Investigation* **35**, 738.
Tai, P. C. and Spry, C. J. F. (1976). Studies on blood eosinophils. I. Patients with a transient eosinophilia. *Clinical and Experimental Immunology* **24**, 415–422.
Tanaka, J. and Torisu, M. (1978). *Anisakis* and eosinophil. I. Detection of a soluble factor selectively chemotactic for eosinophils in the extract from *Anisakis* larvae. *Journal of Immunology* **120**, 745–749.
Tanaka, J., Baba, T. and Torisu, M. (1979). *Ascaris* and eosinophil. II. Isolation and characterization of eosinophil chemotactic factor and neutrophil chemotactic factor of parasite in *Ascaris* antigen. *Journal of Immunology* **122**, 302–308.
Tanaka, K. R., Valentine, W. N. and Fredericks, R. E. (1962). Human leukocyte arylsulfatase activity. *British Journal of Haematology* **8**, 86–92.
Tanner, M. and Weiss, N. (1978). Studies on *Dipetalonema viteae* (Filarioidea). II. Antibody dependent adhesion of peritoneal exudate cells to microfilariae *in vitro*. *Acta Tropica* **35**, 151–160.

Tauber, A. I., Goetzl, E. J. and Babior, B. M. (1979). Unique characteristics of superoxide production by human eosinophils in eosinophilic states. *Inflammation* **3**, 261–270.

Tavares, C. A. P., Soares, R. C., Coelho, P. M. Z. and Gazzinelli, G. (1978). *Schistosoma mansoni*: evidence for a role of serum factors in protecting artificially transformed schistosomula against antibody-mediated killing *in vitro*. *Parasitology* **77**, 225–233.

Tavares, C. A. P., Cordeiro, M. N., Mota-Santos, T. A. and Gazzinelli, G. (1980). Artificially transformed schistosomula of *Schistosoma mansoni*: mechanism of acquisition of protection against antibody-mediated killing. *Parasitology* **80**, 95–104.

Taylor, D. W. and Butterworth, A. E. (1982). Monoclonal antibodies against surface antigens of schistosomula of *Schistosoma mansoni*. *Parasitology* **84**, 65–82.

Taylor, D. W., Hayunga, E. G. and Vannier, W. E. (1981). Surface antigens of *Schistosoma mansoni*. *Molecular and Biochemical Parasitology* **3**, 157–168.

Thompson, J. M., Meola, S. M., Ziprin, R. L. and Jeska, E. L. (1977). An ultrastructural study of the invasion of *Ascaris suum* larvae by neutrophils. *Journal of Invertebrate Pathology* **30**, 181–184.

Thorne, K. J. I. and Blackwell, J. M. (1983). Cell-mediated killing of protozoa. *Advances in Parasitology* **22**, 43–151.

Torisu, M., Iwasaki, K., Tanaka, J., Iino, H. and Yoshida, T. (1983). *Anisakis* and eosinophil: pathogenesis and biologic significance of eosinophilic phlegmon in human anisakiasis. *In* "Immunobiology of the Eosinophil" (T. Yoshida and M. Torisu, eds), pp. 343–367. Elsevier Biomedical, New York.

Torpier, G. and Capron, A. (1980). Differentiation and expression of surface antigens in relation with *Schistosoma mansoni* membrane structure. *In* "The Host Invader Interplay" (H. Van den Bossche, ed.), pp. 143–146. Elsevier/North-Holland Biomedical Press, Amsterdam.

Torpier, G., Capron, M. and Capron, A. (1977). Structural changes of the tegumental membrane complex in relation to developmental stages of *Schistosoma mansoni* (Platyhelminthes: Trematoda). *Journal of Ultrastructural Research* **61**, 309–324.

Torpier, G., Ouaissi, M. A. and Capron, A. (1979). Freeze-fracture study of immune-induced *Schistosoma mansoni* membrane alterations. I. Complement-dependent damage in the presence of antisera to host antigenic determinants. *Journal of Ultrastructural Research* **67**, 276–287.

Turnbull, L. W. and Kay, A. B. (1976). Eosinophils and mediators of anaphylaxis. Histamine and imidazole acetic acid as chemotactic agents for human eosinophil leukocytes. *Immunology* **31**, 797–802.

Vadas, M. A. (1982). Genetic control of eosinophilia in mice: gene(s) expressed in bone marrow-derived cells control high responsiveness. *Journal of Immunology* **128**, 691–695.

Vadas, M. A. (1983). Activation of eosinophils and regulation of eosinophilia. *In* "Immunobiology of the Eosinophil" (T. Yoshida and M. Torisu, eds), pp. 77–95. Elsevier Biomedical, New York.

Vadas, M. A., David, J. R., Butterworth, A. E., Pisani, N. T. and Siongok, T. A. (1979a). A new method for the purification of human eosinophils and neutrophils, and a comparison of the ability of these cells to damage schistosomula of *Schistosoma mansoni*. *Journal of Immunology* **122**, 1228–1236.

Vadas, M. A., David, J., Butterworth, A. E., Houba, V., David, L. and Pisani, N. (1979b). Comparison of the ability of eosinophils and neutrophils and of eosinophils from patients with *S. mansoni* infection and normal individuals, to mediate *in vitro* damage to schistosomula of *S. mansoni*. *Advances in Experimental Medicine and Biology* **114**, 677–682.

Vadas, M. A., Butterworth, A. E., Burakoff, S. and Sher, A. (1979c). Major histocompatibility complex products restrict the adherence of cytolytic T lymphocytes to minor histocompatibility antigens or to trinitrophenyl determinants on schistosomula of *Schistosoma mansoni*. *Proceedings of the National Academy of Sciences, USA* **76**, 1982–1985.

Vadas, M. A., Butterworth, A. E., Sherry, B., Dessein, A., Hogan, M., Bout, D. and David, J. R. (1980a). Interactions between human eosinophils and schistosomula of *Schistosoma mansoni*. I. Stable and irreversible antibody-dependent adherence. *Journal of Immunology* **124**, 1441–1448.

Vadas, M. A., David, J. R., Butterworth, A. E., Houba, V., Sturrock, R. F., David, L., Herson, R., Siongok, T. A. and Kimani, R. (1980b). Functional studies on purified eosinophils and neutrophils from patients with *Schistosoma mansoni* infections. *Clinical and Experimental Immunology* **39**, 683–694.

Vadas, M. A., Dessein, A., Nicola, N. and David, J. R. (1981). *In vitro* enhancement of the helminthotoxic capacity of human blood eosinophils. *Australian Journal of Experimental Biology and Medical Science* **59**, 739–741.

Veith, M. C. and Butterworth, A. E. (1983). Enhancement of human eosinophil-mediated killing of *Schistosoma mansoni* larvae by mononuclear cell products *in vitro*. *Journal of Experimental Medicine* **157**, 1828–1843.

Veith, M. C., Butterworth, A. E. and Boylston, A. W. (1983). The enhancement of eosinophil-mediated killing of schistosomula of *Schistosoma mansoni* by mononuclear cell products. *In* "Immunobiology of the Eosinophil" (T. Yoshida and M. Torisu, eds), pp. 305–325. Elsevier Biomedical, New York.

Vernes, A., Biguet, J., Floch, F. and Tailliez, R. (1972). L'hypersensibilité de type retardé dans la trichinose experimentale: evaluation par les tests d'inhibition de la migration et de l'étalement des macrophages (étude preliminaire). *Bulletin de la Societé de Pathologie Exotique* **65**, 704–713.

Vernes, A., Poulain, D., Prensier, G., Deblock, S. and Biguet, J. (1974). Trichinose experimentale. III. Action "in vitro" des cellules peritoneales sensibilisées sur les larves musculaires de premier stade. Étude preliminaire comparative en microscopie optique et electronique à transmission et à balayage. *Biomedicine* **21**, 140–145.

Verwaerde, C. Grzych, J. M., Bazin, H., Capron, M. and Capron, A. (1979). Production d'anticorps monoclonaux anti *Schistosoma mansoni*. *Comptes Rendus, Académie des Sciences, Paris* **289D**, 725–727.

Vincent, A. L., Ash, L. R., Rodrick, G. E. and Sodeman, W. A. (1980). The lymphatic pathology of *Brugia pahangi* in the Mongolian jird. *Journal of Parasitology* **66**, 613–620.

Voge, M., Sogandares-Bernal, F. and Martin, J. H. (1979). Fine structure of the tegument of *Mesocestoides* tetrathyridia by scanning and transmission electron microscopy. *Journal of Parasitology* **65**, 562–567.

Von Lichtenberg, F. and Ritchie, L. S. (1961). Cellular resistance against schistosomula of *Schistosoma mansoni* in *Macaca mulatta* monkeys following prolonged infection. *American Journal of Tropical Medicine and Hygiene* **10**, 859–869.

Von Lichtenberg, F., Sher, A., Gibbons, N. and Doughty, B. L. (1976). Eosinophil-enriched inflammatory response to schistosomula in the skin of mice immune to *Schistosoma mansoni*. *American Journal of Pathology* **84**, 479–500.

Wakelin, D. (1975a). Genetic control of immune response to parasites: immunity to *Trichuris muris* in inbred and random-bred strains of mice. *Parasitology* **70**, 397–405.

Wakelin, D. (1975b). Genetic control of immune response to parasites: selection for responsiveness and non-responsiveness to *Trichuris muris* in random-bred strains of mice. *Parasitology* **71**, 377–384.

Wakelin, D. (1978). Genetic control of susceptibility and resistance to parasitic infection. *Advances in Parasitology* **16**, 219–308.

Wakelin, D. (1980). Genetic control of immunity to parasites. Infection with *Trichinella spiralis* in inbred and congenic mice showing rapid and slow responses to infection. *Parasite Immunology* **2**, 85–98.

Wakelin, D. and Donachie, A. M. (1980). Genetic control of immunity to parasites: adoptive transfer of immunity between inbred strains of mice characterized by rapid and slow immune expulsion of *Trichinella spiralis*. *Parasite Immunology* **2**, 249–260.

Walls, R. S. (1976). Lymphocytes and specificity of eosinophilia. *South African Medical Journal* **50**, 1313–1318.

Walls, R. S. and Beeson, P. B. (1972). Mechanisms of eosinophilia. 8. Importance of local cellular reactions in stimulating eosinophil production. *Clinical and Experimental Immunology* **12**, 111–119.

Walls, R. S., Basten, A., Leuchars, E. and Davies, A. J. S. (1971). Mechanisms for eosinophilic and neutrophilic leucocytoses. *British Medical Journal* **3**, 157–159.

Walls, R. S., Bass, D. A. and Beeson, P. B. (1974). Mechanism of eosinophilia. X. Evidence for immunologic specificity of the stimulus. *Proceedings of the Society for Experimental Biology and Medicine* **145**, 1240–1242.

Warren, K. S. (1972). The immunopathogenesis of schistosomiasis: a multidisciplinary approach. *Transactions of the Royal Society of Tropical Medicine and Hygiene* **66**, 417–432.

Warren, K. S. (1973). Regulation of the prevalence and intensity of schistosomiasis in man. Immunology or ecology? *Journal of Infectious Diseases* **127**, 595–609.

Warren, K. S. (1982). The secret of the immunopathogenesis of schistosomiasis: *in vivo* models. *Immunological Reviews* **61**, 189–213.

Wasserman, S. I., Goetzl, E. J. and Austen, K. F. (1975). Inactivation of slow reacting substance of anaphylaxis by human eosinophil arylsulfatase. *Journal of Immunology* **114**, 645–649.

Wassom, D. L. and Gleich, G. J. (1979). Damage to *Trichinella* spiralis newborn larvae by eosinophil major basic protein. *American Journal of Tropical Medicine and Hygiene* **28**, 860–863.

Wassom, D., David, C. S. and Gleich, G. J. (1979). Genes within the major histocompatibility complex influence susceptibility to *Trichinella spiralis* in the mouse. *Immunogenetics* **9**, 491–496.

Weiss, J., Elsbach, P., Olsson, I. and Odeberg, H. (1978). Purification and characterization of a potent bactericidal and membrane active protein from the granules of human polymorphonuclear leukocytes. *Journal of Biological Chemistry* **253**, 2664–2672.

Weiss, N. and Tanner, M. (1979). Studies on *Dipetalonema viteae* (Filarioidea). 3. Antibody-dependent cell-mediated destruction of microfilariae *in vivo*. *Tropenmedizin und Parasitologie* **30**, 73–80.

Weller, P. F., Wasserman, S. I. and Austen, K. F. (1980). Selected enzymes preferentially present in the eosinophil. *In* "The Eosinophil in Health and Disease" (A. A. F. Mahmoud and K. F. Austen, eds), pp. 115–130. Grune and Stratton, New York.

Williams, J. F., Picone, J. and Engelkirk, P. (1980). Evasion of immunity by cestodes. *In* "The Host-Invader Interplay" (H. Van den Bossche, ed.), pp. 205–216. Elsevier/North Holland Biomedical Press, Amsterdam.

Willms, K. and Merchant, M. T. (1980). The inflammatory reaction surrounding *Taenia solium* larvae in pig muscle: ultrastructural and light microscopic observations. *Parasite Immunology* **2**, 261–275.

Wing, E. J. and Remington, J. S. (1978). Role for activated macrophages in resistance against *Trichinella spiralis*. *Infection and Immunity* **21**, 398–404.

Wong, L. and Wilson, J. D. (1975). The identification of Fc and C3 receptors on human neutrophils. *Journal of Immunological Methods* **7**, 69–76.

Wong, M. M., Guest, M. F., Lim, K. C. and Sivanandam, S. (1977). Experimental *Brugia malayi* infections in the rhesus monkey. *Southeast Asian Journal of Tropical Medicine and Public Health* **8**, 265–273.

Woodruff, A. W. (1970). Toxocariasis. *British Medical Journal* **ii**, 663–669.

Worms, M. J. and McLaren, D. J. (1982). Macrophage-mediated damage to filarial worms (*Dipetalonema setariosum*) *in vivo*. *Journal of Helminthology* **56**, 235–241.

Yodoi, J., Hirashima, M. and Ishizaka, K. (1980). Regulatory role of IgE-binding factors from rat T lymphocytes. II. Glycoprotein nature and source of IgE-potentiating factor. *Journal of Immunology* **125**, 1436–1441.

Yodoi, J., Hirashima, M. and Ishizaka, K. (1981). Lymphocytes bearing Fc receptors for IgE. V. Effect of tunicamycin on the formation of IgE-potentiating factor and IgE suppressive factor by Con A-activated lymphocytes. *Journal of Immunology* **126**, 877–882.

Zeiger, R. S., Yuram, D. L. and Colten, H. R. (1976). Histamine metabolism. II. Cellular and subcellular localization of the catabolic enzymes, histaminase and histamine methyl transferase, in human leukocytes. *Journal of Allergy and Clinical Immunology* **58**, 172–179.

Zinkernagel, R. M. and Doherty, P. C. (1979). MHC-restricted cytotoxic T cells. Studies on the biological role of polymorphic major transplantation antigens determining T cell restriction-specificity, function and responsiveness. *Advances in Immunology* **27**, 51–59.

Zinkernagel, R. M. and Rosenthal, K. L. (1981). Experiments and speculation on antiviral specificity of T and B cells. *Immunological Reviews* **58**, 131–155.

Ziprin, R. and Jeska, E. L. (1975). Humoral factors affecting mouse peritoneal cell adherence reactions to *Ascaris suum*. *Infection and Immunity* **12**, 499–504.

Zucker-Franklin, D. (1980). Eosinophil structure and maturation. *In* "The Eosinophil in Health and Disease" (A. A. F. Mahmoud and K. F. Austen, eds), pp. 43–59. Grune and Stratton, New York.

# Subject Index

## A

Acetaldehyde dehydrogenase, glycolytic pathway, in, 112
Acetone, 127
Acetylcholine, 2, 15, 16
   neurotransmission, role in, 4
Acetyl-CoA synthetase, glycolytic pathway, in, 112
Actin, 131
Acute pyogenic infections, 148
Adenosine cyclic 3′,5′-monophosphate (cyclic AMP)
   cell regulation in parasitic worms, role in, 14–18, 23–25
   eukaryotic cells, 15–16
   glycolytic enzymes, activation of, 27, 29
   intracellular concentration, receptor activation of, **24**, 25
   protein kinase, activation of, 25–27
   secondary messenger, role as, 19, 27, 29
Adenosine triphosphate (ATP), 12
   -dependent phosphofructokinase activity, 108, 122
   formation from acetyl-CoA, 112, 121
Adenylate cyclase, 15
   hormone-activation of, in higher organisms, 18–20
   guanosine triphosphate function in, 25
   serotonin-activation of, in trematodes, 16–18, 21–23, 29
Agonists
   definition, 20
   serotonin receptors, activation of, 21–23

Albino rats, 195
Alcohol dehydrogenase, glycolytic pathway, in, 112, 124
Aldolase, 124
Alloantigens, 179
*Amblyomma variegatum*, 57
Amino acids, amino proteins, incorporation in, 118–119
Aminotriazole, 158, 170
Amoebae
   aerobic metabolism, 129
   amino acid incorporation, 118–119
   amino acid synthesis, 126
   amylase, 117
   anaerobic metabolism, 129
   associate cells, effect on, 122–124
   axenically grown, 108–116
   bacteria grown, 116–117
   bacterial activities, 123
   bases uptake, 122
   carbohydrate hydrolases, 117
   carbohydrate transport, 122
   cell harvesting, 132
   cell membrane transport, 122
   cholesterol requirements, 124–125
   cloning, 132
   cyst-forming capability, 123, 131
   electron transport, 121, 129
   DNA expression, 123
   enzyme activities in, 108–113, 122–123, 128
      associate organisms, from, 128–129
      induction, 128–129
      regulation, 126–127
   evolution, 127–128, 130
   genome expression alteration, 123
   glucose transport, 122

# SUBJECT INDEX

Amoebae (cont.)
  glycogen metabolism, 128
  glycogen synthesis apparatus, 128
  glycolytic enzymes regulation, 126–127, 128–129
  growth measurement, 133
  growth requirements, 122, 124, 126, 130, 131
  HK9 strain, 118
    Golgi-like apparatus, 127
  homogenization, 132
  host animal hypercholesteraemia and susceptibility to, 123
  hydrogen gas production, 123
  infectivity, 123
  intermediary metabolism, 132
  iron-sulphur complexes in, 125
  Laredo strain, 127
  lipid metabolism, 119–120, 131
  membrane enzymes, 131
  niacin uptake, 122
  nuclear division, 127
  nucleic acid metabolism, 120–121
  nucleosides uptake, 122
  oxygen uptake, 122, 129
  pathogenicity, 120, 123–124, 131
    effect of host heat stress on, 124
    effect of viruses on, 124
  pinocytotic substrate uptake, 122
  protein metabolism, 118–119, 131
  protein synthesis, 126
  purine synthesis, 120
  pyrimidine synthesis, 121
  regulatory mechanisms, 127
  respiration, 129
  riboflavin uptake, 122
  ribosomes, 120
  size measurement, 132
  substrate requirements, 125–126
  substrate transport, 122
  synchronized cultures, 131
  total cell protein, 132
  vitamin requirements, 124–125
  virulence, 123
Amphotericin B, 176
*Amplicaecum robertsi*, metabolism regulation, 10
*Angiostrongylus cantonensis*, 6, 161
*Anisakis*, 166, 197

Anti-eosinophil serum, 186
Anti-epsilon serum, 162, 203
Antigen-antibody phagocytosis, 149
Anti-mouse eosinophil serum, 202
Anti-mouse erythrocyte serum, 177
Antiserotonin agents, 8
Apicomplexa, 96
Arachidonic acid, 150, 161
Arginine, 178
Arylsulphatase B, 149, 150
*Ascaris* sp., 4
  carbohydrate metabolism, 27–28
  effector mechanisms active against, 191–192
  glycolytic enzyme activity in, 27
  muscle functions, 27
  serotonin receptors in, 27–28
*A. suum*
  eosinophil response to, 160, 165
  immune response to, 191
  larvae, 191–192
  metabolism regulation, role of serotonin in, 11
Atropine, 4
Avidin, 124
Axenic amoebae
  *see also* Amoebae
  carbohydrate metabolism, 108–113, 130
  cell cycle, 131
  cytotoxic agents in, 124
  electron transport, 121
  enzyme activities in, 122–123
  enzyme activities not found in, 115–116
  galactose metabolism, 113–114
  glucose metabolism in, 108–113
  glycerolphosphate synthesis, 120, 126
  glycogen cycle, 113
  growth media free amino acid concentrations, 133
  growth requirements, 124–125
  hexose catabolism speculative pathway, 130
  hexose-pentose interconversion, 114
  iron-sulphur proteins, 119
  lipid metabolism, 120
  NIH : 200 strain, 116

## SUBJECT INDEX

Axenic amoebae (cont.)
  nucleotide reduction, 115
  pathogenic properties, 106
  protein metabolism, 119
  substrate requirements, 125–126
  virulence, 123–124
  vitamin requirements, 124–125
Azide, 158, 170

### B

*Babesia*
  development
    erythrocytes, in, 73–74, 96
    sexual stages *in vitro*, 75–83
    species with intralymphocytic schizonts, 69–72
    tick organs, in, 83–89
    vector tick salivary gland, in, 89–95
  differentiated kinetes
    electron micrographs, **82, 86, 90**
    host salivary gland cells penetration, 91
    structure, 89
  gametes fusion, 78
    electron micrographs, **79**
  gamonts, 74
  geographic distribution, 42–44
  history of discovery, 38
  intraerythrocytic stages, 38, 96
  intralymphocytic schizont
    cytomeres, 69
  kinete development, 78, 83, 96
    asexual reproduction, 83, 89
    cytomere differentiation, 83, 89
    electron micrographs, **82, 85, 86, 87**
    host cell invasion, 83
    polymorphic stage subdivision, 83
    tick organs, in, 83, 89
  kinete differentiation, 89
  kinete formation, 78
    diagrammatic representation, **81, 84**
  life cycle, 69–95
  light micrograph differentiation of, **39**

*Babesia* (cont.)
  merozoites
    differentiation, 72
    extracellular, 72
    intraerythrocytic, 74
    penetration of erythrocytes, 72, 74
    reproduction, 74
    structure, 74
  parasitic stages, 89–95
  ray-body
    development, 75, 78, 83
    structure, 75
  sexual stages
    diagrammatic representation, **75**
    tick intestine, in, 75–83
  size in erythrocytes, 42–44
  sporonts, 91
  sporozoites
    concentration in salivary gland host cells, 89
    development, 91, 95, 96
    electron micrographs, **90, 92, 94**
    light micrographs, **94**
    structure, 95
  vector species, 88
  vector tick salivary gland, development in, 89–95
  vectors, 42–44
  vertebrate host, chemotherapy of, 97
  zygote, electron micrograph, **80**
*B. bigemina*
  development, summary of, 88
  intraerythrocytic stages, electron micrograph, **73**
  kinete formation, electron micrographs, **82, 90**
  merozoites, 74
  ray-body, 75
    light micrographs, **80**
  sporozoites, 95
*B. bovis*
  development, summary of, 88
  merozoites, 74
  sporozoites, 95
*B. caballi*
  development, summary of, 88
  merozoites, 74

240    SUBJECT INDEX

*B. canis*
    development, summary of, 88
    gametes fusion, electron
        micrograph, **79**
    gamonts, 74
    intraerythrocytic stages, light
        micrograph, **73**
    life cycle, diagrammatic
        representation, **70**
    merozoites, 74
    ray-body, 75
        electron micrographs, **77, 80**
        light micrographs, **80**
    sporozoite development, 95
        diagrammatic representation,
            **93**
        electron micrographs, **94**
        light micrographs, **94**
*B. divergens*, 69
    merozoites, 74
*B. equi*
    development, summary of, 88
    generic classification, 97
    intraerythrocytic stages
        electron micrograph, **73**
        light micrograph, **39**
    merozoites
        formation, 72
        structure, 74
        tetrad formation (Maltese-
            cross arrangement, 74
    ray-body, 75
    schizonts
        development in lymphocytes,
            69, 97
        electron micrograph, **71**
    sporozoites, 69, 95
        electron micrograph, **90**
    systematic position, 72
*B. microti*
    gamonts, 74
    merozoites, 74
    ray-body, 75
    sporozoites, 69, 95
    systematic position, 72
*B. motasi*, merozoites, 74
*B. ovis*
    development, summary of, 88
    kinete development, electron
        micrographs, **85, 86, 87**

Babesioses, 38
Baboon, helminth infection, 152, 168,
    190
    skin reaction, 199, 200
Bacteria
    adenylate cyclase activation in, 20
    cell regulation, role of cyclic AMP
        in, 14
    granulocyte-mediated damage to,
        148
    macrophage-mediated damage to,
        177
Bacteria-grown amoebae
    *see also* Amoebae
    enzyme activities, 116–117,
        122–123
    protein metabolism, 119
    virulence, 123–124
BCG immunization, 176, 188
Benign African theileriosis, 40
Berenil, 97
Biological assays, cell-mediated
    damage to helminths, 145
Biozzi high-responder mice, 193
Blue bodies *see* Koch's bodies
Bone marrow cultures, eosinophil
    colony development in, 149
*Boophilus* sp., 88
Bovine eosinophils
    basic protein, 157, 186
    helminth damage mediation, role
        in, 192
    morphology, 166
Bovine lymphocytes *see* Lymphocytes
*Breinlia booliati*, 6
2-Bromolysergic acid diethylamide
    (BOL), 2, 8
    parasitic flatworm motility,
        stimulation of, 5
*Brugia* sp., effector mechanisms active
    against, 196
*B. malayi*, 196
*B. pahangi*, 152
    immune responses to, 196

### C

*Caenorhabditis elegans*, 6
Calcium ionophore A23187, 158
Calvin cycle, 114

# SUBJECT INDEX

*Candida albicans*, eosinophil-mediated damage to, 149
*Capillaria hepatica*, 197
Carbohydrate hydrolases, 117
Carbohydrate metabolism regulation in helminths, 10–12
Catalase, 158, 170, 171
Catecholamines, 15
  enzyme activation by, 26, 27
Cattle
  babesian parasites of, 69
  helminth infection, 186, 188
  lungworm, 192
  lymphatic cell attack by theilerian parasites, 67
Cell regulation, cyclic AMP role in, 14–18
Central nervous system, helminth infection of, 161
Cestodes
  effector mechanisms active against, 187–189
  migration behaviour, role of serotonin in, 7–8
  morphology, 187
Cholesterol, 120
  amoebal metabolism, role in, 125–126
Cholinergic antagonists, 4
Chronic granulomatous disease, 170, 203
*Clonorchis sinensis*, serotonin receptors in, 6
Cobra venom factor, 202
Colchicine, 127
Concanavalin A (Con A), 157, 171, 179, 180, 195
Corridor disease, 40
Cortisol, 10
*Corynebacterium parvum*, 176
  macrophage activation by, 177
Cotton rats, 188
Crithidia, in amoebic cultures, 123
Cyanide, 158, 159
Cyclic 3′,5′ nucleotide phosphodiesterase, cyclic AMP regulation by, 27
Cyclic nucleotides, eukaryote development, role in, 15
Cycloheximide, 123
Cyclophyllidea, cell-mediated damage to, 187
Cyproheptadine, 23
Cytolytic T lymphocytes (CTL), 178, 179–180

## D

*Dermacentor nitens*, 88
*D. reticulatus*, 88
Diamidine compounds, 97
*Dictyocaulus* sp., effector mechanisms active against, 192
*D. filaria*, 192
*D. viviparus*, 153
  cell-mediated damage to, 192
  immune responses to, 192
Diethylcarbamazine, 161
Digenea, cell-mediated damage to, 182
Dihydroergotamine, 12
3-(Dimethylaminomethyl)-indole *see* Gramine
*N,N*-Dimethyltryptamine, 21, 22, 28
*Dipetalonema* sp., effector mechanisms active against, 193–195
*D. setariosum*, 195
*D. viteae*, 152
  cell-mediated damage to, 194–195
  immune responses to, 193
  larvae, 193
  microfilariae
    eosinophil-mediated damage to, 153
    immunity against, 193–194
    macrophage-mediated damage to, 173
*Dipylidium caninum*, 6
*Dirofilaria* sp., effector mechanisms active against, 193
*D. immitis*, 152
  immune responses to, 193
  microfilariae, 193
Dog
  babesian parasites of, 69
  helminth infection, 193
Dopamine, 4, 15, 27
  parasitic helminth electrical activity, effect on, 7
Dyes, helminth permeability changes to, 147

## E

East Coast fever, 40
*Echinococcus* sp. cell-mediated damage to, 188
*E. granulosus*, 165–166
*E. multilocularis*
　immune response to, 188
　protoscolices growth suppression, 188
*Eimeria*, sporozoite transmission, 45
Electron microscopical assays
　cell-mediated damage to helminths, 147–148, 187
　eosinophil-mediated damage to helminths, 155–156
Electron micrographs
　babesian parasites
　　gametes fusion, **79**
　　intraerythrocytic stages, **73**
　　kinetes, **82, 85, 86, 87, 90**
　　ray-bodies, **77**
　　schizonts, **71**
　　sporozoites, **90, 92, 94**
　　zygotes, **80**
　theilerian parasites
　　final development, **68**
　　intraerythrocytic stages, **52**
　　kinete development, **66**
　　kinete differentiation, **63, 64**
　　kinete formation, **62**
　　ray-bodies, **56**
　　schizonts, **48**
Embden-Meyerhof pathway, 113
Enoplida, 189
*Entamoeba histolytica*
　axenically grown cells
　　activities in, 108–115, 131
　　activities not present in, 115–116
　bacteria grown cells, 116–117
　　vitamin requirements, 124
　biology, 106
　carbohydrate metabolism, 106, 108–117
　culture studies, 107
　cyst, 106
　deoxyribonucleic acid (DNA) synthesis, 115, 120, 131
　galactose link to glycolysis, 113–114

*Entamoeba histolytica* (cont.)
　glucose catabolism, 108–113
　　steps in, 108, 110–113
　glycogen cycle, 113
　glycogenolysis, 113
　glycolytic pathway, 106
　　allosteric modulation, 126
　　enzyme catalysing steps, 108–113, 128–129
　　regulation of, 126–127, 131
　　schematic representation, **109**
　hexose to pentose interconversion, 114
　intermediary metabolism, 107
　life cycle, 106
　metabolism, 106–133
　　World Health Organization review, 106–107
　morphology, 127
　natural habitat, 127
　nuclear division, 106
　nucleotide reductase, 115
　parasitic existence, evolution of, 128
　pathogenicity, 131
　protein metabolism, 119
　strain designations, 107
　taxonomy, 107
　trophozoite, 106
　vitamin requirements, 124–125
*E. invadens*, 107, 113
*E. moshkovskii*, 107
*E. terrapinae*, 107
Entner-Doudoroff pathway, 116, 117, 129
Enzymes
　carbohydrate metabolism by amoebae, activity in
　　Entner-Doudoroff pathway 116, 117
　　galactose metabolism, 113–114
　　gluconate fermentation, 117
　　glucose fermentation, 108, 117
　　glycolytic pathway, 108–113
　　glycogen cycle, 113
　　hexose–pentose interconversion, 114
　electron transport in amoebae, activity in, 121
　lipid metabolism by amoebae, activity in, 119–120

# SUBJECT INDEX

Enzymes (cont.)
    nucleic acid metabolism by amoebae, activity in, 120–121
    protein metabolism by amoebae, activity in, 118–119
    substrate transport in amoebae, activity in, 122
Eosinophil chemotactic factor of anaphylaxis (ECF-A), 150, 164–165
    chemotactic tetrapeptides, 165, 171
Eosinophil stimulation promoter (ESP), 161, 163–164
Eosinophilic meningitis, 161
Eosinophilopoiesis, 160
Eosinophilopoietic activity mediators, 163
Eosinophilopoietin, 160
Eosinophils, 159
    activation, 163
    antibody-coated spores killing by, 149
    antigen-antibody phagocytosis, 149
    asthma-like reactions, 161
    cationic protein (ECP), 149–150
        toxicity to schistosomula, 156
    colony stimulating factor (CSF), 160
    enzymes, 149
    functional activity advancement, 159–166
        eosinophilia conditions, in, 162–164
        mast cell mediators, by, 164–165
        parasite products, by, 165–166
    functional properties, 150–151
    helminth damage mediation, role in, 152–166, 206
        adherence and degranulation, 157–158
        antibody-coated mammalian target cells killing, 163
        antibody response, 162
        anti-schistosomular antibody, 163
        autofluorescence, 163
        chemotactic responses to casein, 163

Eosinophils (cont)
    colony formation, 160
    colony-stimulating factor (CSF), 160–161, 163
    comparison with neutrophil-mediated damage, 166–168
    cytophilic antibodies, 165
    degranulation, 155, 158, 163
    electron microscopical studies, 155–156
    eosinophilia response, 159–162
    ESP-stimulation, 161
    Fc receptor blocking, 162, 163, 165
    hexose monophosphate shunt activity, 159, 161
    hexose transport, 163
    hydrogen peroxide generation, 158–159
    IgE response, 159–162, 171
    IgG antibodies mediation, 153
    immune effector system, 171
    inhibition of, 158–159
    killing mechanisms, 159
    lectin mediation, 154
    ligand mediation, 153–155
    mechanisms of damage, 155–159
    non-oxidative mechanisms, 158–159
    oxidative mechanisms, 158–159, 163
    oxygen consumption, 159
    respiratory burst activity, 158, 159, 161
    schistosomula killing capacity, 163
    superanion production, 159
    superoxide production, 163
    surface attachment, 153–154
    T cell mediation, 161
    toxic granule content release, 156–157
humoral mediators, 160–161
immunoglobulin (IgE)-dependent mast cell degranulation and, 150–151
major basic protein (MBP), 149, 163

Eosinophils (cont.)
    toxicity to newborn larvae, 157
    toxicity to schistosomula, 156
    mast-cell mediator breakdown, role in, 150–151
    mode of action, 152
    morphological abnormalities, 163
    peroxidase, 158
    properties, 148
    proteins, 149–150
    purification, 152
    species differences, 166
    T lymphocyte dependent, 160
    thymus-dependent increase in numbers of, 151
Epinephrine, 4, 10–12, 15, 26, 27
*Equus burchelli*, 43, 44
Erythrocytes, theilerian merozoite development in, 50, 51–53
Ethidium bromide, 147
*Euglena*, carbohydrate metabolism, 16
Eukaryote cells, 108
    neurotransmitter role in development of, 15–16

## F

*Fasciola* sp., effector mechanisms active against, 186–187
*F. hepatica*
    acetylcholine receptors in, 4
    adenylate cyclase activity, role of serotonin in, 20–21
    carbohydrate metabolism, effect of serotonin on, 10–12, 28
    cell-mediated damage to, *in vitro* biological assay, 145
        isotopic assay, 146
    cell regulation, role of cyclic AMP in, 15
    eosinophil-mediated damage to, 157
    eosinophilic response to, 160
    immune response to, 186
    metabolism regulation, role of serotonin in, 10–12
        carbohydrate metabolism, 22
        glucose metabolism, 10–11
        glycolysis, 12
        lactic acid production, 12

*F. hepatica* (cont.)
    $^{31}$P NMR studies of, 12–14
    metacercariae, 146, 186
    effector mechanisms active against, 186–187
    morphology, 182
    motility, effect of
        BOL on, 5
        LSD on, 5–6
        phosphodiesterase on, 27
        serotonin on, **5**, 28
    neurotransmission, role of serotonin in, 2
    $^{31}$P NMR spectra, **13**
    protein kinase activation, role of serotonin in, 25–26
    serotonin biosynthesis in, 8–9
    serotonin receptors in, 21–23, 29
        desensitization of, 23–25
*Felis sylvestris*, 43, 44
Fermentation, thermodynamic regulation, 126
Ferric ammonium citrate, 125
Ficoll-hypaque, 180
Filarial nematode helminths, neutrophil-mediated damage to, 167
Flukes *see* Trematodes
Fluorescein, 147
Formyl methionyl peptides, 171
Freeze-fracture electron microscopical assays, cell-mediated damage to helminths, 148
Fungi, 14

## G

Galactose oxidase, 180
Glucosamine, 126
Glucose oxidase, 170
Glucose-6-phosphate dehydrogenase, 117
Glucokinase, 126
β-Glucuronidase, 174
Glycerol-3-phosphate dehydrogenase, 116
L-α-Glycerophosphoryl choline, 14
Glycogen phosphorylase, 26
Glycogen synthase, serotonin activation of, 27–28
Glycogenolysis in mammals, 28

Glycogenolytic enzymes, serotonin
  effect on activity of, 27–28
Glycolysis in mammals, 28
Glycolytic pathway, 108–113
  regulation, 126–127
Gordon phenomenon, 150
Gramine, 21
Granulocyte cells, effector mechanism, 149–172
Granulomatous disease, 158
Guanosine triphosphate (GTP), 16
  adenylate cyclase activation by serotonin, regulation of, 25
Guinea pig, major basic protein of, 149, 156

H

*Haemaphysalis leachi*, 88
Haemosporidia, 61, 96
Halofuginone, mode of action, 50–51
Hamster
  amoebae virulence in, 123–124
  helminth infection, 193
    cell-mediated damage to, 194
Helminths, 2
  cell-mediated damage to, *in vitro*, 145–172
    antibody-dependent, 151–152
    biological assay, 145
    criteria defining, 146–147
    electron microscopical assay, 147–148
    eosinophil-mediated, 148, 152–166, 172, 181
    granulocyte-mediated, 149–172
    isotopic assay, 146
    lymphocyte-mediated, 178–182, 206
    macrophage-mediated, 172–178, 181
    microscopical assay, 146–148
    neutrophil-mediated, 166–171, 172
  cellular effector mechanisms
    *in vivo*, 198–205, 206
    accessory cells, role of, 202
    antibody mediation, 201
    complement, role of, 202

Helminths (cont.)
    correlative studies in animals and man, 198–199, 203–205
    eosinophil-mediated, 200
    experimental animal model manipulation, 198, 201–203
    genetic analysis of immunity, 200–201
    histological observations, 198, 199–200
    immune response studies, 200–201, 203–205
    passive transfer studies, 203
    specific acquired immune response, role of, 201
    transfer and depletion studies, 201
  effector mechanisms active against, 144, 148–182, 206
    particular species, 182–198
    summary of, 181–182
  eosinophil-mediated damage to, 152–166, 181
    adherence and degranulation, 157–158
    antibody response, 162
    antigens, 162
    complement enhancement, 154
    electron-microscopical studies, 155–156
    functional activity enhancement, 159–166
    IgE response, 159–162, 194, 206
    ligand mediation, 153–155
    mechanisms of, 155–159
    non-oxidative mechanisms, 158–159
    oxidative mechanisms, 158–159
    peripheral blood eosinophilia, 162
    purified eosinophils, effects of, 152–153
    toxic granule content release, 156–157
  granulocyte-mediated damage to, 149–172

Helminths (cont.)
 early demonstrations of, 151–152
 human immunity mechanism, 144
 lymphocyte-mediated damage to, 178–181, 182, 206
  cytolytic T lymphocytes, 179–180
  K cells and NK cells, 180
 macrophage-mediated damage to, 172–178, 181
  antibody-dependent, 172–176
  antibody-independent, 176–178
  IgE response, 173–174, 194, 206
  IgG response, 175–176
  mechanisms of damage, 177–178
  stimulation after IgE-antigen binding, 174–175
  T lymphocyte-dependent activation mechanisms, 176
 neutrophil-mediated damage to, 166–171, 182, 206
  functional activity enhancement, 171
  ligand mediation, 168
  mechanisms of, 168–170
 non-phagocytosable surface, 144, 155
 surface structure, 144, 206
Heparin, 156
Hexosamine, 126
Histaminase, 149, 150
Histamine, 27, 150
Human alveolar macrophages, 174
Human eosinophils, 152, 162
 C3b receptors, 168
 Fc receptors for IgE, 165
 helminth damage mediation, role in, 153, 190, 195
  degranulation, 155
  functional activity enhancement by mast-cell mediators, 164–165
  MBP toxicity to schistosomula, 156
 morphology, 166

Human erythrocyte transport, 122
Human leukocytes, 170
Human lymphocytes, 180
Human monocytes
 helminth damage mediation, role in, 153
 IgE receptors, 174
Human neutrophils, 167
 helminth damage mediation, role in, 168, 190
 interaction with schistosomula, 169, 173
Human placental conditioned medium (HPCM), 160, 163
Human SRS-A, 150
*Hyalomma anatolicum*, 57
Hyaluronidase, 117
Hydatid cyst, 161
Hydrocortisone, 10
Hydrogen gas, production by amoebae, 123
Hydrogenase, activity in amoebae, 116, 123, 129
5-Hydroxy-*N,N*-dimethyltryptamine, 21
Hydroxyeicosatetraenoic acid, 150
5-Hydroxy-*N*-methyltryptamine, 21, 22
5-Hydroxytryptamine *see* Serotonin
5-Hydroxytryptophan, 8
 decarboxylase, 9
*Hymenolepis diminuta*, 6, 189
 metabolism regulation, 11
  lactic acid production, 12
 migration behaviour, 7–8
 migration behaviour, role of serotonin in, 7–8
 serotonin biosynthesis in, 9
*H. nana*, 9
Hyperimmune rabbit serum, 168

I

Imidazole acetic acid, 150
Immunofluorescence detection, eosinophil MBP toxicity to schistosomula, 156
Indoleamines, 2, 28
Iron, role in amoebal metabolism, 125
Isobutylmethylxanthine (IBMX), 27
Isolated organ bath system, parasitic worm motility studies in, 4

# SUBJECT INDEX

Isotopic assays, cell-mediated damage to helminths, 146
Ixodid ticks, 45

## K

Ketanserin, 23
Killer (K) cells, 178, 180
Koch's bodies, 45
Koch'sche Kugeln *see* Koch's bodies

## L

Lactate dehydrogenase, 115, 117
Lectins, 154, 180
*Leishmania*, 172
Leprosy, 204
Leukotrienes, 150
Light micrographs
    babesian parasites
        intraerythrocytic stages, **39, 73**
        ray-bodies, **80**
        schizonts, **71**
        sporozoites, **94**
    theilerian parasites
        differentiated kinetes, **64**
        intraerythrocytic stages, **39**
        schizonts, **48**
Light microscopical assays, cell-mediated damage to helminths 146–147, 187
*Listeria monocytogenes*, 172
    macrophage-mediated damage to, 176
*Litomosoides* sp., effector mechanisms active against, 195
*L. carinii*
    cell-mediated damage to, *in vitro*
        eosinophil-mediated, 153, 197
        granulocyte-mediated, 152
        isotopic assay, 146
        macrophage-mediated, 173
    microfilariae, 146, 195
        cell-mediated damage to, 195
        neutrophil-mediated damage to, 171
Liver fluke *see Fasciola hepatica*
Lymphocytes
    cytotoxic effector properties, 178
    helminth damage mediation, role in, 178–181

Lymphocytes (cont.)
    cytolytic T lymphocytes, 179–180
    killer (K) cells, 178, 180
    killing mechanisms, 180–181
    lectin-induced cytotoxicity, 180
    natural killer (NK) cells, 178, 180
    target cell zeiosis, 181
    toxic mediators release, 180
    theilerian sporozoite development in, 45–51
D-Lysergic acid diethylamide, 28
    2-bromo derivative, 2, 22, 23
    parasitic flatworm motility, stimulation of 5–6, 22
    psychotomimetic action on man, 5
    receptor binding studies, use in, 22–23
    serotonin-activated adenylate cyclase, antagonization of, 21–23
    structure, 2
Lysophospholipase, 149

## M

Macrophages
    cell-surface receptors, 172
    IgE receptors, 173–174
    functional properties, 172
    helminth damage mediation, role in, 172–178
        activation, 172–173, 176–177
        antibody-dependent, 173–176
        antibody-independent, 176–178
        arginase-dependent mechanisms, 178
        catalase inhibition, 175
        cytochrome *c* inhibition, 175
        glucosamine incorporation, 175
        hydrogen peroxide generation, 175, 177, 178
        IgE-dependent killing, 173–174
            mechanisms, 175
        IgG-dependent killing, 175–176
        killing mechanisms, 175

Macrophages (cont.)
  lysosomal enzymes release, 175
  macrophage cyclic GMP, intracellular level increase, 175
  mechanisms of damage, 177–178
  non-specific antibody-independent effect, 177
  oxygen-dependent mechanisms, 178
  oxygen metabolite generation, 173
  peptide inhibition, 175
  plasminogen activator secretion, 175
  pre-incubation of macrophages, effect of, 174
  protease peptone stimulation, 176
  stimulation after IgE-antigen complex binding, 174–175
  superoxide anion generation, 175
  T cell mediation, 177
  thioglycollate stimulation, 176
 properties, 149
Macroschizonts *see* Koch's bodies
Major histocompatibility complex (MHC), 178, 179
Malaria parasites, 96
 ookinete, 61
Malignant ovine and caprine theileriosis, 41
Mammals
 adenylate cyclase communicated hormone signals in, 18
 brain receptors, serotonin antagonist binding studies, 23
 glycogenolysis control mechanisms, 25–26, 28
Man
 eosinophil cationic protein (ECP), 149–150
 eosinophil major basic protein (MBP), 149
 helminth infection

Man (cont.)
 correlative studies, 204–205
 immune response, 144, 198, 203–206
 large-scale field studies on immunity, 205
 skin reaction, 204
Mast-cell mediators, role in eosinophil activity, 151–152, 164–165
Mediterranean theileriosis, 40
Menoctone, 50
*Meriones libycus*, 195
*Mesocestoides corti*, 189
 serotonin receptors in, 6
Metal ions, role in amoebal metabolism, 124–125
Metergoline, 23
Methiothepin, 23
5-Methoxy-$N,N$-dimethyltryptamine, 21, 22
5-Methoxytryptamine, 21, 22, 28
Methylene blue, 147, 180
α-Methylmannoside, 158
$N$-Methyltryptamine, 21, 22
Methysergide, 8, 12
Metrizamide, 152, 180
Metrizoate, 152
Mianserin, 23
Microgamont *see* Ray-body
Molluscs, neurotransmission in, 2
Mouse
 glycogenolysis stimulation by serotonin in, 28
 eosinophils, 166
 helminth infection, 145, 151
  antibody-mediated immunity, 201
  eosinophil-mediated damage to, 154, 189, 193
  IgE responses, 162
  immune response, 202
  lymphocyte-mediated damage to, 180
  macrophage-mediated damage to, 172, 176–178, 188, 193
  metabolism behaviour in, 10
  skin reaction, 199
  T lymphocyte response, 160
Murine peritoneal exudate cells, 191
Murine toxocariasis, 197

# SUBJECT INDEX

*Mycobacterium tuberculosis*, 172
Myeloperoxidase, 170

## N

NADH-linked diaphorase, 129
NADPH flavin oxidoreductase, 129
Natural cord factor, 176
Natural killer (NK) cells, 178, 180
Nematodes
    effector mechanisms active
        against, 189–198, 206
    morphology, 189
    serotonin receptors in, 27–28
*Nematospiroides* sp., effector
    mechanisms active against, 193
*N. dubius*
    cell-mediated damage to, *in vitro*
        biological assay, 145
    immune responses to, 193
    larvae, 145, 193
Neuraminidase, 180
Neurohumoral transmitters, 4
Neuromuscular regulators, 4
*Neurospora*, 14, 20
Neurotransmitters, role in eukaryote
    development, 15–16
Neutrophils
    C3 receptors, 168
    helminth damage mediation, role
        in, 157, 166–171, 206
        adherence, 157–158
        comparison with eosinophil-
            mediated damage, 166–168
        degranulation, 169
        Fc receptor blocking, 168
        functional activity
            enhancement, 171
        fusion, 169
        hydrogen peroxide generation,
            158, 170
        inhibition of, 170
        killing mechanisms, 171
        ligand mediation, 168
        mechanisms of damage,
            168–171
        mode of action, 152
        oxidative killing mechanisms,
            170
        respiratory burst activity, 170

Neutrophils (cont.)
        schistosomulum membrane,
            fusion between, 169
        superoxide production, 170
        toxic cationic protein release,
            170
    myeloperoxidase, 158
    properties, 149
    species differences, 171
    surface receptors, 168
*Nicollia*, 72
Nicotinamide adenine dinucleotide
    (NAD), 112, 115, 121
Nicotine, 4
*Nippostrongylus* sp., effector
    mechanisms active against, 191
*N. brasiliensis*, 162
    cell-mediated damage to, 191
    eosinophil chemotactic activity,
        166
    immune response to, 191
Nitroblue tetrazolium reduction, 170,
    191
Non-phagocytosable helminths, 173
Norepinephrine, 10, 11, 15, 27
Nuclear magnetic resonance (NMR)
    spectroscopy, use in metabolism
    regulation studies, 12–14
Nuclear magnetic resonance (NMR)
    spectrum, *Fasciola hepatica*, **13**
*Nuttallia*, 69, 72

## O

Octopamine, 27
OKT8 marker, 180
Oligomeric aldolase, 127
*Onchocerca* sp., effector mechanisms
    active against, 197
*O. gibsoni*, 197
*O. volvulus*, 153
Onchocerciasis, 161
Opsonizing antibodies, 194, 195
Oriental theileriosis, 41
*Ostertagia circumcinta*, metabolism
    regulation, 10

## P

Panmede, 120
*Panthera leo*, 43, 44

Parasites, role of serotonin in
 metabolism of, 10–12
Parasitic flatworms, motility, 4–8
Parasitic helminths
 acetylcholine receptors in, 4
 adenylate cyclase activity, role of
  serotonin in, 16–18
 cell regulation, role of cyclic AMP
  in, 14–18
 electrical activity in, 7
 metabolism regulation, role of
  serotonin in, 10–12
 migratory behaviour, 3
 motility, 3–4
  effect of serotonin on, 4–7
 musculature, 3
 nervous system, 3
 neuroanatomy, 3
 sensory receptors, 3
Parasitic worms, serotonin receptors
 in, 2–29
Partial agonist, definition, 22
Parvaquone (993c) *see* Menoctone
Peanut agglutinin, 180
Percoll, 152
Peripheral blood eosinophilia, 162
Peroxidase-labelled anti-IgE, 174
Phagocytotic protozoa, 116
Phase-contrast cinemicrography
 studies, eosinophil-mediated
 damage to helminths, 155
Phorbol myristate acetate, 158
Phosphate acetyltransferase, 116
Phosphate transacetylase, 123
Phosphoarginine, 14
Phosphocreatine, 14
Phosphodiesterase, inhibition, 27
Phosphoenolpyruvate (PEP) carboxy-
 phosphotransferase, activity in
 glycolytic pathway of amoebae,
 112
Phosphofructokinase
 activation of, 26
 glycolytic pathway of amoebae,
  activity in, 108, 122, 126–127,
  129
6-Phosphofructokinase adenosine
 triphosphate, 116
Phosphoglycerate kinase, activity in
 glycolytic pathway of amoebae, 108

Phospholipase D, 150
*Physarum polycephalum*, 15
Phytohaemagglutinin, 171, 181, 195
Pig, helminth infection, 188
Pinocytotic protozoa, 116
Piroplasms
 genera, 38
 group classification, 96
 life cycle, 38–97
  phases, 96
 sexual stages, 38–97
 sporozoite transmission, 45
Planarians, hormone receptor
 amplification, 18
*Plasmodium*, 53
*P. gallinaceum*, 78
Platyhelminthes, effector mechanisms
 active against, 182–189
Polyarginine, 156–157
Progesterone, 10
Prokaryotes, 108
*Propionibacterium shermanii*, 108, 112
Prostaglandins, 150
Protamine, 156–157
Protein kinase, serotonin activation of,
 25–26
Protozoa, damage to
 granulocyte-mediated, 148
 lymphocyte-mediated, 178
 macrophage-mediated, 177
Pyridoxal phosphate, 9
Pyruvate decarboxylase, 116
Pyruvate dehydrogenase, 116
Pyruvate kinase, 116, 126
Pyruvate oxidase, activity in amoebae,
 116
Pyruvate-phosphate dikinase, 112
Pyruvate synthase, activity in glycolytic
 pathway of amoebae, 112, 126, 130

R

Rabbit, eosinophil proteins, 150
Radioisotope release measurement, in
 helminth viability assays, 145, 146
Rat
 alveolar macrophages, 174
 cestode migration behaviour in,
  7–8
 eosinophils, 166, 190

# SUBJECT INDEX

Rat (cont.)
  glycogenolyis stimulation by
    serotonin in, 28
  helminth infection, 151, 193
    anaphylactic antibodies, 165
    antibody mediation, 153, 187,
      188
    eosinophil-mediated damage
      to, 154, 155, 165, 190, 204
    IgE response, 162, 194, 203
    immune response, 204
    infection immunity, 186
    macrophage-mediated damage
      to, 175, 204
    neutrophil-mediated damage
      to, 167, 168, 173, 190
    T lymphocyte response, 160,
      162
  peritoneal macrophages, 174
Ray-body
  discovery, 53–54, 75
  babesian parasites
    development, 75, 78, 83
    diagrammatic representation,
      **76**
    electron micrographs, **77, 80**
  theilerian parasites
    diagrammatic representation,
      **54, 55**
    electron micrographs, **56**
Rhabditida, 189
Rhesus monkey, helminth infection, 152
  skin reaction, 199
*Rhipicephalus appendiculatus*, 57
*R. bursa*, 88
*R. sanguineus*, 88
Rodents, babesian parasites, 69
Ruminants, babesian parasites, 69

## S

*Sarcocystis*, sporozoite transmission, 45
Scanning electron microscopical
  assays, cell-mediated damage to
  helminths, 188
*Schistosoma* spp.
  effector mechanisms active
    against, 182–186
  serotonin inactivating enzymes, 9
  skin response to schistosomula of,
    199

*S. haematobium*, 9
*S. japonicum*, 9, 166
  effector mechanisms active against,
    182
  skin reaction to, 199
*S. mansoni*
  adenylate cyclase activity, role of
    serotonin in, 16–18
  cell-mediated damage to, *in vitro*,
    182–186
    antibody-dependent, 151
    biological assay, 145
    eggs, 185–186
    isotopic assay, 146
    light microscopical assay, 146
  cercariae, 16
  early phase immunity to, 202
  effector mechanisms active against,
    144, 182
    adult worms, 185
    eggs, 185–186, 201
    T lymphocyte dependent, 201
  eosinophil chemotactic activity, 166
  immune response to, 182,
    201, 202, 204
  metabolism regulation, role of
    serotonin in, 11, 12
  monoamine oxidase activity in, 9–10
  morphology, 182
  ova, 145
  schistosomula, 16, 18, 146
    antibody-dependent cell-
      mediated damage to,
      151–152, 157
    cationic polypeptide damage
      to, 156
    ECP damage to, 156
    eosinophil-mediated damage
      to, 148, 153, 154, 155–166
    lymphocyte-mediated damage
      to, 179, 180
    macrophage-mediated damage
      to, 173, 174, 175, 176
    maturation capacity
      measurement, 145
    MBP damage to, 156
    neutrophil-mediated damage
      to, 157, 167, 170, 171
    preparation of, 151
    skin reaction to, 199

S. *mansoni* (cont.)
  toxic granule content release damage to, 156
  serotonin biosynthesis in, 9
  serotonin receptors in, 6–7, 29
Schistosome eggs
  effector mechanisms active against, 185–186
    granuloma pathogenesis, 203
  eosinophil-mediated damage to 185–186
Schistosomes, effector mechanisms active against, 182–186
  adult worms, 185
  eggs, 185–186
  schistosomula, 183–185
Schistosomiasis, 160
  immune response to, 204, 206
  prevalence of infection, 204
  vaccine development, 206
Schistosomula
  adenylate cyclase activity in, 18
  cell-mediated damage to, *in vitro*, 182–186
    antibody production, 183
    antibody-binding capacity, 184
    host molecule uptake, 184
    immune attack susceptibility, 184
    protective antigens, 183–184
    tegumental membrane lipid bilayer formation, 156, 183
  effector mechanism active against, 183–185
  maturation, 184
    skin penetration, 183
  skin response to, 199–200
Sea urchin, 15, 16
Serotonin, 14, 15
  adenylate cyclase communicated hormone signals, role in, 18–20
  analogues, adenylate cyclase activity, effect on, 21
  antagonists, 8
  carbohydrate metabolism regulation, role in, 10–12
  cestode migration behaviour, role in, 7–8
  degradation, metabolic reactions in, 8

Serotonin (cont.)
  distribution, 2
  eukaryote development, role in, 15–16
  glycogenolysis and glycolysis in mammals, role in, 28
  glycolytic enzyme activity, role in, 27–28
  identification in tissue, 8
  inactivation, 8
  metabolic reactions in synthesis, 8
  metabolism regulation, role in, 10–12
  $^{31}P$ NMR studies of, 12–14
  neurotransmitter, role as, 2, 4
  parasitic flatworm motility, stimulation of, 4–7, 22
  protein kinase activation, role in, 25–26
  receptors in mammals, 28
  receptors in nematodes, 21–23, 27–29
    desensitization of, 23–25
  structure, 2
  trematodes, occurrence in, 8–10
Shaffer-Frye culture system, 117, 120, 122, 123
Sheep, helminth infection, 10, 186, 192
Slime moulds, 14
Slow-reacting substance of anaphylaxis (SRS-A), 150
Sodium oxalate, 125
Southern blot technique, 120
Spiroperidol, 23
Spirurida, 189
Sporozoa, 96
*Staphylococcus aureus*, protein A (SPA), 157
Strahlenkörper *see* Ray-bodies
*Strongyloides* sp., effector mechanisms active against, 192
*S. ratti*, 192
Succinate dehydrogenase, 129
Superoxide dismutase, 171
*Syncerus caffer*, 40
Synephrine, 27

T

*Taenia* spp., cell-mediated damage to, 188
*T. crassiceps*, 188

# SUBJECT INDEX

*T. pisiformis*, serotonin receptors, 6
*T. saginata*, 188
*T. solium*, 188
*T. taeniaeformis*, 188
   immune response to, 200
*Taurotragus oryx*, 40
Testosterone, 10
*Tetrahymena pisiformis*, carbohydrate metabolism, 16
*Theileria*
   differentiated kinetes
      electron micrograph, **64**
      light micrographs, **64**
      structure, 61, 64–65
   erythrocytopaenia induction by, 53
   gamete fusion, 58, 61
      diagrammatic representation, **58**
      electron micrographs, **59**
   history of discovery, 38
   intraerythrocytic stages, 38, **39**, **52**, 96
   kinete development, 65–69, 96
      electron micrographs, **66**
      host cell cytoplasm ingestion, 65–66
      light micrograph, **66**
   kinete differentiation, 64–65, 89
   kinete formation, 61, 64–65
      diagrammatic representation, **60**, **81**
      electron micrographs, **62**, **63**
   life cycle, 45–69
      babesian parasites, comparison with, 96–97
      development in erythrocytes, 51–53
      development in vector tick salivary gland, 65–69
      diagrammatic representation, **46**
      schizogony inside lymphocytes, 45–51, 96
      sexual stages in tick intestine, 53–65
      sexual stages *in vitro*, 53–65
      summary of, 57
   merozoites
      development in erythrocytes, 50, 51–53

*Theileria* (cont.)
      formation, 47–51
      pathogenic effects, 50
      stages of, 51, 53
      structure, 50
   ovoid form, release from erythrocytes, 55
   ray-body
      development, 54–55, 58
      diagrammatic representation, **54**, **55**
      electron micrographs, **56**
      structure, 55, 58
   schizonts
      culture *in vitro*, 47
      development, 47
      drug mode of action on, 50
      electron micrographs, **48**
      formation, 45
      light micrographs, **48**
      morphology, 47
   species
      diseases caused, 40–41
      geographic distribution, 40–41
      light micrograph differentiation of, **39**
      vectors, 40–41
      vertebrate hosts, 42–44
   sporozoites
      concentration in salivary gland host cells, 67, 69
      electron micrographs, **68**
      formation, 67, 96
      light micrographs, **68**
      transmission, 45
   vector species, 57
   vector tick salivary gland, development in, 65–69
      electron micrographs, **66**
      light micrograph, **66**
   vertebrate host infection, 45, **46**
*T. annulata*
   gamete fusion, electron micrograph, **59**
   intraerythrocytic stages, **52**
   kinete formation, electron micrographs, **62**, **63**, **64**
   merozoites
      comma-shaped form, 51

*T. annulata* (cont.)
  development in erythrocytes, 50, 51
  division by binary fission, 53
  formation, 47
  metabolism, 53
  spherical stage, 53
  stages of, 51
  tetrad formation (Maltese-cross arrangement), 53
  ray-body
    diagrammatic representation, **54**
    electron micrographs, **56**
  schizonts, **48**
  summary of development, 57
*T. hira*, merozoites, 51
*T. mutans*
  merozoites, 51
  tetrad formation, 53
  summary of development, 57
*T. orientalis*, merozoites, 53
*T. ovis*
  kinete development
    electron micrographs, **66**
    light micrograph, **66**
  kinete formation, 61
    electron micrographs, **62, 63**
  merozoites, 51
  sporozoite formation, **68**
*T. parva*
  gamete fusion, electron micrograph, **59**
  kinete formation, 65
    electron micrographs, **63, 66**
  merozoites
    development in erythrocytes, 50, 51
    formation, 47
    stages of, 51
  ray-body
    diagrammatic representation, **55**
    electron micrographs, **56**
  schizonts, **48**
  sporozoite formation, 67, 69
    electron micrograph, **68**
    light micrograph, **68**
  summary of development, 57

*T. taurotragi*, summary of development, 57
*T. velifera*, summary of development, 57
Theilerioses, 38
Thyroxine, 10
Tick
  haemolymph, 65
  intestine, parasite sexual stages development in
    babesian parasites, 75–83, 96
    theilerian parasites, 53–65, 96
  salivary glands, parasite kinetes development in
    babesian parasites, 89–95
    theilerian parasites, 65–69
Toluidine blue, 147
*Toxocara canis*, 161, 197
*Toxoplasma* sp., sporozoite transmission, 45
*T. gondii*, 172, 176
  macrophage-mediated damage to, 176, 177
Transaldolase, 116
Transmission electron microscopical assays,
  cell-mediated damage to helminths, 148, 169
Trematodes
  acetylcholine receptors in, 4
  effector mechanisms active against, 182–186, 206
  serotonin occurrence in, 8–10
Tricarboxylic acid cycle, 127
*Trichinella* sp., effector mechanisms active against, 190–191
*T. spiralis*
  cell-mediated damage to, *in vitro* biological assay, 145
    gross disruption of, 146–147
  effector mechanisms active against, 144, 189, 190–191
  eosinophil-mediated damage to, 148, 153, 156–157
  eosinophilic response to, 160, 186
  granulocyte-mediated damage to, 152
  IgE response to, 162
  immune response to, 191, 200, 201, 202, 203

# SUBJECT INDEX

*T. spiralis* (cont.)
  larvae
    encystation capacity measurement, 145
    macrophage-mediated damage to, 176
    neutrophil-mediated damage to, 168, 170
*Trichostrongylus colubriformis*, 197
*Trichuris muris*, 200
Trinitrophenyl, 179
Triosephosphate dehydrogenase, 117
*Tritrichomonas foetus*, 112
Tropical eosinophilia, 161
Tropical theileriosis, 40
Trypan blue, 147
*Trypanosoma cruzi*, 173
  epimastigotes, 178
Tryptamine, 21, 28
Trypticase, 120
Tryptophan, 8
Tuberculosis, 177
Tubocurarine, 4
Tumour cells, damage to
  cytotoxic T lymphocyte mediated, 145, 178
  macrophage mediated, 177

## U

Ubiquinone, role in amoebal metabolism, 124
UDP glucose-hexose-1-phosphate uridyltransferase, 116

## V

Viruses
  amoebal pathogenicity, effect on, 124
  lymphocyte-mediated damage to, 148
Vitamins, requirements in amoebal metabolism, 124–125
*Vulpes vulpes*, 43, 44

## W

*Wuchereria* spp., effector mechanisms active against, 196–197
*W. bancrofti*, 196

## X

Xanthine oxidase, 170